Advances in Spatial Science

Springer

Berlin
Heidelberg
New York
Hong Kong
London
Milan
Paris
Tokyo

Titles in the Series

C. S. Bertuglia, M. M. Fischer and *G. Preto* (Eds.)
Technological Change, Economic Development and Space
XVI, 354 pages. 1995. ISBN 3-540-59288-1
(out of print)

H. Coccossis and *P. Nijkamp* (Eds.)
Overcoming Isolation
VII, 272 pages. 1995. ISBN 3-540-59423-X
(out of print)

L. Anselin and *R. J. G. M. Florax* (Eds.)
New Directions in Spatial Econometrics
XVX, 420 pages. 1995. ISBN 3-540-60020-5
(out of print)

H. Eskelinen and *F. Snickars* (Eds.)
Competitive European Peripheries
VIII, 271 pages. 1995. ISBN 3-540-60211-9

J. C. J. M. van den Bergh, P. Nijkamp and *P. Rietveld* (Eds.)
Recent Advances in Spatial Equilibrium Modelling
VIII, 392 pages. 1996. ISBN 3-540-60708-0

P. Nijkamp, G. Pepping and *D. Banister* (Eds.)
Telematics and Transport Behaviour
XII, 227 pages. 1996. ISBN 3-540-60919-9

D. F. Batten and *C. Karlsson* (Eds.)
Infrastructure and the Complexity of Economic Development
VIII, 298 pages. 1996. ISBN 3-540-61333-1
(out of print)

Y. Leung
Intelligent Spatial Decision Support Systems
XV, 470 pages. 1997. ISBN 3-540-62518-6

C. S. Bertuglia, S. Lombardo and *P. Nijkamp* (Eds.)
Innovative Behaviour in Space and Time
X, 437 pages. 1997. ISBN 3-540-62542-9

A. Nagurney and *S. Siokos*
Financial Networks
XVI, 492 pages. 1997. ISBN 3-540-63116-X

M. M. Fischer and *A. Getis* (Eds.)
Recent Developments in Spatial Analysis
X, 434 pages. 1997. ISBN 3-540-63180-1

R. H. M. Emmerink
Information and Pricing in Road Transportation
XVI, 294 pages. 1998. ISBN 3-540-64088-6

P. Rietveld and *F. Bruinsma*
Is Transport Infrastructure Effective?
XVI, 384 pages. 1998. ISBN 3-540-64542-X

P. McCann
The Economics of Industrial Location
XII, 228 pages. 1998. ISBN 3-540-64586-1

L. Lundqvist, L.-G. Mattsson and *T. J. Kim* (Eds.)
Network Infrastructure and the Urban Environment
IX, 414 pages. 1998. ISBN 3-540-64585-3

R. Capello, P. Nijkamp and *G. Pepping* (Eds.)
Sustainable Cities and Energy Policies
XI, 282 pages. 1999. ISBN 3-540-64805-4

M. M. Fischer and *P. Nijkamp* (Eds.)
Spatial Dynamics of European integration
XII, 367 pages. 1999. ISBN 3-540-65817-3

M. M. Fischer, L. Suarez-Villa and *M. Steiner* (Eds.)
Innovation, Networks and Localities
XI, 336 pages. 1999. ISBN 3-540-65853-X

J. Stillwell, S. Geertman and *S. Openshaw* (Eds.)
Geographical Information and Planning
X, 454 pages. 1999. ISBN 3-540-65902-1

G. J. D. Hewings, M. Sonis, M. Madden and *Y. Kimura* (Eds.)
Understanding and Interpreting Economic Structure
X, 365 pages. 1999. ISBN 3-540-66045-3

A. Reggiani (Ed.)
Spatial Economic Science
XII, 457 pages. 2000. ISBN 3-540-67493-4

D. G. Janelle and *D. C. Hodge* (Eds.)
Information, Place, and Cyberspace
XII, 381 pages. 2000. ISBN 3-540-67492-6

P. W. J. Batey and *P. Friedrich* (Eds.)
Regional Competition
VIII, 290 pages. 2000. ISBN 3-540-67548-5

B. Johansson, Ch. Karlsson and *R. R. Stough* (Eds.)
Theories of Endogenous Regional Growth
IX, 428 pages. 2001. ISBN 3-540-67988-X
(out of print)

G. Clarke and *M. Madden* (Eds.)
Regional Science in Business
VIII, 363 pages. 2001. ISBN 3-540-41780-X

M. M. Fischer and *Y. Leung* (Eds.)
GeoComputational Modelling
XII, 279 pages. 2001. ISBN 3-540-41968-3

For information about further volumes please see end of the book

Yasuhide Okuyama · Stephanie E. Chang
Editors

Modeling Spatial and Economic Impacts of Disasters

With 72 Figures
and 40 Tables

Springer

Dr. Yasuhide Okuyama
Regional Research Institute, West Virginia University
511 North High Street, Morgantown, WV, 26506-6825, USA
yaokuyama@mail.wvu.edu

Professor Stephanie E. Chang
School of Community and Regional Planning
#433-6333 Memorial Road
University of British Columbia
Vancouver, BC V6T 1Z2, Canada
stephanie.chang@ubc.ca

ISBN 3-540-21449-6 Springer-Verlag Berlin Heidelberg New York

Cataloging-in-Publication Data applied for

A catalog record for this book is available from the Library of Congress.

Bibliographic information published by Die Deutsche Bibliothek
Die Deutsche Bibliothek lists this publication in the Deutsche Nationalbibliografie;
detailed bibliographic data available in the internet at *http://dnb.ddb.de*

Springer-Verlag Berlin Heidelberg New York
is a part of Springer Science+Business Media
springeronline.com

Printed in Germany

Cover design: Erich Kirchner, Heidelberg

SPIN 10998540 42/3130 - 5 4 3 2 1 0 - Printed on acid-free paper

In memory of Barclay G. Jones

Preface

This volume is dedicated to the memory of Barclay G. Jones, Professor of City and Regional Planning and Regional Science at Cornell University. Over a decade ago, Barclay took on a fledgling area of study – economic modeling of disasters – and nurtured its early development. He served as the social science program director at the National Center for Earthquake Engineering Research (NCEER), a university consortium sponsored by the National Science Foundation and the Federal Emergency Management Agency of the United States. In this capacity, Barclay shepherded and attracted a number of regional scientists to the study of disasters. He organized a conference, held in the ill-fated World Trade Center in September 1995, on "The Economic Consequences of Earthquakes: Preparing for the Unexpected." He persistently advocated the importance of social science research in an establishment dominated by less-than-sympathetic natural scientists and engineers.

In 1993, Barclay organized the first of a series of sessions on "Measuring Regional Economic Effects of Unscheduled Events" at the North American Meetings of the Regional Science Association International (RSAI). This unusual nomenclature brought attention to the challenge that disasters – largely unanticipated, often sudden, and always disorderly – pose to the regional science modeling tradition. The sessions provided an annual forum for a growing coalition of researchers, where previously the literature had been fragmentary, scattered, and episodic. Since Barclay's unexpected passing in 1997, we have continued this effort in his tradition.

Many of the papers in this volume were originally presented in these sessions at recent RSAI meetings and have subsequently been revised and updated for publication. Several other authors who were unable to attend one of the recent sessions nonetheless contributed chapters. The wide spectrum of modeling approaches in this volume is a reflection of Barclay's vision for new and creative methods of dealing with the losses from disasters.

Over the last decade, a series of urban disasters throughout the world has drawn attention to the timeliness and importance of the topic. In terms of earthquakes alone, major disasters struck Los Angeles in 1994, Kobe (Japan) in 1995, Turkey and Taiwan in 1999, and India in 2001. These and other natural disasters, as well as the terrorist events of September 11, 2001, vividly demonstrate the need to better understand, prepare for, and mitigate the spatial and economic dimensions of disaster losses.

We would like to thank Geoffrey J. D. Hewings for his encouragement and guidance to this project. Mary Lou Myer of the Regional Research Institute has assisted in the formatting of the chapters. Katharina Wetzel-Vandai and Marianne Bopp at Springer have been extremely supportive and patient. Finally, during the production process of this book, each of us underwent some life-altering events, some scheduled and others unscheduled. We are grateful to our respective families for their continuous support. They have made the production of this book especially meaningful to us.

Yasuhide Okuyama, Morgantown, WV
Stephanie E. Chang, Vancouver, BC

March 2004

Table of Contents

Part III: Integrative Models

1 Introduction

Yasuhide Okuyama[1] and Stephanie E. Chang[2]

[1] Regional Research Institute, West Virginia University, USA
email: yaokuyama@mail.wvu.edu
[2] School of Community and Regional Planning, and Institute for Resources, Environment, and Sustainability, University of British Columbia, Canada
email: stephanie.chang@ubc.ca

1.1 Overview

The problem of disasters has gained considerable interest in the last decade. The 1990s was designated as the International Decade for Natural Disaster Reduction (IDNDR) by the United Nations General Assembly (resolution 42/169), and parallel national efforts were undertaken across the globe. Disaster losses nonetheless continued their rapid growth (Mileti, 1999; Red Cross, 2002). During this decade, a series of major natural and human-induced disasters struck large urban areas and plainly demonstrated the need to better understand, anticipate, and prepare for such calamities. For example, in the United States, record losses were caused by Hurricane Andrew which struck the Miami region in 1992, the Great Midwest Floods of 1993, and the 1994 Northridge earthquake in Los Angeles (NRC, 1999). By one estimate, the Northridge earthquake inflicted some $40 billion in damage (Eguchi *et al.*, 1998). In Japan, the 1995 Great Hanshin earthquake struck directly beneath the modern industrialized urban area of Kobe. It killed more than 6,000 people and resulted in an estimated $100 billion in damages, or about 2 percent of Japan's gross national product (Scawthorn *et al.*, 1997). The 1999 Chi-Chi earthquake in Taiwan caused an estimated $8 billion in loss and, significantly, caused sizable spatial repercussions as semiconductor prices spiked worldwide (Chang, 2000). The terrorist attacks of September 11, 2001, in New York City killed close to 3,000 people and caused some $83 billion in direct and indirect loss to the city's economy. Adjusting for insurance and federal government reimbursements, the net loss may have amounted to some $16 billion (NYC Partnership, 2001). Each of these events demonstrated, moreover, how disasters not only inflict human and physical damages but can also cause considerable economic disruption to vulnerable cities and regions. For developing countries, disasters can have devastating consequences for their development and growth (Red Cross, 2002).

It is useful to clarify the concept of disaster in order to provide a common reference for this volume. The terms "hazard" and "disaster" are often used interchangeably in the literature, leading to some confusion. Within the social science literature, numerous definitions have been proposed, debated, and reconceptualized over the last several decades (see Tierney *et al.*, 2002, for an excellent review). For example, the functionalist approach of the classic sociological literature defined "disasters" as events that exceed the capacities of communities or social units to perform necessary functions, whereas early geographers in the field considered "natural hazards" to be extreme natural events that, by definition, intersect with human choices and adjustments. Some more recent definitions reflect social constructionist, political economy, or ecological vulnerability perspectives. While regional economists and regional scientists have written little on conceptually defining disasters, they commonly consider these to be extreme events, whether natural or human-induced in origin, that shock an economic system and produce measurable impacts.

Here, we make a distinction between hazards and disasters. John Whittow (1979) distinguished between these two terms as follows: "*A hazard is a perceived natural event which threatens both life and property—a disaster is the realization of this hazard* (p.19)." This definition can be extended to other types of threats such as technological accidents, terrorism events, riots, and even wars. In his way, *hazard* is the occurrence of the physical event per se, such as an earthquake, and *disaster* is its consequence. Not all hazardous events necessarily lead to disaster; disasters result when the physical event intersects with vulnerable built and socio-economic environments. This suggests that while research on hazards should involve the mechanism and/or prediction of occurrence of these events, research on disasters should incorporate the societal impacts, reconstruction, and recovery from them.

From the perspective of spatial economic modeling, disasters caused by natural and human-induced events reflect many similarities, although a discussion of their differences is useful. A major difference is that in contrast to natural disasters, many human-induced disasters are designed intentionally to cause destructions. Consequently, the resulting damages are often concentrated on facilities or areas that are of great significance to the well-being of the impacted society, whether strategically, economically, or symbolically. In addition, within the scope of natural hazards, certain types of hazards create some specific implications for modeling. The effects of sudden-onset hazards, such as earthquakes, hurricanes, and floods, are more readily captured in spatial economic models than slow-onset disasters such as droughts and epidemics. In the context of scientific and technological advances, a further distinction can be made between predictable hazards, such as hurricane and floods, and unpredictable ones, such as earthquakes. Predictions and predictability affect human behavior, trigger preparedness actions, and affect mitigation options that can modify the impact of an event when it occurs. While different types of hazards lead to different features in their consequences and impacts, to some extent, many common attributes extend across disasters. These include the occurrence of direct business interruption and consequent indirect losses, reconstruction and recovery processes, etc. In many cases, therefore, modeling and analysis can apply similar frameworks across disaster types. Detailed study of the transferability of models across disaster types remains, however, an area needing further research.

In the recent years, there has been a growing recognition that disasters cannot be adequately handled within the framework of conventional spatial economic models. While Econometrics, Input-Output, Social Accounting, and other types of economic models have been applied to study the impacts of disasters, they are based on a number of assumptions that are questionable in catastrophes. In their excellent summary of issues and modeling strategies for disaster impact studies, West and Lenze (1994) noted that researchers need to address issues such as double-counting, the response of households, and the evaluation of financial situations. A recent report by the National Research Council (1999) argued the main problem as follows:

> *The core of the problem with statistically based regional models is that the historical relationships embodied in these models are likely to be disrupted in a natural disaster. ...In short, regional economic models have been developed over time primarily to forecast future economic conditions or to estimate the effects of a permanent change (e.g., the opening or closing of a manufacturing plant). The abruptness, impermanence, and often unprecedented intensity of a natural disaster do not fit the event pattern upon which most regional economic models are based. The models are thus inappropriate for simulating natural disaster losses.* (p.40)

This problem can be recast as a series of challenges that disasters pose to conventional modeling. These challenges concern: (a) data availability; (b) scope; (c) broad influences; (d) uncertainty; and (e) stability and dynamic process:

- Data on disasters are sparse and difficult to interpret. Not only are disasters rare, but also their consequences are unique to each event. Empirically, it may be difficult to distinguish the effects of the disaster from changes that would have occurred in any case. Moreover, disaster data are mostly available at the physical and localized levels, providing little information on socio-economic damages or spatial effects.
- Conventional models usually reflect stylized and limited aspects of society. The damages and losses caused by disasters reflect complex linkages between the natural, built, and socio-economic environments. Thus, disasters cross disciplinary boundaries.
- In addition, disasters expose the underlying vulnerability or resiliency of socio-economic agents and groups which may not be realized in the pre-disaster period, so that impacts are influenced by broad factors other than physical damage. Issues of financial assistance greatly influence disaster outcomes.
- Most conventional models are based on the assumptions of perfect information and foresight. Many disasters occur suddenly and with little warning or anticipation. Risk information is also imperfect. Risk perception and related behaviors, such as purchasing insurance, influence disaster impacts.
- Conventional models assume incremental and steady changes and are equilibrium-oriented. Disasters often create chaos and economic disequilibrium, and can also cause unexpected long-term, structural changes. In addition, the speed of adjustments is a critical dimension of impacts and recovery. Dynamic interactions among agents and behavioral adjustments in a disaster are very complex.

Notable progress has been made in recent years in tackling many of these challenges. Researchers have developed increasingly sophisticated models of the spatial economic impacts of urban disasters. This volume brings together a comprehensive cross-section of the state-of-the-art in socio-economic disaster modeling. Many of the chapters have been motivated by recent catastrophes and the need to better understand and prepare for future events. Each of the chapters addresses one or more of the challenges noted above, often proposing refinements to conventional methods or innovative new modeling frameworks. Collectively, they indicate that while issues still remain, significant inroads have been made.

The chapters are organized into three parts. Chapters in Part I elaborate on the discussion of conceptual and modeling issues. Part II proposes major refinements for modeling disasters in the context of conventional economic frameworks. Part III expands upon the economic approaches by adopting integrative strategies. These focus on linking economic models with the models of physical environment and disaster damage.

1.2 Conceptual and Modeling Issues

The three chapters in Part I explore issues and challenges for modeling the spatial economic impacts of disasters. Rose (chapter 2) sets the stage by discussing foundational concepts and identifying current issues and research priorities. By clarifying economic principles and key terms, such as issues of double-counting, stocks vs. flows, and direct vs. higher-order effects, he addresses some important sources of current confusion. Rose also identifies current issues in the literature and makes the case for others requiring attention. Recent research has raised such questions as non-market effects, the dynamic nature of impacts, and economic resiliency. At the same time, some topics that have long been neglected in disaster studies, like distributional impacts and sustainability, remain essential for public policy. Rose also provides an unusually candid discussion of data availability and model validation, and concludes that the greatest research need for hazard loss modeling is to improve its empirical basis. This, he argues, is necessary to broaden popular acceptance of the idea of economic impacts among researchers and policy-makers alike.

The following chapter by Cochrane (chapter 3) complements the discussion of direct and higher-order effects. Cochrane broadens the conceptualization of higher-order (indirect) effects to include not only ripple effects from production disruptions, but also systemic risk in the form of disaster-induced shortages of credit and insurance. He first discusses HAZUS, a landmark disaster loss model developed in the late 1990s by the U.S. Federal Emergency Management Agency (FEMA). As the principal architect of HAZUS' indirect economic loss module, Cochrane provides a useful summary of its methodology and generalized statements about economic loss patterns. He then extends the discussion of indirect losses to issues of systemic risk. While the financial market may have significant influence on disaster mitigation, reconstruction, and recovery, it has been largely neglected in the literature with the exception of a few studies on disaster relief and the role of insurance (*e.g.*, Kunreuther and Roth, 1998). Cochrane addresses this gap by examining the question of disaster-induced systemic loss in the insurance arena, emphasizing recent cases in California and Florida, as well as New York following the World Trade Center attacks in 2001.

The economic impacts of the events of September 11, 2001, on the economy of New York City are thoroughly analyzed in chapter 4 by Bram *et al.* Terrorist incidents represent another type of disaster or unscheduled event. While differing in some respects from natural disasters (*e.g.*, intentional causation by human beings, selection of targets, and potential avoidance), they also exhibit many similarities (*e.g.*, uncertainty of consequences, fear of repeated occurrences, and stages of reconstruction and recovery). Bram *et al.* find that while the attack caused a sharp temporary disruption, the New York economy is likely to keep its previous pace of economic growth over the medium term. However, they warn that much depends upon economic policy responses. Based on their research, the authors concur with a number of issues discussed by Rose and Cochrane in the previous chapters and raise the importance of others for reconstruction and recovery, including fiscal impacts on local government, the effects on industry and population (labor) structures, and intergovernmental relationships.

Many of the issues discussed in Part I are addressed and integrated in the modeling frameworks presented in Parts II and III.

1.3 Economic Models

The chapters in Part II propose a variety of approaches to modeling the spatial economic impact of disasters from the perspective of conventional economic methodologies. Since the pioneering work of Dacy and Kunreuther (1969) and Cochrane (1975), many economic modeling frameworks have been proposed for analyzing natural disasters. Each of the recent approaches presented in Part II challenges different aspects of disaster modeling; for instance, temporal impacts, insurance, resiliency, long-term evolution, and risk perception. At the same time, the models cover many common features, such as uncertainty and macroeconomic consequences.

The first three chapters of Part II respectively employ the conventional frameworks of Input-Output, Social Accounting, and Computable General Equilibrium modeling. Their contributions entail developing novel modification schemes for dealing with the unique aspects of disasters.

As Rose notes in Part I (chapter 2), the Input-Output (I-O) model has been employed in many studies on economic impacts of disasters. Criticisms against I-O pertain mainly to the inflexibility of the model's fixed coefficients, its static nature, and its equilibrium orientation, in which the market always clears regardless of what disturbances have occurred. Many researchers have used the I-O model for the preliminary estimation of the total economic impacts of a disaster; however, some studies have modified the I-O framework to reflect the particular aspects of disasters and the recovery process[1]. Okuyama *et al.* (chapter 5) highlight the spatial and temporal distributions of the economic impacts of a disaster and overcome some of the drawbacks of the I-O model by utilizing an interregional I-O table within the Sequential Interindustry Model (SIM) framework. With this modification, a static I-O table is extended to a dynamic system with spatial dimensions. It is thus able to deal with certain aspects of disasters such as uncertainty, demand-supply mismatch, and technology replacement.

As discussed in the previous section, the modeling literature has largely neglected the important issue of insurance and, more generally, disaster protection. Building on the Social Accounting Matrix (SAM) framework, Cole (chapter 6) develops an

[1] In recent examples, Rose *et al.* (1997) combined I-O modeling with linear programming to analyze optimal rationing policies of electric power after a disaster, while Okuyama *et al.* (1999) used the Leontief-Strout-Wilson version of the multi-regional I-O model to assess the interregional economic impacts of hypothetical earthquakes in the central U.S.

insurance accounting matrix to assess the ways that disasters affect social agents and propagate through the economy. He provides some insightful simulation results for evaluating the effects of insurance under a series of repeated disasters. Cole's approach is unique in recognizing that uncertainties are pervasive in normal production and daily survival. Agents invest in insurance to ward against these uncertainties, and disasters occur when major events exceed their coping capacities. Moreover, Cole's work is unique also in considering both formal and informal types of insurance, an example of the latter being the development of social networks. The approach thus directly addresses issues relevant to developing countries, which are often characterized by a dual economic structure consisting of modern and traditional sectors.

Computable General Equilibrium (CGE) models, like I-O and SAM models, are based on the general principles of a multi-sector general equilibrium-oriented framework. A relatively new modeling approach, they are distinguished by the inclusion of more sophisticated dynamic processes for price-quantity (supply-demand) adjustment, input substitution, and trade relationships. While CGE models have been proposed for disaster modeling in concept, Rose and Guha (chapter 7) present one of the few attempts to implement a CGE for an actual regional economy. Their case study of Memphis, Tennessee, focuses on the impact of electric power disruption in disasters. Simulation results reveal that in contrast to static I-O models, which overstate the rigidities in an economy, CGE models tend to exaggerate an economy's flexibility and resiliency. CGE models will thus understate economic losses unless properly adjusted. This analysis is accompanied by a discussion of how different types of models may be appropriate for capturing different stages of disaster recovery, from the emergency response period to the long run.

The long-term impact of a catastrophic event on an economy represents a critical research need that is difficult to address within the framework of traditional economic models. Reggiani and Nijkamp (chapter 8) approach this topic through the recently introduced framework of Self-Organized Criticality (SOC). In this framework, a major exogenous force (*i.e.*, a disaster) drives the economy (at the macro-level) to a critical state, together with strong localized interactions between individual elements of the system (at the micro-level). To illustrate the application of SOC to a disaster situation, Reggiani and Nijkamp assess the fall of the Iron Curtain as a socio-economic crisis, analyzing major shifts or jumps in regional labor markets in West Germany during the 1990s. Their findings suggest that the SOC concept may also be useful for explaining the dynamics and dislocations of natural disasters.

In contrast to the preceding chapters, which largely concern *ex post* analyses of disaster impacts, Tatano *et al.* (chapter 9) address from an *ex ante* perspective the spatial economic implications of risk and risk perception. This study is more traditional in the sense that it employs a purely theoretical approach using mathematical modeling techniques from neo-classical economics. It explores the effects of information provision regarding local vulnerability in potential future disasters. Tatano *et al.* show that, with imperfect risk information, perception gaps may lead to a divergence between equilibrium land use patterns and optimal ones. They conclude that appropriate planning is needed for achieving optimal land-use. This chapter thus addresses a new dimension in the spatial economic modeling of disasters, the effects of risk information and risk perception. This is another crucial public policy issue that needs to be studied thoroughly.

1.4 Integrative Models

The spatial economic consequences of disasters derive in a complex manner from the interactions between physical damages and the regional economy. To a much greater extent than the chapters in Part II, those in Part III focus on the integration of damage estimation and economic modeling. In many cases, this involves engineering as well as economic analyses. Such comprehensive and typically large-scale modeling approaches have gained notable interest in recent years with increasing efforts in multi-disciplinary research, methodological advances, and improvements in computational power.

A major challenge in modeling the spatial economic impacts of disasters is the need to accommodate non-economic information. This includes inputs on the direct social and physical damages caused by disasters, such as loss of human lives and damage to the built environment, as well as information regarding potential future disasters such as geologic and meteorological data. Because no economic model can deal with these physical data directly, they must be interpreted in ways that allow economic models to treat them as inputs. This task is critical, as economic estimates are only as good as the quality of the disaster-related inputs; however, it is daunting, due to differences in such factors as units of analysis and sensitivity between physical data and economic models (West and Lenze, 1994). Each of the chapters in Part III propose novel approaches to integrating physical damage and economic models for disasters.

The dynamic processes of disaster recovery present a major challenge for conventional economic models and an important void in the literature. Chang and Miles (chapter 10) address the complexities of recovery through an innovative simulation model. Building on the empirical literature, they first present a comprehensive yet parsimonious conceptual framework for disaster recovery. In contrast to conventional approaches, this framework emphasizes interactions between households, businesses, infrastructure systems, neighborhoods, and community government. This work is also distinctive in its attention to urban spatial disparities in recovery. Chang and Miles use Object Modeling Techniques (OMT) to simulate the disaster recovery process for individual agents and an earthquake-stricken community as a whole. Their findings reveal how interactions between agents and across scales may magnify and complicate the impacts of public sector decisions regarding mitigation investment, emergency response, and recovery. The rapid restoration of transportation systems, for example, is the key to facilitating urban disaster recovery.

The economic importance of elevated highway systems in particular, which carry substantial volumes of commodities and people on a daily basis, has been strikingly demonstrated in recent earthquake disasters. Highway disruption was particularly noteworthy in the 1989 Loma Prieta earthquake that struck the San Francisco Bay Area, the 1994 Northridge earthquake in Los Angeles, and the 1995 earthquake that struck Kobe, Japan. Inspired by these events, Gordon *et al.* (chapter 11) develop an integrated, operational model for Los Angeles that emphasizes how damages to transportation and to industrial production capacity affect the metropolitan economy. This model links a spatial Input-Output model with a transportation network and earthquake damage model. Their ultimate goal is to present a full-cost approach to estimating disaster impacts that can help policy makers evaluate the costs and benefits of earthquake retrofit and reconstruction strategies. In this chapter, Gordon *et al.*

estimate that a magnitude 7.1 scenario earthquake could cost the region – as a lower bound estimate – in excess of $94 billion. This includes not only structural damage and business interruption, but also the repair costs, increased freight and travel costs, and changes in travel behavior associated with highway system damage.

From a regional perspective, the Kobe earthquake also demonstrated that the impacts of a catastrophic disaster can spread through interregional trade to other regions. Sohn *et al.* (chapter 12) employ and calibrate an interregional commodity flow model for analyzing how transportation damage in a hypothetical earthquake may cause economic impacts across the United States. The model evaluates transport cost increases as well as reduced final demands due to the disruption of transportation networks. It is innovative in being able to deal with both highway and railroad networks in an interregional context and in allowing the partial closure of damaged bridges and links. As with the Gordon *et al.* model, it can be utilized to assess the economic impacts of a disaster and to support cost-benefit analysis of retrofit strategies by policy makers and planners.

As with transportation systems, other infrastructure lifelines such as electricity and water networks also provide essential services in a disaster situation. From a largely engineering perspective, Shumuta (chapter 13) develops a benefit-cost methodology for renewal planning of electric power systems. A case study is presented for an actual system in Japan. Shumuta's approach is distinctive in considering disaster risk not as an isolated problem, but rather within the context of infrastructure renewal planning. It also differs from others in this volume by evaluating system performance in the face of multiple hazards, including earthquakes, typhoons, and environmental factors associated with equipment deterioration. Moreover, Shumuta pays particular attention to assessing uncertainties related to such factors such as earthquake occurrence probabilities.

The following chapter by Shinozuka and Chang (chapter 14) also analyzes electric power systems, but places greater emphasis on measuring the social and economic impacts of power loss in disasters. This chapter introduces the concept of seismic resilience of infrastructure systems in general and of utility power systems in particular. A comprehensive methodology is developed for analyzing disaster resilience, from equipment damage to system functionality to social and economic impacts. The model is applied to the Los Angeles Department of Water and Power's system after the 1994 Northridge earthquake. Based on the analysis, a useful set of data is derived, and the robustness and rapidity for restoration of electric power networks and grids during and after a catastrophic event are investigated. Results are represented in the novel form of risk curves. The risk curves enable analysts to compare expected electric power performance against criteria of acceptable risk, as well as to evaluate the benefits of seismic retrofit investments.

1.5 Final Thoughts

The papers presented in this volume demonstrate the vast strides that have recently been made in modeling the spatial economic impact of disasters. Particularly important advances have been made in methods that address economic resiliency, dynamic adjustment processes, and integration of the physical and economic consequences of disasters. Collectively, researchers have addressed many of the shortcomings of traditional models in the disaster context that were outlined earlier in this chapter, while several challenges still remain; for example, more research is needed on long-term effects, distributional impacts, and common modeling frameworks.

Many of the models developed in this volume are applied to a specific type of disaster, and most can, to some extent, be generalized to become applicable to other types of disasters. It should be recognized, however, that each disaster poses a different set of threats and potentially entails different types of impact. For example, the duration of hazards such as earthquakes and hurricanes is relatively short, whereas floods can linger for months. The extent, to which modeling strategies are transferable across natural disasters, or between natural disasters and other major dislocations such as terrorism events or sudden economic change, remains to be explored. Recent efforts toward the development of a multi-hazard loss estimation model by the U.S. Federal Emergency Management Agency (FEMA) indicate the currency of this issue.

Better models are needed to improve the effectiveness of disaster reduction efforts. In a recent synopsis of the state of natural hazards research, Mileti (1999) claimed that:

> *Many disaster losses – rather than stemming from unexpected events – are the predictable result of interactions among three major systems: the physical environment, which includes hazardous events; the social and demographic characteristics of the communities that experience them; and the buildings, roads, bridges, and other components of the constructed environment. Growing losses result partly from the fact that the nation's capital stock is expanding, but they also stem from the fact that all these systems – and their interactions – are becoming more complex with each passing year.* (p.3)

While various disciplines—including engineering, economics, geography, sociology, and emergency management—have long histories of disaster research, multidisciplinary efforts that comprehensively investigate disasters are still at an early stage. Moreover, until recently, there have been few efforts to model how disasters affect urban and regional economies. It is our hope that by enhancing the understanding of disasters and encouraging multidisciplinary collaboration across fields, this collection may ultimately contribute to reducing the toll of disasters.

References

Chang, S.E. 2000. "Economic Impacts," chapter 9 in Lee, George C. and Chin-Hsiung Loh, eds., *The Chi-Chi, Taiwan, Earthquake of September 21, 1999: Reconnaissance Report, Technical Report MCEER-00-0003*. Buffalo, NY; Multidisciplinary Center for Earthquake Engineering Research: 101-114.

Cochrane, H. 1974. "Predicting the Economic Impact of Earthquakes," in H.C. Cochrane, *et al.* eds. Social Science Perspectives on the Coming San Francisco Earthquake, National Hazards Research Paper, No. 25, NHRAIC. Boulder, CO; U. of Colorado.

Cochrane, H. 1997. "Forecasting the Economic Impact of a Midwest Earthquake," in Barclay G. Jones ed. *Economic Consequences of Earthquakes: Preparing for the Unexpected*. Buffalo, NY; National Center for Earthquake Engineering Research: 223-247.

Dacy, D.C. and H. Kunreuther. 1969. *The Economics of Natural Disasters: Implications for Federal Policy*. New York, NY; the Free Press.

Eguchi, R.T., J.D. Goltz, C.E. Taylor, S.E. Chang, P.J. Flores, L.A. Johnson, H.A. Seligson, and N.C. Blais. 1998. "Direct Economic Losses in the Northridge Earthquake: A Three-Year Post-Event Perspective," *Earthquake Spectra*, 14 (2): 245-264.

International Federation of Red Cross and Red Crescent Societies (Red Cross). 2002. *World Disasters Report 2002*. Geneva, Switzerland.

Kunreuther, H. and R.J. Roth, Sr. eds. 1998. *Paying the Price: The Status and Role of Insurance Against Natural Disasters in the United States*. Washington, DC; Joseph Henry Press.

Mileti, D.S. 1999. *Disasters by Design: A Reassessment of Natural Hazards in the United States*. Washington, DC; Joseph Henry Press.

National Research Council (NRC), Committee on Assessing the Costs of Natural Disasters. 1999. *The Impacts of Natural Disasters: A Framework for Loss Estimation*. Washington, DC; National Academy Press.

New York City Partnership and Chamber of Commerce (NYC Partnership). 2001. *Working Together to Accelerate New York's Recovery: Economic Impact Analysis of the September 11th Attack on New York City*. http://www.nycp.org.

Okuyama, Y., G.J.D. Hewings, T.J. Kim, D.E. Boyce, H. Ham, and J. Sohn. 1999. "Economic Impacts of an Earthquake in the New Madrid Seismic Zone: A Multiregional Analysis," in William M. Elliott and Peter McDonough eds. *Optimizing Post-Earthquake Lifeline System Reliability*, Technical Council on Lifeline Earthquake Engineering, Monograph No.16: 592-601.

Rose, A., J. Benavides, S.E. Chang, P. Szczesniak, and D. Lim. 1997. "The Regional Economic Impact of an Earthquake: Direct and Indirect Effects of Electricity Lifeline Disruptions," *Journal of Regional Science*, 37 (3): 437-458.

Scawthorn, C., B. Lashkari, and A. Naseer. 1997. "What Happened in Kobe and What if it Happened Here?" in Barclay G. Jones ed. *Economic Consequences of Earthquakes: Preparing for the Unexpected*. Buffalo, NY; National Center for Earthquake Engineering Research: 15-49.

Tierney, K.J., M.K. Lindell, and R.W. Perry. 2001. *Facing the Unexpected: Disaster Preparedness and Response in the United States*. Washington, D.C.: John Henry Press.

West, C.T. and D.G. Lenze. 1994. "Modeling the Regional Impact of Natural Disaster and Recovery: A General Framework and an Application to Hurricane Andrew," *International Regional Science Review*, 17 (2): 121-150.

Whittow, J. (1979) *Disasters: The Anatomy of Environmental Hazards*. Athens, GA; University of Georgia Press.

Part I: Conceptual and Modeling Issues

2 Economic Principles, Issues, and Research Priorities in Hazard Loss Estimation*

Adam Rose

Department of Geography and Natural Hazards Research Center, The Pennsylvania State University, USA
e-mail: azr1@psu.edu

2.1 Introduction

The quantification of economic losses from natural and manmade hazards is necessary to gauge individual and community vulnerability, evaluate the worthiness of mitigation, determine the appropriate level of disaster assistance, improve recovery decisions, and inform insurers of their potential liability. Several notable studies dealing with hazard loss estimation have recently been undertaken. These include chapters in surveys by the National Research Council (NRC, 1999; Mileti, 1999) and Heinz Center (2000), as well as various case studies (see, *e.g.*, Cole, 1995; Tierney, 1997; Shinozuka *et al.*, 1998; Gordon *et al.*, 1998; Chang *et al.*, 2001).

The purpose of this chapter is to identify major issues surrounding conceptual and empirical aspects of natural hazard loss estimation and to specify future research needed to resolve them. This includes clarifying basic economic principles of loss estimation, such as the need to consider both property damage and business interruption, the distinction between direct and indirect losses, property damage and business interruption losses, and real resource costs and transfers. It emphasizes the importance of the spatial and temporal context in which a natural hazard takes place, and the fact that hazard losses are highly variable because of business/consumer resiliency and public policy. The chapter also summarizes major modeling approaches and suggests that they are outpacing much needed data gathering. Finally, we explore traditional areas that have been neglected far too long, such as distributional impacts, as well as new areas especially worthy of attention, such as sustainability. Although the emphasis is on natural hazards, the chapter is applicable to manmade disasters as well.

---

* The author is grateful to the editors and to an anonymous reviewer for helpful suggestions for improving the chapter. However, the author is solely responsible for any remaining errors and omissions. Also, the opinions presented are solely those of the author and do not necessarily represent those of the institution with which he is affiliated.

2.2 Basic Principles

Welfare economics, the scientific basis for economic policy-making (see, *e.g.*, Boadway and Bruce, 1985), provides a starting point for an analysis of economic losses from natural and manmade hazards. A major point is that cost should be measured in terms of the value of resources used (or destroyed) and at prices that represent their efficient allocation, and not necessarily at market prices, which often do not account for inefficiencies or may not even exist in cases such as environmental resources. This provides a guide for avoiding double-counting and covering all resources, including non-market ones. Business interruption losses represent a proxy for the ideal resource valuation because of measurement problems and because businesses, insurers, and governments typically make decisions on the basis of such metrics as lost sales or profits.

Economists distinguish between gross output, the total value of production or sales, including the production of intermediate goods (industrial goods used to produce other goods), and net output, the value of final products. On the income side, net output is equivalent to the return to primary factors of production (labor, capital, natural resources, in the form of wages, profits and royalties). This is sometimes confusing because gross national product is really a net measure, except that it includes depreciation. When depreciation is subtracted, the quantity is referred to as net national product. Business interruption losses are in gross terms if measured by lost production or sales, and they are in net terms if measured by lost wages, profits, and royalties. The matter is further complicated when what economists call "welfare" (well-being) measures are calculated, typically using the concepts of producer and consumer surplus (see Zerbe and Dively, 1994). The former is equivalent to economic profits, or net returns of business (including deducting a market rate of return on investment and deducting depreciation). The latter includes consumer satisfaction from goods and services in excess of their market price, a concept very difficult to measure. It is no wonder that concepts like sales are used as a proxy in everyday decision-making.

2.2.1 Stocks versus Flows

One of the fundamental distinctions recognized in economics is between stocks and flows. Stocks refer to a quantity at a single point in time, whereas flows refer to the services or outputs of stocks over time. Property damage represents a decline in stock value and usually leads to a decrease in service flows. Business interruption losses are a flow measure, but emanate only in part from a company's own property damage.

Though property damage estimates have dominated loss reporting until recently, flow measures are superior to stock measures in many ways. First, direct business interruption losses can take place even in the absence of property damage, and hence represent broader coverage of the scope of losses. For example, a factory may be unscathed by an earthquake, but may be forced to shut down if its electricity supply is cut off due to earthquake-induced damage to power stations, substations, transmission lines, or distribution lines.

The value of an asset is the discounted flow of net future returns from its operation. Hence, for ordinary property damage the stock and flow measures represent the same things, and, at first pass, including both would involve double-counting. The situation is, however, complicated in the case of natural hazards. It would be straightforward if a machine with a one-year lifespan were destroyed, and if it were not replaced for a year. Here we could either take the machine value or the value of lost production, though again not both. (Note the ideal machine value is an imputed measure of worth and not necessarily equal to original purchase cost or replacement cost both adjusted for depreciation.) The more likely situation is one where the machine has a useful life of ten years, and is out of service for only three weeks. Repair cost and the value of lost production are not likely to be equal, and the latter is the preferable measure. The analyst might be tempted also to just settle for the larger of the two—property damage in this case. That would double-count the direct flow loss; hence it should be subtracted. This is equivalent to adjusting an asset value for the depreciation that would have taken place for the relevant period of production.[1]

A third reason flow measures are superior is that they are more consistent with indices of individual wellbeing, such as business profits and consumer satisfaction, or with aggregate measures, such as gross national (or regional) product. In this regard, property damage measures can exaggerate losses because only a portion of the property value translates into service flows in any one year. A final reason flow measures are superior is that they are more readily linked to indirect effects.

2.2.2 Double-Counting

In addition to some stock/flow overlaps, care should be taken to avoid other types of double-counting of hazard losses. Many goods and services have quite diverse attributes, and all of those damaged/interrupted should be counted (*e.g.*, a hydroelectric dam provides electricity, recreational opportunities in the reservoir behind it, and flood control). It is important, however, to remember that some goods and services cannot yield all of these attributes to their maximum simultaneously, and that only one or the other, or some balance of the two, should only be counted (*e.g.*, a river can provide services to swimmers or it can be a repository for waste but not both at the same time).

Double-counting can be avoided by not attributing losses to more than one entity in the case of private goods, as in the case of avoiding counting retail store sales as a loss to both the storeowner and its customers. Just as important, however, is the inclusion of all relevant losing entities or stakeholders. Caution must be exercised here because of the regional character of most hazards and the inclination just to consider those living within its boundaries. Tourism associated with natural environments is an excellent case in point. Loss of environmental value should not just be gauged by local residents but by all potential users. This problem arises most readily when losses are measured by a questionnaire administered to those on site.

A closely related consideration pertains to the distinction between costs and transfers. If the expenditures needed to build a hydroelectric dam, in part for flood protection or to repair earthquake damage to a bridge, are \$10 million, and 5% of the costs were various types of taxes (sales, import tariffs, property, etc.), tax expenditures do not reflect the use of resources and are not real costs to society. In general, such

[1] This is a controversial subject. I am in agreement with analysts who suggest it is appropriate to include both the stock and flow measures in the case of damaged property, where the latter represents the opportunity costs of *delays* in restoring production.

taxes are important to individual businesses, but simply represent a shifting of dollars from one entity to another. The complication that arises here, however, pertains to the spatial delineation of the affected group. Local property or sales taxes within a region are transfers, but payments of federal income tax do represent an outflow and can be legitimately included in the regional cost estimates. Of course, there is the danger of being too provincial in such assessments.[2]

2.2.3 Direct versus Higher-Order Effects

The distinction between direct and indirect effects has been the subject of great confusion from the outset. For example, the characterization that direct loss pertains to property damage and indirect loss pertains to business interruption (see, *e.g.*, ATC, 1991; Heinz Center, 2000) is not helpful because both have direct and indirect components.[3] While total business interruption losses are the bottom line, distinguishing components helps ensure everything is counted and provides more precise information for decision-making (*e.g.*, as illustrated below, direct effects usually pertain to private concerns of individual businesses, while indirect effects raise additional public policy issues).

Direct flow losses pertain to production in businesses damaged by the hazard itself, or what the NRC (1999, p. 15) study refers to as the "consequences" of physical destruction, though without distinguishing direct vs. indirect components as does Mileti (1999; p. 98). They have also come to include lost production stemming from direct loss of public utility and infrastructure services (Rose *et al.*, 1997). For example, earthquake-induced disruptions of water supplies may force the closing of a high-rise office building for fire safety reasons (fire engine hoses can only reach the first several floors, and the remainder of fire control is dependent on internal sprinkling systems). A factory may have to shut down because the bridge that its suppliers and employers use to reach it is damaged. Again, the office building and factory may not suffer any direct physical damage.

The extent of business interruption does not stop here, but sets off a chain reaction. A factory shutdown will reduce supplies to its customers, who may be forced to curtail their production for lack of critical inputs. In turn, their customers may be forced to do the same, as will the customers of these customers, and so on. These types of effects are called downstream, forward, or supply-side linkages. A set of counterparts refers to upstream, backward linkage, or demand-side indirect effects. The factory shutdown will also reduce orders for its inputs. Its suppliers will then have to reduce their production and hence cancel orders for their inputs. The suppliers of the suppliers will follow suit, and so forth. The sum total of all of these indirect

---

[2] Some taxes, such as property taxes, do reflect an indirect payment for services, such as water and sewer, but tariffs and sales taxes do not. Property taxes would only be included in the resource cost tabulation if the water and sewer services were actually used in the construction of the hydroelectric dam and then only at a level commensurate with the service costs.

[3] Indirect effects can also be associated with stock losses or property damage (e.g., damage from fires caused by earthquakes, traffic accidents due to failed traffic signals, and buildings made more vulnerable to subsequent weather damage). However, except in extreme cases, such as the San Francisco Earthquake of 1906, these direct stock effects are likely to be relatively small when compared with the flow-induced indirect losses. NRC (1999) makes a useful distinction by referring to such instances as "secondary direct" losses, which helps distinguish them from indirect flow losses.

effects is a multiple of the direct effects; hence, the concept of a "multiplier" is often applied to their estimation (Rose and Miernyk, 1989; and FEMA, 2001).[4]

An important distinction is a terminological one relating to the term "indirect effects," which has come to be known as all economic impacts beyond direct (including intangibles). A problem arises because much modeling of region-wide losses for natural hazards has been undertaken with the use of input-output (I-O) models, which have their own long-established terminology (see, *e.g.*, Miller and Blair, 1985; and Rose and Miernyk, 1989). In I-O parlance, indirect effects refer only to interactions between businesses, while the interactions between consumers (through lost wage and profit income and consequent reduced spending) and businesses are referred to as "induced" effects, though the NRC (1999) study utilizes this term only to refer to the income side (p. 17). Usually, in I-O parlance, the sum of the two effects is referred to as "second-order" or "higher-order" effects, to avoid using "indirect" in two different ways. All of this is further complicated by the fact that an advanced modeling approach, computable general equilibrium (CGE) analysis, is gaining prominence in hazard impact analysis (see, *e.g.*, Boisvert, 1992; and Rose and Guha, 2004). It is able to model a broader range of higher-order impacts, typically referred to as "general equilibrium" effects, which, rather than being confined to economic interdependence (based solely on quantities of inputs and outputs), capture responses to price changes in factor and product markets (Shoven and Whalley, 1992; and Rose, 1995).

Thus, I propose the use of the term "higher-order effects" to cover all flow losses beyond those associated with the curtailment of output as a result of hazard-induced property damage in the producing facility itself. This term overcomes the misuse of I-O parlance and also is intended to be general enough to include both I-O quantity interdependence and general equilibrium price interdependence effects.

Many analysts are hesitant to measure higher-order losses for various reasons. First, they cannot be as readily verified as direct losses. Second, modeling them requires utilizing simple economic models carefully, or, more recently, utilizing quite sophisticated economic models. Third, the size of higher-order effects can be quite variable depending on the resiliency of the economy and the pace of recovery (see, *e.g.*, Rose *et al.*, 1997, as well as the discussions and illustrations below). Fourth is the danger of manipulating these effects for political purposes (*e.g.*, it is not unusual in the context of economic development for promoters to inflate multipliers). However, none of these reasons undercut the importance of higher-order effects, especially if one considers their likely size (see, *e.g.*, Cochrane, 1997; Webb *et al.*, 2000; Bram *et al.*, 2002).

[4] Some further clarification is in order. First, the current line of demarcation between direct and indirect effects is somewhat arbitrary, specifically, the convention of counting business losses due to cut-off from utility lifelines as direct effects. There is equal justification for considering these to be first-round indirect effects. The advantage to the distinction made here is that it emphasizes the key role of utilities and infrastructure in the economy, and emphasizes their prominent role in contributing to losses. Also, it helps ensure that these effects will be taken into account, because most analysts are not able to or do not bother to consider indirect effects.

2.3 Current Issues

2.3.1 Non-Market Effects

Hazard researchers are becoming increasingly aware of the ever-broader scope of disaster losses. The recent Heinz Center Report (1999) does an excellent job of enumerating their extent, including categories of Social, Health and Safety, and Eco-System costs. Most of the losses in the latter category, as well as a significant portion of losses to one of the other two categories identified in the Heinz Report—the Built Environment—are characterized by economists as "non-market." This means they are not bought or sold and hence do not readily have a price tag. However, just because something does not have a price does not mean it does not have value; it simply means a "market failure" has occurred. In this case, a market has failed to form, and absence of prices will cause resources to be misallocated.

The major area of attention to non-market aspects of natural hazards to date has been on one part of the built environment—public infrastructure, such as highways/bridges and utility lifelines (electricity, gas, and water). Non-market effects arise here primarily because the former category is publicly (rather than privately) owned, and hence services are provided free of charge, or because both categories have features of decreasing cost activities (natural monopolies), and appropriate pricing is made difficult (see also Howe and Cochrane, 1993).[5]

The various flow impacts of natural hazards on the public sector built environment have been termed "infrastructure user costs" (see Rose *et al.*, 1998). For the case of a highway washed away by a flood, there is no direct production loss measure, *e.g.*, no lost public highway "sales," except in the case of toll roads, where the toll is not necessarily an accurate measure of lost value in any case. Direct losses would, according to the convention noted earlier, best be represented by lost revenue of businesses that are required to shut down because their employees could not get to work, inputs could not be accessed, or outputs could not be delivered.

Several other non-market direct impacts take place, however, as do conventional market and unconventional non-market higher-order impacts. Commuters are

---

[5] Both ecosystem losses and public infrastructure losses arise due to what economists call "public goods," which have the characteristics that two or more people can utilize the services of the good simultaneously without detracting completely from one another, and from which people cannot be excluded because it is technologically impossible, socially unacceptable, or economically impractical. Major examples of public goods are national defense, television broadcasting signals, national parks, and environmental resources in general. This is in contrast to more typical "private goods," which are utilized by one person at a time and for which a price can readily be extracted (e.g., clothing, restaurant meals, etc.). Not all public goods are provided by government; some are provided by the private sector under the right circumstances, and most environmental goods are provided by nature. There is considerable momentum to reduce the number of goods and services provided by government, even for what were previously thought to be public goods. This typically involves enhancing the "excludability" characteristics so that a user fee (not be to confused with user cost) can be charged. This is not necessarily simple since efficient pricing would actually require that different users be assessed different changes, according to their marginal willingness to pay. Another complication is that some goods have different values and different degrees of "publicness" at different times (a classic example is a road, which can accommodate traffic at zero cost during normal hours, but that is subject to congestion, which imposes costs on all users during peak hours). Several remedies to this situation have been proposed, as well as for the more complicated situation where periods of congestion (and hence increasing costs) exist. All of these remedies require a careful scrutiny to make sure that the price charged represents true valuation of the resources used.

adversely impacted by transportation outages through loss of time due to congestion (even the subsequently decreased leisure time has a value); however, there are no multiplier effects associated with this activity. On the other hand, the loss of productivity to producers or transportation companies results in cost increases that have price multiplier effects first (a form of "cost-push" inflation) and output multiplier effects subsequently. Consumers may also curtail their shopping trips due to bridge or highway outages. These decreases in direct consumption also generate higher-order effects (see, *e.g.*, Gordon *et al.*, 1998).

For the case of utility lifelines, direct and indirect production losses are likely to be the major loss category. Also, productivity losses are likely to occur due to downtime or declines in product quality and will spawn multiplier effects as in the transportation example. Decreases in household activity (reduced showers, reading time, cooking) are not part of economic indices, but they should be considered in broader measures of well-being, though multiplier effects are not applicable.[6] The consumer side is important but lifeline disruptions will have little effect on shopping over and above that attributable to business operation itself. For example, if a power outage causing the closure of a department store were listed as a direct output (sales) loss for the producer, it would be double-counting if included as a consumption loss as well.

The largest potential area of non-market losses pertains to the natural environment, ranging from conventionally marketed economic activity, such as agriculture and forestry, but extending to damages to the environment in general, even including "option value" (in part, the value one places on potential access to the resource in the future). An extensive literature on non-market valuation exists (see, *e.g.*, Freeman, 1993) but has been virtually unnoticed by hazard researchers (the one exception is Howe and Cochrane, 1993), though it is a major focus of the closely related area of research on climate change (see, *e.g.*, Fisher *et al.*, 2000; Oladosu, 2000). "Eco-System" costs, however, are beyond the scope of this chapter. Note that while climate change is usually characterized by long-term warming, it also gives rise to short-term climate variability, which manifests itself in an increasing number of hurricanes and other types of severe storms that can lead to direct or indirect losses through water or wind damage (see IPCC 2000; Rose *et al.* 1999).

2.3.2 Timing and Recovery

Another reason flow measures are superior to stock measures is that the former include a time dimension. Stock measures pertain simply to the value of an asset at a single point in time. The typical measure of damage (purchase or replacement cost) is thus invariant to how long the asset is out of service. For example, if the roof of a factory is blown off by a hurricane, there is a tendency to specify the loss in fixed terms, irrespective of whether production is shut down for a week or a year awaiting repairs.

Attention to flow losses represents a major shift in the focus of hazard loss estimation—that losses are not a definite or set amount but are highly variable depending on the length of the "economic disruption," typically synonymous with the recovery plus reconstruction periods. This also brings home the point that disaster

[6] Property damage to residential structures also has a flow counterpart, termed the "imputed rental value of owner-occupied dwellings." This non-market cost might be measured as well; it has no higher-order effects, except those associated with payments for temporary shelter.

losses are not simply determined by the strength of the stimulus (coupled with initial vulnerability), but also highly dependent on human ingenuity, will, and resources. Caution should be exercised, however, before rushing toward minimizing losses without consideration of the increased recovery costs incurred. The broader objective is to minimize the joint cost of impacts and recovery/reconstruction. Fortunately, economists have recently identified a set of costless, or near costless, measures to greatly reduce losses during the recovery period (see, *e.g.*, Rose *et al.*, 1997). These include both market (private sector) and non-market (public policy responses) to be discussed in the following two subsections.

2.3.3 Resiliency

Another aspect of post-disaster response is "resiliency," which researchers have long found to be prevalent in dire situations surrounding natural hazards (see, *e.g.*, White and Haas, 1975). Resiliency can be roughly defined as the ability to cushion or mute potential losses from a natural hazard (see Tierney, 2001, for an in-depth analysis).[7] It applies to individual firms, households, and institutions, and also to the community or economy as a whole (Bruneau *et al.*, 2002). At the level of the individual firm, for example, it relates to some ordinary business practices, such as the ability to substitute one type of fuel for another or to obtain inputs from alternative suppliers (from within or outside the impacted region) in the event that a long-standing supplier is temporarily out of business. In addition to these inherent possibilities are additional adjustments that reflect adaptive behavior under stress (see also Comfort, 1999, for examples within organizations). For example, enterprises are inspired to ingenuity in substituting inputs or to extra effort at conserving inputs in short supply.

Rose and Liao (2002) have defined firm level resiliency as the difference between actual direct output loss and some projection of potential loss. An example of the latter would be the linear proportional loss of output equivalent to the loss of a critical input. For example, if a chemical company sustains a 40% loss of electricity, this would translate into a 40% reduction in production for the disruption period. The result would exclude the possibility of resiliency responses noted about. Of course, it is somewhat arbitrary, and one can make the case that a severe earthquake would cause business disorganization beyond that associated with the reduction in the critical input, thereby making a non-linear baseline appropriate.

Studies have shown that individual business resiliency can be sizable. For example, Tierney (1995), through a questionnaire survey, found that an 8.3 percent loss of electricity at the peak of the Northridge earthquake resulted in only 1.9 percent loss of direct output in average for firms in the affected area. Using a simulation

---

[7] Resiliency is closely related to the concept of "adaptation," or how society copes with a disaster after it takes place. Adaptation is often thought of as a more passive "suffering through it" response but in fact has come to include more active coping measures, such as those examples of resiliency presented earlier in the text. Even passive adaptation may involve an implicit (non-market) cost. For example, living in a warmer atmosphere detracts from the quality of our life, though at some point many species are able to survive and thrive as before and eventually get used to changing conditions, at which point adaptation damages cease. At the same time, it should be emphasized that adaptation of living species best leads to such positive outcomes only if a sufficiently long period transpires, a period very much longer than is likely associated with current projections of global warming. "Resiliency" as used in this chapter might best be thought of as short-run adaptation.

model, Rose and Lim (2002) estimated this output loss to be only 0.42 percent, or a 95 percent reduction over the linear baseline.

Resiliency is much more complex at the aggregate level (see, *e.g.*, Tierney, 2001). Here we confine out attention to the regional economy because broader aspects of the social fabric are beyond the scope of this chapter. One definition would be the macroeconomic analog to that applied to individual firms above (see also Rose and Liao, 2002). This would be the difference between region-wide output taking into account higher-order (indirect, induced, general equilibrium, etc.) effects and those projected by a linear model such as input-output analysis (see also section 2.4.2 below). Typically, for a major metropolitan region, this involves a multiplier between 2 and 3 times the direct loss. Simulation analyses utilizing computable general equilibrium models, however, typically estimate the higher-order losses to be closer to 1.5 (see, *e.g.*, Rose and Guha, 2004). The difference between the two can be attributed to regional economic resiliency, based on adjustments through markets and prices that induce a more efficient allocation of limited resources in the effected economy.

2.3.4 Public Decision-Making

Another major consideration affecting losses is public policy, which impacts not only on the public sector (in terms of resiliency) but on businesses and households as well. Losses can be minimized by a host of policy measures including: substitution of public services for private ones, maintaining civil order, providing health services, providing financial assistance for recovery and reconstruction, and providing information to supplement or substitute for market processes.

A major example is put forth by Rose and Benavides (1998; 1999), pertaining to the reallocation of electricity resources across sectors and sub-regions in the aftermath of an earthquake. For example, a simulation of an electricity disruption following a major New Madrid Earthquake indicated that losses could be decreased by more than 70% if electricity were rerouted to those sectors contributing most to gross regional product in terms of direct and indirect use of electricity inputs. Improvements of another 10% would be forthcoming if the order of restoration of substation services were made on the basis of regional economic considerations rather than simple repair cost considerations. Such resource allocations can be accomplished by government decree or through private and public sector cooperation in establishing a clearinghouse for information about the availability of and demand for goods and services (*e.g.*, one which matches suppliers without their traditional customers to customers without their traditional suppliers). Public sector involvement need not be heavy-handed and resemble central planning. A similar outcome can be achieved by using market mechanisms, such as non-interruptible service premia, which are already prevalent in the electric utility industry. The work of Rose and Benavides (1999), however, indicates that a market failure may currently exist in this institution because premia are based on private business considerations and do not typically take into account indirect effects on suppliers and customers, thereby understating potential losses from electricity interruption. A similar market failure exists with respect to contagion effects, *e.g.*, where protection against terrorism by one apartment owner is undercut by failure of neighboring owners to do so (see, *e.g.*, Kunreuther and Heal, 2002).

Again, we emphasize that such institutions or policies are near costless to administer. Not taking advantage of them represents a missed opportunity to reduce losses and is as devastating as if the hazard in question had actually toppled the buildings in which the potentially avoidable lost production would have originated.

2.4 Loss Estimation Model Assessment and Validation

2.4.1 Basic Approaches

There are several broad categories of hazard loss estimation methodologies (see, *e.g.*, section 2.1). Though no formal assessment and comparison of the accuracies of all the methodologies has been undertaken, it is generally considered that the sounder the data, the more reliable the results. In general, approaches based on primary data are more applicable to direct loss estimation, and other methods more applicable to higher-order loss estimation. Also, loss estimation is used both to evaluate losses after they have taken place (retrospective) and to predict potential losses (prospective). Primary data techniques are typically being better suited to the former and simulation techniques better suited to the latter. Here, we will summarize several approaches and then discuss the more comprehensive approaches to hazard loss estimation in detail in the following subsections.

Many firms collect primary data on hazard losses both for their own internal assessments and as the basis for applications for external aid. Uniform procedures do not exist for the collection of these data, and hence these compilations suffer somewhat from inconsistencies. Also, they typically do not distinguish direct vs. higher-order losses. Moreover, findings are often considered proprietary and not readily available to analysts or policy-makers. The use of "self-collected" primary data on loss estimation has thus far been limited. However, the establishment of standardized definitions and reporting procedures, as well as confidentiality guarantees, could make its use widespread.

Primary data are also collected through questionnaire, interview, and telephone surveys. Responses are sometimes taken from internal reporting documents, but questionnaires can raise inquiries about broader matters as well, such as high-order losses (see, *e.g.*, Tierney, 1997). Use of the questionnaire approach has been somewhat limited because of its cost, but it has the potential for widespread use. This approach is also frequently used to measure environmental impacts by eliciting information on willingness to pay for environmental resources, as in contingent valuation studies (see, *e.g.*, Mitchell and Carson, 1989; Hammitt *et al.*, 2001).

Secondary data compilations refer to typically published data by government agencies, philanthropic organizations (*e.g.*, Red Cross), private companies or associations (*e.g.*, insurance industry) and independent researchers (see, *e.g.*, West and Lenze, 1994). Many are based on primary data, but the aggregate tabulation poses some limitation on use. Moreover, the analyst is not always informed about definitional, measurement, or other complications that may bias the data when using them.

Data adaptation/transfer refers to using information about one hazard event to provide a quick estimate of losses from another (see, *e.g.*, Boyle and Bergstrom, 1992). It is often based on ad hoc procedures or mere rules of thumb, and at the very least should correct for differences in physical magnitude of a hazard, economic character of the region, and population size. This can readily be formally accomplished by using statistical techniques.

Statistical analysis represents an attempt to establish patterns in primary and secondary data. It is an inferential approach to loss estimation that fills in gaps in knowledge and provides greater insight into the event. It includes both exploratory analyses, such as correlation, and causal analysis, such as regression. The latter explicitly specifies explanatory variables that can help in prediction or quick loss estimation of other hazard events in a more formal "data transfer function" (see, *e.g.*, Cochrane, 1997; Kirchoff *et al.*, 1997). Macroeconometric models and more elaborate versions of statistical models include sets of simultaneous equations characterizing the entire economy and are capable of forecasting future economic growth and departures to it stemming from external shocks (see, *e.g.*, Ellson *et al.*, 1984).

Deterministic simulation analysis involves the use of a set of ad hoc formulas or consistent mathematical models based on formal theories and specific data. Examples at the microeconomic level include engineering-economic analysis (see, *e.g.*, White *et al.*, 1997; Newnan *et al.*, 2001), whereas examples at the macroeconomic level include input-output analysis, linear programming, and computable general equilibrium analysis (to be discussed in more detail below). These can be as extensive as the integrated assessment models involving stochastic estimation. Major limitations pertain to numerous simplifications inherent in this modeling approach and the fact the results are often presented as point estimates, unless sensitivity tests are performed.

The final category, stochastic simulation analysis, refers to the application of Monte Carlo or various other methods to compensate for a lack of data or data uncertainty (see, *e.g.*, Taylor *et al.*, 2001). Examples include decision analysis (see, *e.g.*, Resnick, 1987; Howard and Matheson, 1989) and uncertainty analysis (see, *e.g.*, Law and Kelton, 1991; Porter *et al.*, 2002). The approach can be applied to various aspects of the hazard loss process, including the original physical stimulus (*e.g.*, ground shaking), the workings of the built environment, vulnerability analysis, direct physical loss estimation, direct economic loss estimation, and less so to higher-order economic loss estimation. It often provides a range of possible loss estimates, which are preferable to a point estimate from a statistical standpoint, though at much greater cost from the standpoint of data and computing effort.

2.4.2 General Issues

In the following sub-section, we examine the validation of economic loss estimation by three comprehensive analytical modeling approaches: input-output, computable general equilibrium, and econometric. In the two former approaches, models are typically constructed by applying various "data reduction" or "down-scaling" techniques to secondary data because of the prohibitive expense of constructing them from primary data. More direct survey approaches might first appear superior to the abstraction of modeling higher-order losses. For example, a survey might be circulated simply asking firms to estimate their indirect losses from an earthquake, *i.e.*, those not associated with direct physical damage to their facilities or to direct lifeline curtailments. This would seem to obviate the expense of constructing a model, but it has two shortcomings. First, higher-order effects are so extensive and so often many steps removed from direct losses that it would be impossible for respondents to answer this question accurately. The situation is difficult enough for the case of input or output reductions and virtually impossible for responses to price changes. Second, the question is useful in hindsight for a one-shot assessment but not as useful as a model for analysis, prediction, and policy.

The accuracy of each loss estimation model depends on the context of the application of the model. To evaluate the potential severity of an earthquake or flood in a region, it may only be necessary to have a general idea of potential higher-order impacts, with a reasonable error of as much as 50 percent. For the sake of evaluating mitigation options, the confidence interval must be closed significantly. For the purpose of re-allocating resources during the immediate aftermath of a disaster or during near-term recovery, high degrees of accuracy are needed, because of the political sensitivity of this type of policy.

2.4.3 Summary of Comprehensive Methodologies

2.4.3.1 Input-Output Models

The input-output (I-O) model is the most widely used tool of regional economic impact analysis, and its application to natural hazards dates back to the seminal work of Cochrane (1974).[8] I-O is a static, linear model of all purchases and sales between sectors of an economy, based on the technical relations of production (Miller and Blair, 1985; Rose and Miernyk, 1989). Its focus on production interdependencies makes it especially well suited to examining how damage in some sectors can ripple throughout the economy. The Indirect Economic Loss Module of the HAZUS Loss Estimation Methodology, for example, is based on this approach (FEMA, 2001).

Other advantages of I-O are that it is: an excellent organizational framework for data collection and display, able to provide a transparent view of the structure of an economy, and capable of readily accommodating engineering data. Disadvantages of the basic I-O model include its linearity, lack of behavioral content, lack of interdependence between price and output, lack of explicit resource constraints, and lack of input and import substitution possibilities. In fact, the simplicity of I-O is sometimes misleading, and many of the inaccuracies associated with it are due to the inability to use it correctly, rather than the shortcomings of the model itself. Several decades of refinement of this approach have been able to overcome many of these problems (Rose and Miernyk, 1989). Any two or three of these refinements are manageable, but the comprehensive model that overcomes them all can be especially unwieldy, and this is one of the reasons for the accelerated interest in CGE models to be discussed below (Rose, 1995).

Several refinements of the I-O methodology have been suggested that improve accuracy, not of the underlying model, but of its application to hazard loss estimation. For example, Boisvert (1992) and Cochrane (1997) have developed methodologies for more flexible treatment of imports, which typically rise to compensate for shortages of regionally produced inputs in the aftermath of an earthquake (the latter has been incorporated into HAZUS). Cole (1988) has improved the assessment of the timeframe of regional impacts in general, and Cole (1998) and Okuyama (1999a and 1999b) have extended the area of coverage to account for economic ripple effects in

---

[8] Application of I-O to manmade hazards dates back to strategic bombing studies during World War II and extends to the recent World Trade Center attacks.

regions adjacent and non-adjacent to where the hazard occurred. Additional refinement and application of I-O has been undertaken for optimal recovery (Rose, 1980; Rose *et al.*, 1997; Cole, 1998; and Rose and Benavides, 1999), transportation impacts (Gordon *et al.*, 1998; and Cho *et al.*, 2001), lifeline impacts (Rose *et al.*, 1997), overall impacts (Hewings and Mahidhara, 1996; and Cochrane, 1997), incorporation of resiliency and time-phasing of impacts (Okuyama *et al.*, 2000; and Rose and Lim, 1997), and integrated system modeling (Shinozuka *et al.*, 1998).

A related consideration pertains to the point that not only is it desirable to have an accurate model but also to apply it in an appropriate way. A good case in point is the approach of using standard I-O multipliers irrespective of the length of the recovery period. For example, full multiplier effects are likely to take place only in short duration hazard situations, such as the Northridge earthquake, where all power was restored within 36 hours. Here the economy does not have much opportunity to adapt, and the full brunt of hazard impacts, except those likely to be mollified by inventories of non-electricity inputs (electricity cannot be stored), are likely to be felt. On the other hand, hazards of longer duration allow for more resiliency options, including the rearrangement of trading patterns. Application of ordinary multiplier analysis would lead to estimates of various types of higher-order effects that would further reduce electricity demand, even below its availability (an obvious waste of resources). However, a factory that was cut off from supply of a critical input because its producer was without electric power is likely to re-contract with another supplier (either inside or outside the region) of that input, whose customer could not absorb it due to loss of its electricity or other reasons for business interruption. In this context, higher-order effects could actually be near zero. An exception would be if one of the sectors of the economy that was much more excessively damaged than others caused a supply "bottleneck," so that even re-contracting would not allow full utilization of available resources (Rose *et al.*, 1997).

There are several approaches to incorporating uncertainty into I-O models. The first is to perform scenario analyses, essentially contingent forecasts, depending on a variety of conditions (*e.g.*, resource availabilities, resiliency, speed of recovery). Second is to alter major parameters of the model (*e.g.*, input and import coefficients) and to perform sensitivity tests of their influence on the results. Third is to include a probability distribution of key parameter values. The fourth is to perform a Monte Carlo simulation relating to the distribution of these parameter values (West, 1986). All of these approaches shift the attention from validating a point estimate result to providing a reasonable range of estimates as is becoming more prevalent in hazard impact analysis (see, *e.g.*, Cochrane, 1997).

All of these methods are somewhat ad hoc, since no definitive rules have been established for their application to the idiosyncrasy of I-O analysis. Still, some of them, such as Monte Carlo simulations, have well-established protocols in general. These approaches have a distinct advantage over a purely deterministic approach, as they provide a range of outcomes within which there is some chance of being correct. Point-estimates exaggerate the certainty of the analysis and will almost assuredly be incorrect.

Another way to assess the validity of model results is to examine inherent biases of the model itself. The two major ones for I-O are linearity and infinity elasticities of supply. The first permeates every facet of the workings of the model and implies no economies or diseconomies of scale and no input substitution. The absence of the latter will lead to upper-bound results, *e.g.*, a twenty percent decrease in electricity available in any sector would lead to a twenty percent reduction in that sector's output. Of course, input coefficients can be adjusted for some types of resiliency (*e.g.*, based upon empirical evidence on conservation to moderate the extreme outcomes), but explicit input substitution possibilities (*e.g.*, switching from piped to trucked water), are difficult to accommodate. The inflexibility of anything other than constant returns to scale, however, might lead to underestimating losses (as scale decreases due to capital stock damage or input curtailment, unit cost increases would not be reflected). Infinite supply elasticities overlook real world capacity limitations and are difficult to modify without transforming the I-O model into a linear programming format. This is not likely to lead to a problem in the basic aspects of loss estimation, since the impact activity is on the down-side. However, it will, in some cases, lead to an overstatement of positive impacts of resource spending because of capacity limitations in construction and related industries, which could spur inflationary pressures and subsequent dampening of production throughout the economy.[9]

Very few examples of actually testing the accuracy of I-O based indirect losses exist. One exception is Rose and Lim's (2002) analysis of the electricity lifeline impacts of the Northridge Earthquake, which yielded aggregate estimates reasonably close to those of a business interruption survey undertaken by Tierney (1997). Of course, accuracy of the contingent ("if-then") projections of any model will always be limited because of the many factors that are held constant but that change in the real world.

2.4.3.2 Computable General Equilibrium Models

The use of CGE models for impact analysis in general is rapidly increasing, especially at the regional level (see Partridge and Rickman, 1998). CGE is a multi-market simulation model based on the simultaneous optimizing behavior of individual consumers and firms in response to price signals, subject to economic account balances and resource constraints (see, *e.g.*, Shoven and Whalley, 1992). This approach is not so much a replacement for I-O as a more mature cousin or extension, and it retains many of the latter's advantages and overcomes most of its disadvantages (Rose, 1995). For example, CGE retains the multi-sector characteristics and emphasis on interdependence, but also incorporates input/import substitution, increasing or decreasing-returns-to-scale, behavioral content (in response to prices and changes in taste or preferences), workings of markets (both factor and product) and non-infinite supply elasticities (including explicit resource constraints). Moreover, the empirical core of most CGE models is an I-O table extended to include disaggregated institutional accounts, *i.e.*, a social accounting matrix (SAM).

At the same time, CGE models do have shortcomings, the major ones being the assumption that all decision-makers optimize and that the economy is always in equilibrium. The latter is not a problem when the period of analysis is more than one year and the external shock is small, but natural hazards have the opposite characteristics.

[9] The shortage of goods and services in the aftermath of a major disaster would seem to be ripe for price increases and, in fact, price gouging. However, for several reasons (beyond the scope of this analysis) there is little actual experience with systematic price increases after such events (reasons including the fear of short-term and long-term reprisals, economic resiliency, etc.).

Until recently, all applications of CGE models to hazards have been experiments with synthetic models (see Boisvert, 1992; and Brookshire and McKee, 1992). More realistic applications have been undertaken by Rose and Guha (2004) and Rose and Liao (2002) to impacts of utility lifeline disruptions in the aftermath of earthquakes. Potential application to other aspects of hazard loss estimation, however, is unlimited given the capabilities of this approach.

Just as I-O has inherent biases, so does CGE, and typically at the other extreme. I-O models are overly rigid and exaggerate hazard impacts, but the typical CGE model is overly flexible and understates them (Rose and Guha, 2004). Most input substitution elasticities are between .5 and 1.5, which would greatly understate the cost penalty or overstate the technical possibilities of input substitution (*e.g.*, of replacing natural gas with fuel oil in an industrial boiler in a period of just a few weeks). While dual-fired boilers exist, in most cases retrofitting would have to take place and fuel-oil delivery established. While basic I-O models indicate that a twenty percent curtailment in lifeline service would lead to at least a twenty percent disruption in sectoral output, experiments with standard CGE models indicate a drop in sectoral and regional output of only a couple of percentage points at most in relation to a twenty percent input curtailment, owing to the inherent ease of substitution of inputs and of imports for the shortage of regionally produced goods (Rose and Guha, 2004). It is imperative that modelers at the very least begin utilizing elasticities applicable to the "short-run", or, even better, "very short-run" context in performing any simulations with this modeling approach.

Another reason that CGE models provide lower input estimates than I-O models is, however, legitimate.[10] This pertains to the fact that not all causation in CGE models is unidirectional, *i.e.*, functional relationships often offset each other, the most vivid example being the relationship between price change and output changes, which is short-circuited in all but the most sophisticated I-O models.[11] The curtailment of an input will raise its price, further reducing its demand directly, as well as some indirectly related demands. At the same time, the initial reduction in these indirect demands will lower prices, thereby providing a demand stimulus. Therefore, the rules of thumb of multiplier impacts presented in the previous section do not apply, except for representing extreme upper-bounds in the context of CGE models.

A new way of improving the accuracy of CGE models for hazard impact analysis is to utilize survey data on resiliency and direct impacts. Rose and Liao (2002) have demonstrated how these data can be used to revise individual production function parameters of CGE models. The individually adjusted production functions are inserted and the entire model is recalibrated. It is then used to simulate the region-wide impacts of direct flow or stock damages, the difference between total impacts and direct impacts representing the higher-order (including general equilibrium) impacts.

Again care must be exercised in applying CGE models properly. I-O models, with an adjustment for inventories, are probably better suited to recovery periods of

[10] We should note that CGE models also potentially have an upward bias on loss estimates due to the fact that they often assume businesses are running at full capacity when, in fact, excess capacity often exists. However, input bottlenecks are more likely to be a problem than production capacity considerations. Moreover with some reasonably minor adjustments, CGE models can allow for excess capacity. Also, making up lost production following an event by working over-time hours, for example, needs to be taken into account in any impact analysis, CGE or otherwise.

[11] This relationship is especially useful in accurately estimating the extent of the oft-found "reconstruction boom" from the infusion of insurance payments and government aid. The strain on capacity in construction and related industries could "heat-up" the economy, setting in motion a series of dampening mechanisms.

less than one week, but CGE models are better suited to all other cases, except possibility where martial law is declared and resource reallocation is undertaken by centralized administration and not through market signals (in this case a multi-sector mathematical programming model would suffice, as proposed by Rose, 1981). CGE models can be adjusted for a greater range of resiliency options than I-O, though the adjustment process is more complex. Also, care needs to be taken that the production function recalibration does not violate any inherent properties of the underlying production function. Note, however, that the business recontracting issue applicable to I-O is not relevant here because of market mechanism automatically takes into account substitution possibilities.

There are several approaches to incorporate uncertainty in CGE models, all basically the same as noted for I-O: scenario analysis, sensitivity tests on parameters, probability distributions for parameters, and stochastic simulations. Since there are many more parameters in a CGE than an I-O model, sensitivity tests on the parameters are especially important.

2.4.3.3 Econometric Models

A third alternative is macroeconometric models, which are statistically estimated simultaneous equation representations of the aggregate workings of an economy (Glickman, 1971). These models have only rarely been used in regional economic loss estimation because of their expense, huge data demands, and difficulty in distinguishing direct and higher-order effects (see, *e.g.*, Ellson *et al.*, 1984; and Guimaraes *et al.*, 1993). The statistical rigor of these models requires time series data with at least ten observations (typically years) and preferably many more. Data needed are not usually available at the regional level for this purpose, so various data reduction strategies have been developed, as in the case of I-O and CGE models (see, *e.g.*, Treyz, 1993). Moreover, the historical experience upon which these models are based is unlikely to be representative of experience and behavior during a disaster situation. Also, appropriate adjustments for this are much more difficult than in I-O and CGE models.

Still, the potential application of these models to hazard loss estimation is great, since neither I-O nor CGE models have forecasting capabilities, which are especially useful in examining potential impacts of a future earthquake or in distinguishing the actual activity of an economy from what it would have been like in the absence of the shock (*i.e.*, establishing a baseline). Also, econometric models have their own, well-established set of criteria for model validation, which are much more well-established and rigorous than the previous two models we have discussed.

Econometric models have no major biases, as do I-O and to a lesser extent CGE models, but do contain a number of more subtle biases (see, *e.g.*, Leamer, 1983). A major limitation is the lack of explicit behavioral content other than straightforward optimization. Expectations have been incorporated into macroeconometric models, but short-run adaptive behavior generally has not.

Uncertainty is inherently incorporated into econometric models by virtue of their stochastic estimation. Specific goodness of fit measures exist for individual parameters and for a model as a whole. Additional diagnostics of uncertainty can be performed by way of sensitivity tests and scenario analysis.

2.5 Broadening the Focus in Future Research

The top priority of future research is to improve the empirical basis of hazard loss estimation. This includes the establishment of protocols for defining and collecting data that is comparable across disasters and across impacted regions. It also includes testing major hypotheses regarding determinants of losses, strength of resiliency, and effectiveness of public policy. Finally, it means incorporating these advances into hazard loss estimation models. Otherwise, the major priorities are to broaden the scope of hazard loss estimation in terms of socioeconomic and temporal analysis. These two areas are somewhat related and are discussed in more detail below.

2.5.1 Distributional Impacts

The distribution of hazard impacts is often omitted because of lack of models or data, but is especially important for evaluating equity considerations (Cochrane, 1975) and in communicating risk to stakeholders, thereby facilitating their input into the policy process (Rose *et al.*, 1989). Distributional impacts are likely to be more controversial than aggregate ones but no less important. For example, achieving accuracy is more difficult for subsets of a region. Also, there is likely to be a mismatch between those who may have to incur the costs of mitigation or post-disaster recovery and those who benefit from their implementation. Still, accurate distributional estimates are a useful supplement to the aggregate numbers used in most cost-benefit analyses (CBA). Ordinary CBA implicitly justifies decisions on the basis of how the community is impacted as a whole. It works well in the context of a single, custodial decision-maker (increasingly less the case these days), or, in the case of public participation if people are entirely altruistic (also unlikely). Distributional information, on the other hand, can help affected parties to see what stake they have in dealing with natural hazards. At the very least, this will help make potential impacts more poignant and generate greater interest in the problem.

I-O, and CGE models as well, are well-suited to performing distributional analysis. They disaggregate the economy into sectors, providing insight to the inherent unevenness of direct and higher-order impacts across industries and between industries, households, government, and other institutions. Extended I-O models allow for further analysis of socioeconomic or institutional accounts by disaggregating income, consumption, and trade flows (Batey and Rose, 1990).[12] This modeling is reasonably straightforward, including calculation of short-cut distributional multipliers, *e.g.*, how a direct change in income to one socioeconomic group affects all others directly and indirectly (for a recent application to hazards see Okuyama *et al.*, 1999a). The major limitation here is data, especially mapping of income flows from sectors to socioeconomic groups. Still some useful data reduction and adaptation techniques exist here as well (see Rose *et al.*, 1988; and Li *et al.*, 1999), so that this area of application is considered to be reasonably accurate, though not as much as aggregate impact estimation.

---

[12] The number of income brackets is limited by the availability of data but typically involves as many as ten separate classes. Extensive disaggregations of occupational groups are possible because of the long-standing work of the U.S. Bureau of Labor Statistics. Disaggregations according to racial/ethnic groups are more difficult, but a good deal of data is generally available from the U.S. Bureau of Census.

CGE models are especially well suited to distributional impact analysis. They have a multi-sector character, though, for the sake of manageability, are usually more aggregated than I-O models, which can readily accommodate four hundred plus sectors, as opposed to fifty or fewer in the typical CGE model. Moreover, socioeconomic impacts are facilitated by the existence of a social accounting matrix underlying CGE models to trace income and consumption flows. Also, CGE models usually contain a formal labor market component, often including occupational data by sector. Formulation and application of CGE models for distributional analysis is fairly advanced (see, *e.g.*, Hanson and Rose, 1997), though none to date have been applied to loss estimation from natural hazards. At the same time, this area cries out for application since previous analyses, using a variety of methods and models, indicate that low income households are relatively more vulnerable to hazards and that extreme hazards render many people poor (see, *e.g.*, Mileti, 1999).

In addition to providing valuable information to stakeholders, distributional impact analysis also addresses the more recently prominent issue of "environmental justice," which has typically been applied to evaluating differential impacts of environmental impacts of public policy across racial/ethnic groups. This topic is becoming important for reasons of fairness, but also for pragmatic reasons relating to lawsuits brought by minority group members when they have felt an unequal burden of pollution (and potentially natural hazard damage) or felt they were incurring a disproportionate percentage of the cost of mitigation or remediation.

2.5.2 Sustainability

The topic of sustainability has recently been introduced to hazard analysis in a major, and even all encompassing, manner (see, *e.g.*, Mileti, 1999; and Burby, 1999). For many years, "sustainability" was typically used in the context of economic development, and was often viewed as a buzzword emanating from "green" initiatives (see, *e.g.*, Brundtland, 1984). Over time, it has come to represent a worthy, and now well-established, principle, if not a major strategy, applicable to a broad range of subjects, including hazards. Sustainability originally referred to the harmonious pursuit of economic development and enhancement of environmental quality. Its major principles called for utilizing natural resources in such a way as to not undercut the economic development opportunities of future generations, and, as such, focuses on "intergenerational equity." It emphasizes that natural (including environmental) resources are indispensable to future economic growth and quality of life. It also emphasizes opportunities and decisions that can promote the harmonious pursuit of these two objectives simultaneously (in contrast to major perspectives of the previous two decades that typically had these two objectives at odds). These principles are applicable to hazards and focus attention on the need to improve the inclusion of environmental losses in hazard loss estimation.

Another major theme is that a country or region be self-reliant in its future quests. In addition to the growth/environment features, this principle is especially pertinent to hazards, where a major way of dealing with losses has been dependency on outside aid. Adoption of measures that reduce this dependency are paramount. Prime examples are to stop perpetuating past mistakes, such as rebuilding in flood earthquake zones. Other important tactics include finding ways to enhance individual and community resiliency in general. For example, this has a bearing on how recovery and reconstruction impact future losses.

Sustainability shifts the focus to a time dimension, *e.g.*, how do decisions today affect the long-term vulnerability of the community to hazards? Thus hazard loss estimation principles and methods must be extended in this direction. Benefit-cost analysis represents a start because it requires the time discounting of future considerations. Otherwise, much of the hazard literature has been rather weak on the time dimension. One area that has received attention is the issue of whether there are long-term effects of disasters (see, *e.g.*, Friesema, 1979; and Chang, 2001). However, the topic presents some inherent difficulties, the most notable being the prediction of future baselines from which to gauge the hazard loss. It would appear that econometric modeling approaches would be better suited to analyzing these issues than would be I-O, CGE, or the less comprehensive (from a community standpoint) approaches noted earlier.

2.6 Conclusions

A sound conceptual base is a prerequisite for the accurate empirical estimation of losses from natural hazards. This chapter has endeavored to set forth the necessary economic principles for this purpose. Most of them are well known to economists, though not necessarily familiar to those in other disciplines, while others are novel in order to capture the uniqueness of the field of hazard loss estimation.

This chapter has also indicated how complicated the conceptualization of hazard losses can be. Still there is no pretense that it will clear away all of the obstacles to the empirical application of hazard loss methodologies. That aspect of hazard loss estimation is most difficult, given the idiosyncrasies of individual contexts, limits on resources, and time pressures. Conceptual refinements do not always readily translate into real world counterparts, but are an important first step in comprehensive and accurate natural hazard loss estimation.

Many economic losses, especially direct and indirect business interruption, are not as readily apparent as property damage, and their validity is met by the skepticism of many engineers, some policy-makers, and even some economists. Improved methods to refine and validate relevant models and the estimates from their application are crucial to the acceptance of this important type of impact. Perhaps most important is the collection of more empirical data and improvements in formulating and testing hypotheses about direct and higher-order losses, especially the offsetting effects of resiliency. The latter is especially difficult because of the inability to perform controlled experiments and the difficulty of establishing a reference base (predicting an economic baseline or growth path in the absence of the disaster).

Finally, I emphasize the importance of not perceiving of hazard loss estimation as a passive pursuit, but one with the major objective of actively reducing negative impacts either through mitigation or post-disaster private decisions and public policies. Economists have identified a number of relatively costless mechanisms for doing so, including the reallocation of scarce utility lifeline services in order to minimize regional employment or output losses by way of central administration or market mechanisms. It bears repeating that not taking advantage of such opportunities results in outcomes as devastating as if the disaster had actually toppled the buildings in which the lost production would have originated.

References

Applied Technology Council. 1991. *ATC-25: Seismic Vulnerability and Impact of Disruption on Lifelines in the Conterminous United States*. Federal Emergency Management Agency, Applied Technology Council, CA.

Batey, P. and A. Rose. 1990. "Extended Input-Output Models: Progress and Potential," *International Regional Science Review*, 13: 27-49.

Boadway, R. and N. Bruce. 1985. *Welfare Economics*. Oxford, UK; Basil Blackwell.

Boisvert, R. 1992. "Direct and Indirect Economic Losses from Lifeline Damage," in *Indirect Economics Consequences of a Catastrophic Earthquake*, Final Report by Development Technologies to the Federal Emergency Management Agency.

Boisvert, R. 1995. "Computer General Equilibrium Modeling Approach to Natural Hazard Loss Estimation," Final Report to FEMA, Department of Agricultural Economics, Cornell University, Ithaca, NY.

Boyle, K. and J. Bergstrom. 1992. "Benefit Transfer Studies: Myths, Pragmatism, and Idealism," *Water Resources Research*, 29: 657-63.

Bram, J., J. Orr, and C. Rappaport. 2002. "The Impact of the World Trade Center Attack on New York City: Where Do We Stand?" Federal Reserve Bank of New York, New York, NY.

Brooke, A., D. Kendrick, and A. Meeraus. 1992. *GAMS: A User's Guide*. San Francisco, CA; Scientific Press.

Brookshire, D. and M. McKee. 1992. "Other Indirect Costs and Losses from Earthquakes: Issues and Estimation," in *Indirect Consequences of a Catastrophic Earthquake*, Final Report by Development Technologies to the Federal Emergency Management Agency.

Brookshire, D., S.E. Chang, H. Cochrane, R. Olson, A. Rose, and J. Steenson. 1997. "Direct and Indirect Economic Losses from Earthquake Damage," *Earthquake Spectra*, 13: 683-701.

Brundtland Commission. 1984. *Our Common Future*. Cambridge, UK; Cambridge University Press.

Bruneau, M. *et al.* 2002. "A Framework to Quantitatively Assess and Enhance Seismic Resilience of Communities," MCEER, Buffalo, NY.

Burby, R. ed. 1999. *Cooperating with Nature: Confronting Natural Hazards with Land Use Planning for Sustainable Communities*. Washington, DC; Joseph Henry Press.

Chang, S., A. Rose, M. Shinozuka, and K. Tierney. 2001. "Modeling Earthquake Impact on Urban Lifeline Systems: Advances in Integration," in *B.* Spencey and Y. Hu eds. *Earthquake Engineering Frontiers in the New Millennium*, Lisse, The Netherlands; Balkema.

Chang, S.E. 1998. "Direct Economic Impact," in M. Shinozuka *et al.* eds. *Engineering and Socioeconomic Impacts of Earthquakes: An Analysis of Electricity Lifeline Disruptions in the New Madrid Area*. Buffalo, NY; MCEER.

Chang, S.E. 2001. Structural Change in Urban Economies: Recovery and Long-Term Impacts in the 1995 Kobe Earthquake," *The Kokumin Keizai Zasshi* (*Journal of Economics and Business Administration*), 183: 47-66.

Chang, S.E., H.A. Seligson, and R.T. Eguchi. 1996. *Estimation of the Economic Impact of Multiple Lifeline Disruption Memphis Light, Gas, and Water Division Case Study*. Buffalo, NY; NCEER.

Cho, S., P. Gordon, J. Moore, H. Richardson, M. Shinozuka, and S.E. Chang. 2001. "Integrating Transportation Network and Regional Economic Models to Estimate the Cost of a Large Urban Earthquake," *Journal of Regional Science*, 41: 39-65.

Cochrane, H. 1974. "Predicting the Economic Impact of Earthquakes," in H. Cochrane *et al.* eds. *Social Science Perspectives on the Coming San Francisco Earthquake*, Natural Hazards Research Paper No. 25, NHRAIC, Boulder, CO; University of Colorado.

Cochrane, H. 1975. *Natural Hazards and Their Distributive Effects*. Boulder, CO; Natural Hazards Research Applications Information Center.

Cochrane, H. 1997. "Forecasting the Economic Impact of a Mid-West Earthquake," in B. Jones ed. *Economic Consequences of Earthquakes: Preparing for the Unexpected*. Buffalo, NY; MCEER.

Cochrane, H., M. Laub, and J. Barth. 1992. "The Effect of Earthquakes on Banking and Financial Markets," in *Indirect Economics Consequences of a Catastrophic Earthquake*, Final Report by Development Technologies to the Federal Emergency Management Agency.

Cole, S. 1988. "The Delayed Impacts of Plant Closures in a Reformulated Leontief Model," *Papers of the Regional Science Association*, 65: 135-149.

Cole, S. 1995. "Lifelines and Livelihood: A Social Accounting Matrix Approach to Calamity Preparedness," *Journal of Contingencies and Crisis Management*, 3: 1-11.

Cole, S. 1998. "Decision Support for Calamity Preparedness: Socioeconomic and Interregional Impacts," in M. Shinozuka *et al.* eds. *Engineering and Socioeconomic Impacts of Earthquakes: An Analysis of Electricity Lifeline Disruptions in the New Madrid Area*. Buffalo, NY; MCEER.

Cole, S. 1999. "The Phantom of the Matrix: Inverting the Case on Closure in Cole's Model," *Papers in Regional Science*, 78: 429-36.

Comfort, L. 1999. *Shared Risk: Complex Systems is Seismic Response*, New York, NY; Pergamon.

Dacy, E. and H. Kunreuther. 1969. *The Economics of Natural Disasters: Implications for Federal Policy*. New York, NY; The Free Press.

Davis, H.C. and E.L. Salkin. 1984. "Alternative Approaches to the Estimation of Economic Impacts Resulting from Supply Constraints," *Annals of Regional Science*, 18: 25-34.

Eguchi, R. and S. Pelmulder. 1992. "Indirect Economic Impacts of Energy Network," in *Indirect Economic Consequences of a Catastrophic Earthquake*, Final Report by Development Technologies to the Federal Emergency Management Agency.

Ellson, R., J. Milliman, and R. Roberts. 1984. "Measuring the Regional Economic Effects of Earthquakes and Earthquake Prediction," *Journal of Regional Science*, 24: 559-79.

Federal Emergency Management Agency. 2001. *Earthquake Loss Estimation Methodology (HAZUS)*. Washington, DC; National Institute of Building Sciences.

Fisher, A. *et al.* 2000. *Preparing for a Changing Climate: The Potential Consequences of Climate Variability and Change in the Mid-Atlantic Region*. University Park, PA; The Pennsylvania State University.

Freeman, A.M. 1993. *The Measurement of Environmental and Resource Values: Theory and Methods*. Washington, DC; Resources for the Future.

Friesema, P. *et al.* 1979. *Aftermath: Communities and Natural Disasters*. Beverly Hills, CA; Sage.

Giarratani, F., and R. Garhart. 1991. "Simulation Techniques in the Evaluation of Regional Input-Output Models," in J. Dewhurst, G. Hewings, and R. Jensen eds. *Regional Input-Output Modeling: New Developments and Interpretations*. Aldershot, UK; Avebury.

Glickman, N. 1971. "An Economic Forecasting Model for the Philadelphia Region," *Journal of Regional Science*, 11: 15-32.

Gordon, P., H. Richardson, and B. Davis. 1998. "Transport-Related Impacts of the Northridge Earthquake," *Journal of Transportation and Statistics*, 1: 22-36.

Guimaraes, P., F. Hefner, and D. Woodward. 1993. "Wealth and Income Effects of Natural Disasters: An Econometric Analysis of Hurricane Hugo," *Review of Regional Studies*, 23: 97-114.

Hammitt, J., J.T. Liu, and J.L. Liu. 2001. "Contingent Valuation of a Taiwanese Wetlands," *Environment and Development Economics*, 6: 259-68.

Hanson, K. and A. Rose. 1997. "Factor Productivity and Income Inequality: A General Equilibrium Analysis," *Applied Economics,* 29: 1061-71.

Heinz Center for Science, Economics and the Environment. 2000. *The Hidden Costs of Coastal Hazards: Implications for Risk Assessment and Mitigation.* Washington, DC; Island Press.

Hewings, G.J.D. and R. Mahidhara. 1996. "Economic Impacts: Lost Income, Ripple Effects, and Recovery," in S. Changnon ed. *The Great Flood of 1993.* Boulder, CO; Westview Press.

Howard, R. and J. Matheson. 1989. *Readings on the Principles and Applications of Decision Analysis.* Menlo Park, CA; Strategic Decisions Group.

Howe, C., and H. Cochrane. 1993. *Guidelines for the Uniform Definition, Identification and Measurement of Economic Damages from Natural Hazard Events.* Program on Environment and Behavior Special Publication No. 28, Institute of Behavioral Science, University of Colorado, Boulder, CO.

Intergovernmental Panel on Climate Change. 2000. *Climate Change 2000.* New York, NY; Cambridge University Press.

Jensen, R. 1980. "The Concept of Accuracy in Regional Input-Output Models," *International Regional Science Review,* 5: 139-54.

Kawashima, K. and T. Kanoh. 1990. "Evaluation of Indirect Economic Effects Caused by the 1983 Nihonkai-Chubu, Japan, Earthquake," *Earthquake Spectra,* 6: 739-56.

Kirchhoff, S., B. Colby, and J. LaFrance. 1997. "Evaluating the Performance of Benefit Transfer: An Empirical Inquiry," *Journal of Environmental Economics and Management,* 33: 75-93.

Kunreuther, H. and G. Heal. 2002. "Interdependent Security: The Case of Identical Agents," Columbia University.

Kunreuther, H. *et al.* 2001. "Using Cost-Benefit Analysis to Evaluate Mitigation for Lifeline Systems," *Research Progress and Accomplishments,* Buffalo, NY: MCEER.

Lahr, M. 2001. "A Strategy for Producing Hybrid Regional Input-Output Tables," in M. Lahr and E. Dietzenbacher eds. *Input-Output Analysis: Frontiers and Extensions.* London, UK; Palgrave.

Law, A. and W. D. Kelton. 1991. *Simulation Modeling and Analysis.* New York, NY; McGraw-Hill.

Leamer, E. 1983. "Let's Take the Con Out of Econometrics," *American Economic Review,* 73: 31-43.

Li, P.C., A. Rose, and B. Eduardo. 1999. "Construction of an Input-Output Income Distribution Matrix for the U.S.," in G.J.D. Hewings *et al.* eds. *Understanding and Interpreting Economic Structure,* Heidelberg, Germany; Springer.

Mileti, D. 1999. *Disasters by Design: A Reassessment of Natural Hazards in the United States.* Washington, DC; Joseph Henry Press.

Miller, R.E. and P.D. Blair. 1985. *Input-Output Analysis: Foundations and Extensions.* Englewood Cliffs, NJ; Prentice-Hall.

Minnesota IMPLAN Group (MIG). 1999. *Micro IMPLAN: A User's Guide,* Stillwater, MN; MIG.

Mitchell, R. and R. Carson. 1989. *Using Surveys to Value Public Goods: The Contingent Valuation Method.* Washington, DC; Resources for the Future.

National Research Council. 1999. *The Impacts of Natural Disasters: A Framework for Loss Estimation.* Washington, DC; National Academy of Sciences Press.

Newnan, D., J. Lavelle, and T. Eschenbach. 2001. *Essentials of Engineering Economic Analysis, Second Edition.* New York, NY; Oxford University Press.

Okuyama, Y., G.J.D. Hewings, and M. Sonis. 1999a. "Economic Impacts of an Unscheduled, Disruptive Event: A Miyazawa Multiplier Analysis," in G.J.D. Hewings *et al.* eds. *Understanding and Interpreting Economic Structure.* Berlin, Germany: Springer-Verlag.

Okuyama, Y., G.J.D. Hewings, and M. Sonis. 2000. "Sequential Interindustry Model (SIM) and Impact Analysis: Applications for Measuring Economic Impact of Unscheduled Events," Regional Economics Applications Laboratory, University of Illinois, Urbana, IL.

Okuyama, Y., G.J.D. Hewings, T. Kim, D. Boyce, H. Ham, and J. Sohn. 1999b. "Economic Impacts of an Earthquake in the New Madrid Seismic Zone: A Multiregional Analysis," in W. Elliott and P. McDonough eds. *Optimizing Post-Earthquake Lifeline System Reliability*.

Oladosu, G. 2000. "A Non-Market Computable General Equilibrium Model for Evaluating the Economic Impacts of Climate Change in the Susquehanna River Basin," Ph.D. Thesis, The Pennsylvania State University, University Park, PA.

Partridge, M.D. and D.S. Rickman. 1998. "Regional Computable General Equilibrium Modeling: A Survey and Critical Appraisal," *International Regional Science Review*, Vol. 21, No. 3: 205-248.

Perroni, C. and T. Rutherford. 1993. "Regular Flexibility of Nested CES Functions," *European Economic Review*, 39: 335-43.

Porter, K., J. Beck, H. Seligson, C. Scawthorn, L. Tobin, and T. Boyd. 2002. *Improving Loss Estimation for Woodframe Buildings*. Consortium of Universities for Research in Earthquake Engineering, Richmond, CA.

Resnick, M. 1987. *Choices: An Introduction to Decision Theory*, Minneapolis, MN; University of Minnesota Press.

Rose, A. 1981. "Utility Lifelines and Economic Activity in the Context of Earthquakes," in J. Isenberg ed. *Social and Economic Impact of Earthquakes on Utility Lifelines*. New York, NY; American Society of Civil Engineers.

Rose, A. 1995. "Input-Output Economics and Computable General Equilibrium Models," *Structural Change and Economic Dynamics*, 6: 295-304.

Rose, A. 2000. "Natural Hazard Loss Estimation: Fundamentals and Complications," Department of Energy, Environmental, and Mineral Economics, the Pennsylvania State University, University Park, PA.

Rose, A. and D. Lim. 2002. "Business Interruption Losses from Natural Hazards: Conceptual and Methodological Issues in the Case of the Northridge Earthquake," *Environmental Hazards*, 4: 1-14.

Rose, A. and G. Guha. 2004. "Computable General Equilibrium Modeling of Electric Utility Lifeline Losses from Earthquakes," in this volume.

Rose, A. and J. Benavides. 1998. "Regional Economic Impacts," in M. Shinozuka *et al.* eds., *Engineering and Socioeconomic Impacts of Earthquakes: An Analysis of Electricity Lifeline Disruptions in the New Madrid Area*. Buffalo, NY; MCEER.

Rose, A. and J. Benavides. 1999. "Optimal Allocation of Electricity After Major Earthquakes: Market Mechanisms Versus Rationing," in K. Lawrence *et al.* eds. *Advances in Mathematical Programming and Financial Planning*. Greenwich, CT; JAI Press.

Rose, A. and S. Liao. 2002. "Modeling Regional Economic Resiliency to Earthquakes: A Computable General Equilibrium Analysis of Water Service Disruptions," in *Proceedings of the 7th National Conference on Earthquake Engineering*, Oakland, CA; EERI.

Rose, A. and W. Miernyk. 1989. "Input-Output Analysis: The First Fifty Years," *Economic Systems Research*, 1: 229-71.

Rose, A., B.K. Stevens, and G. Davis. 1988. *Natural Resource Policy and Income Distribution*. Baltimore, MD; Johns Hopkins University Press.

Rose, A., B.K. Stevens, and G. Davis. 1989. "Assessing Who Gains and Who Loses From Natural Resource Policy: Distributional Information and the Public Participation Process," *Resources Policy*, 15: 282-91.

Rose, A., G. Guha, S. Y. Liao and R. Ranjan. 2002. "Recalibrating CGE Models on the Basis of Empirical Data: An Application to Estimating Losses from Natural Hazards," Department of Energy, Environmental, and Mineral Economics, The Pennsylvania State University, University Park, PA.

Rose, A., G. Oladosu, and D. Salvino. 2003. "Regional Economic Impacts of Electricity Outages in California: A Computable General Equilibrium Analysis," Working Paper, The Pennsylvania State University, University Park, PA.

Rose, A., J. Benavides, S.E. Chang, P. Szczesniak, and D. Lim. 1997. "The Regional Economic Impact of an Earthquake: Direct and Indirect Effects of Electricity Lifeline Disruptions," *Journal of Regional Science*, 37: 437-58.

Rose, A., J. Carmichael, G. Oladosu, and D. Abler. 1999. "Modeling the Economics of Natural Hazard Impacts and Policy Responses Using Computable General Equilibrium Analysis," Department of Energy, Environmental, and Mineral Economics, The Pennsylvania State University, University Park, PA.

Rose, A., S.E. Chang, and M. Shinozuka. 1998. "Direct and Indirect Infrastructure User Costs of Natural Hazards," in *Proceedings of the Structural Engineers World Congress*, New York, NY: Elsevier.

Shinozuka, M., A. Rose, and R. Eguchi eds.. 1998. *Engineering and Socioeconomic Impacts of Earthquakes: An Analysis of Electricity Lifeline Disruptions in the New Madrid Area.* Buffalo, NY; MCEER.

Shoven, J. and J. Whalley. 1992. *Applying General Equilibrium.* New York, NY; Cambridge University Press.

Taylor, C., S. Werner, and S. Jakubowski. 2001. "The Walkthrough Method for Catastrophe Decision-Making," Natural Hazards Management, Inc., Torrance, CA.

Tierney, K. 1997. "Impacts of Recent Disasters on Businesses: The 1993 Midwest Floods and the 1994 Northridge Earthquake," in Barclay Jones ed. *Economic Consequences of Earthquakes: Preparing for the Unexpected.* Buffalo, NY; NCEER.

Tierney, K. and J. Dahlhamer. 1998. "Earthquake Vulnerability and Emergency Preparedness Among Businesses," in M. Shinozuka *et al.* eds., *Engineering and Socioeconomic Impacts of Earthquakes: An Analysis of Electricity Lifeline Disruptions in the New Madrid Area.* Buffalo, NY; MCEER.

Tierney, K., J. Nigg, and J. Dahlhamer. 1996. "The Impact of the 1993 Midwest Floods: Business Vulnerability and Disruption in Des Moines," in R. T. Sylves and W.L. Waugh, Jr. eds., *Cities and Disaster: North American Studies in Emergency Management.* Springfield, MA; Charles C. Thomas.

Treyz, G. 1993. *Regional Economic Modeling: A Systematic Approach to Economic Forecasting and Policy Analysis.* Dordrecht, The Netherlands; Kluwer.

Webb, G., K. Tierney, and J. Dahlhamer. 2000. "Business and Disasters: Empirical Patterns and Unanswered Questions," *Natural Hazards Review,* 1: 83-90.

West, C.T. and D.G. Lenze. 1994. "Modeling the Regional Impact of Natural Disaster and Recovery: A General Framework and an Application to Hurricane Andrew," *International Regional Science Review,* 17: 121-150.

West, G.R. 1986. "A Stochastic Analysis of an Input-Output Model," *Econometrica*, 54: 363-74.

White, G. and J. E. Haas. 1975. *Assessment of Research on Natural Hazards.* Cambridge, MA; MIT Press.

White, J., K. Case, D. Pratt, and M. Agee. 1997. *Principles of Engineering Economic Analysis, 4th Edition.* New York, NY; John Wiley & Sons.

Zerbe, R. and D. Dively. 1994. *Benefit-Cost Analysis Theory and Practice.* New York, NY; HarperCollins.

3 Indirect Losses from Natural Disasters: Measurement and Myth

Harold C. Cochrane

Department of Economics, Colorado State University, USA
e-mail: cochrane@lamar.colostate.edu

3.1 Introduction

Early studies of indirect loss framed the problem narrowly and focused on secondary ripple effects triggered by supply bottlenecks, *i.e.*, the shortages that cause undamaged producers and service providers to shut down (Cochrane, 1974). Since then, the concept has been refined and the analytical tools for estimating such losses have advanced markedly (RMS, 1996; Rose *et al.*, 1995). Yet despite such advancements, indirect loss remains a somewhat amorphous concept. In contrast to what was once believed, it does not readily yield simple rules of thumb, *e.g.*, "indirect loss is equivalent in magnitude to direct loss." In fact, it has since been learned that indirect loss can vary dramatically, and under some circumstances may even prove to be negative implying regional net gains (negative losses). This Chapter provides: a conceptual framework for analyzing indirect loss, a brief description of an algorithm for calculating such loss, a discussion of the range of post disaster outcomes that might be expected, and a brief review of the evidence. These elements are then used to evaluate claims various agencies have made regarding the economic fallout from the World Trade Center attack.

3.2 What are Indirect Damages?

Terminologically, indirect loss is any loss other than direct loss. Direct loss is a loss linked *directly* to disaster. It includes all damages, plus employment losses due directly to the closure of damaged facilities. Indirect losses are anything else. For example, factories relying on just-in-time inventory systems may be forced to close if a critical parts supplier cannot meet its obligations. Income lost as a result would be considered indirect damage. In some instances indirect gains may materialize. This is likely if sufficient excess capacity existed prior to the event, and an infusion of outside aid and insurance payments stimulated a reconstruction boom.

Indirect loss is easily confused with change in gross regional income or product. They are not the same however. This is because of the way indirect loss was defined, *i.e.*, losses other than those produced as a *direct consequence* of a disaster. Unemployment occurring in commercial ventures suffering sufficient damage to terminate operations is considered a direct loss. So, changes in regional income stem from *both* direct and indirect movements in regional employment

3.3 Indirect Damage and Systemic Risk

It is somewhat ironic that while regional economists were struggling to understand how supply disruptions rippled to interconnected economic sectors, monetary theorists were addressing the same issue. They focused not on shortage of product but on credit and insurance rationing, referring to this type of dislocation as *systemic risk*. Clearly theses two concepts are similar in nature. They do differ in one important way however; systemic risk embraces the possibility (many believe the likelihood) that economic contraction might lead financial intermediaries to ration credit to a far greater extent than warranted by economic fundamentals. In short, systemic risk is the potential for financial panic arising from "irrational" fear. Indirect loss makes no allowance for this psychological component. Barring this difference, indirect loss and systemic risk are similar enough to warrant using a single term to address both. From here on, both phenomena are referred to as indirect losses. All shocks, insurance, banking, and manufacturing/lifelines, lend themselves to the same type of analysis. The difference between total and direct loss is *indirect*, regardless of what the triggering mechanism might be: damage to factories and lifelines (*real ripple effects*), banking which causes uncertainty and a subsequent tightening of overall credit conditions (*bank systemic risk*), or insurance where companies limit coverage thereby triggering subsequent economic dislocations (*insurance systemic risk*). See figure 3.1.

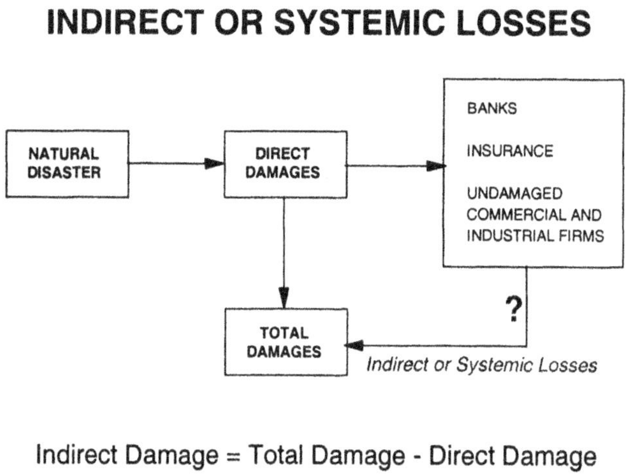

Figure 3.1 Indirect or Systemic Losses

3.4 Modeling Indirect Damage

As just indicated, *real ripple effects* stem from supply shortages or sudden reductions in demand. The relative size of the ripple hinges on the geographic area of focus, and the time frame under scrutiny. Since losses can be displaced geographically and temporally, the narrower the time-space focus, the greater the likelihood that losses will be underestimated. As will be shown below, ripples must be negative for the nation, but can, with the receipt of outside assistance, prove positive for the impacted region.

Indirect losses are a result of dislocations suffered by economic sectors not sustaining direct damage. Activities that are either forward-linked (rely on regional markets for their output) or backward-linked (rely on a regional source of supply) could experience interruptions in their operations. The extent of these losses depends upon such factors as the availability of alternative sources of supply and markets for products, the length of the production disturbance, and deferability of production. Indirect damages occur because: 1) direct damage to production facilities and inventories cause supply shortages for firms needing these supplies (forward linked indirect damages); 2) because damaged production facilities reduce their demand for inputs from other producers (backward linked indirect damage); or 3) because of a disaster caused reduction in government, investment, or export spending. The extent to which these economic linkages tighten to produce spillover losses for economic sectors not suffering direct damage hinges on the degree to which demands and supplies might be augmented. Inventories can be adjusted, alternate supplies imported, alternate markets served (exports), and idle undamaged capacity drawn into use. What this means is that the preexisting state of the economy has much to do with the degree of indirect loss observed after an event, a point that is addressed further in the following discussion.

3.5 A Brief Description of the HAZUS[1] Rebalancing Algorithm

In 1995 the National Institute of Building Sciences began work on an ambitious project that incorporated numerous databases into a commercially available geographic information system. The resultant computer model named HAZUS enabled emergency management personnel and planners to simulate the direct and indirect losses that could be expected from an earthquake anywhere in the U.S. The package has since been refined and extended to other hazards, specifically wind and flood. The Indirect Economic Loss Model (IELM) component of HAZUS was designed to determine how damage to a region's economic sectors would impact employment and income. Since disasters typically produce damage patterns that are uneven, traditional

[1] HAZUS is software developed by the Federal Emergency Management Agency and the National Institute of Building Sciences. It is designed around a geographic information system and is intended to simulate how a specific earthquake, flood or wind storm will effect: the number of buildings damaged, casualties, transportation outages, disruption of electrical and water utilities, homelessness, the cost of repair, and indirect loss.

inter-industry techniques had to be abandoned in favor of an algorithm[2] for tracking and correcting imbalances in regional trading patterns. The IELM is based on published inter-industry transactions tables (IMPLAN[3]), but permits differing assumptions with regard to imports, exports, excess capacity (unemployment and vacancy rates), and the amount of outside assistance received.

The algorithm takes the post-disaster surviving capacity as a starting point for recalculating inter-industry supplies and demands. Direct damage is introduced as a percent of a sector's production surviving the disaster. The algorithm then utilizes this potential post-disaster output to calculate the effect on inter-industry shipments. It does so by first determining the degree to which a sector's purchases are impacted. See figure 3.2 for an illustration of how a single sector's interindustry trade flows are impacted.

The interindustry inputs (transactions matrix columns) are multiplied by the percentage of the sector's post disaster capacity. The result is a measure of inter-industry sectoral demands. The alorithm then calculates the degree to which sectoral shipments are effected; the surving capacity is multiplied by the shipments (transactions matrix rows). This provides a measure of excess inter-industy supplies or demands. The algorithm adds the predisaster final demands (household, government, and exports) to arrive at a complete measure of excess supply or demand by sector.

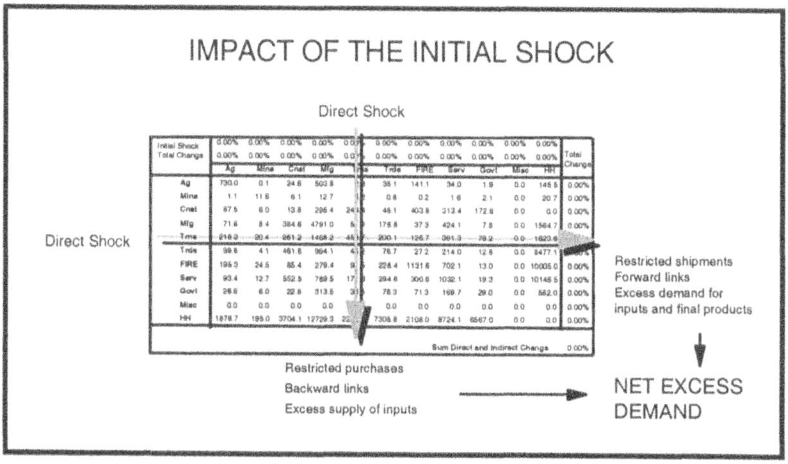

Figure 3.2 Impact of the Initial Shock

In the event excess demand is detected, the algorithm searches for a way of satisfying that demand, specifically by adjusting sectoral capacity to account for

[2] The interest reader is referred to Chapter 16 in the HAZUS technical manual for a more detailed description as to how the model is constructed.
[3] IMPLAN is one of a several commercially available state and county input output tables.

unemployed resources in the region[4], and by importing from other regions. Since these are user defined, the ensuing adjustment can be rapid and eliminate all backward linked losses, or it could be quite slow and severely constrain production in undamaged sectors. Which possibility emerges depends upon the unique characteristics of the economy under study. This adjustment is applied to all sectors except for one, the construction sector. Its output was assumed to be perfectly price elastic, implying that construction talent could be brought in from surrounding regions without any significant escalation in cost.

If excess supplies are detected, the algorithm searches for alternative means of disposing those supplies, specifically through export. The model adjusts potential outputs iteratively until all net excesses are eliminated.

3.6 How the Algorithm Treats the Rebuilding Process

Reconstruction of damaged sectors is presumed to occur in a fashion mimicking a logistics function.[5] The algorithm tracks the amount of rebuilding occurring each quarter, and adjusts both construction demands and sectoral capacity accordingly. To the extent that rebuilding is financed by either insurance or disaster assistance, the normal[6] pattern and magnitude of household spending is held constant. However, in the event that the recovery is internally financed, through the use of savings or borrowed funds, then household demands are adjusted downward[7], reflecting the burden posed by additional debt. This procedure presumes that households behave according to the life-cycle theory of consumption. That is, any temporary shock to wealth is absorbed over the household's life-cycle. The present value of the stream of indirect losses (or gains) is then computed.

3.7 Other Issues and Explanations

The IELM algorithm is not a linear program, nor is it a Computable General Equilibrium (CGE) model. It assumes that household spending is exogenous (hence its Type I flavor), and it takes a regional accounting stance. The reasons for abandoning the more sophisticated and complex approach of a CGE, and why a Type II analysis was eliminated from the model deserve explanation.

Product substitutions: although the algorithm does adjust trading patterns[8], it does not allow for product and input substitutions driven by relative price changes. Such

---

[4] Unemployed resources are determined by the unemployment rate assumed to prevail at the time of the disaster. This rate is converted into a measure of idle capacity, which in turn is used to adjust the supplies available from the surviving capital stock.

[5] The user sets the speed of recovery.

[6] Normal spending excludes outlays for rebuilding and restoring lost consumer durables.

[7] The adjustment applied by sector is assumed to be the amortized payment on the borrowed funds (reduced savings is treated as borrowed funds) times the pre disaster pattern of household demands.

[8] The import-export pattern is variable, but inter-industry trading is assumed fixed.

flexibility is the hallmark of Computable General Equilibrium models. Relative price changes were left out of the process for two reasons. First, it appears that relative prices have not changed dramatically after disaster.[9] If this holds, then the post disaster adjustment process is shaped primarily by infusion of aid, changes in wealth, and supply constraints, and not by product substitution. Second, CGE models contain a host of ad hoc assumptions shaping product substitutions, assumptions that would make the inner workings of the model less transparent, and the results difficult to interpret.

Type I or Type II analysis: as indicated earlier, the algorithm treats households as if they behaved according to the life-cycle model of consumption. By electing this orientation, the level of household consumption is disconnected from regional income. It is assumed that temporary bouts of unemployment lead households to dip into savings or borrow.[10] A disaster induced economic slump is clearly different from a long run change in regional demand. The latter does warrant a Type II multiplier analysis where changing household income will alter spending. Admittedly, an adherence to a Type I analysis is complicated by the fact that reconstruction will lead to a temporary influx of construction talent. Some of the income earned by these temporary residents will be spent within the region. So, a Type I analysis will somewhat understate the effects of the disaster. Whether making household spending endogenous or not then hinges on judgment – it was decided that short run shocks were better represented by a Type I analysis.

3.8 Generalizations about Direct and Indirect Loss

Indirect loss stems from supply shortages or a sudden reduction in demand. The magnitude of real indirect loss hinges on four factors: 1) where the economic boundary is drawn (city, standard metropolitan statistical area, state, or the nation); 2) degree of economic integration; 3) pre-disaster economic conditions (level of unemployment and the ability to substitute imports for regionally produced goods and services); and 4) the amount of outside assistance received (insurance pay-outs and FEMA grants). It is precisely because these factors vary so much by region and over time that it is highly unlikely that a simple rule of thumb connecting indirect loss to direct damage will be discovered. To test this possibility, Cochrane (1997) developed a Monte Carlo simulation that exercised the HAZUS model thousands of times, varying user inputs and damage patterns. No reliable rule of thumb emerged from an econometric analysis of these results, but the following tendencies were noted. Indirect loss:

---

[9] A review of price indices for Dade and Los Angeles Counties after Hurricane Andrew and the Northridge Earthquake suggests that neither event caused a shift in the relative price of construction materials. Despite the persistence of anecdotal evidence to the contrary, a review of published research on the subject revealed no evidence of substantial post disaster price fluctuations.

[10] As indicated earlier, the algorithm treats self-financed reconstruction expenditures as a debt, and then adjusts future spending accordingly. Indebtedness can arise from unemployment as well, since income would be less than spending. At this stage of the model's development this indebtedness is not tracked. Or, the reverse (savings might accumulate) if rebuilding lowered the region's unemployment rate. These fine points are currently ignored.

- *is absent* if all sectors of the economy are damaged to the same degree. In this case the economy simply implodes in a proportionate fashion. All losses are considered to be direct.
- *is the predominant loss* if a small critical sector is destroyed and no substitutes exist (no imports, inventories or excess capacity). Given these circumstances, direct loss to the critical activity will appear small relative to the predominantly forwardly linked losses incurred by industries relying on the critical sector.
- *is negative (implying regional economic gains)* if there is ample excess capacity in the region and reconstruction is financed by outside sources. These circumstances produce a typical demand driven recovery, where regional product grows in proportion to reconstruction spending (*i.e.*, the construction output multiplier times reconstruction expenditures).
- *is temporarily negative (implying temporary gains)* if reconstruction is financed by savings or borrowing. In this case, incomes rise during the period of reconstruction, but fall as repayment lessens aggregate demand in later time periods. The net effect may be positive or negative for the region. Which occurs depends on the net effect of the demand change filtered through the appropriate multiplier.
- *is less sensitive to economic structure (manufacturing dominated or service dominated economy) than to damage pattern, degree of integration (size), preexisting conditions, and who is financing the recovery*

3.9 Accounting Stance National vs. Regional

From a *national accounting stance*, indirect loss is measured by adjusting regional loss for the liability the federal government incurs in providing disaster relief, and for offsetting interregional trade effects (e.g., lost market within the region translate into new opportunities outside the region). The positive effects of outside aid produced for the region are to some degree negated by the liability caused by debt financed federal relief. Direct damage and subsequent indirect loss is transmitted to other regions via altered trading patterns (imports and exports). If the region produces critical materials for which no ready substitutes are available, then local direct and indirect disruptions can spill over to affect the national economy. Hence, it is possible for indirect loss to exceed that sustained by the region alone.

The national economy is also impacted in that external aid has to be financed. As a result, indirect gains disappear; the costs incurred outside the region negate local gains. So, from the vantage point of the nation as a whole, *indirect loss must exist.*

3.10 What is the Evidence? What is Known about Indirect Loss?

A number of case studies have been performed to assess the economic impact of disaster. These have typically focused on the total economic loss (direct and indirect) incurred by the stricken region. Much less is known about the systemic form of indirect loss, that is the loss stemming from dislocations in banking, financial markets, and insurance. A brief review of the evidence follows; particular attention is paid to the issue of systemic loss, the presence and magnitude of which is not well understood, and is subject to considerable debate.

Indirect losses from damages sustained when commercial ventures and lifelines are damaged. — Careful empirical experiments to detect the presence of forward and backward linked losses are a time consuming and expensive undertaking, involving the development of some combination of econometric and programming models. However, some evidence of the overall economic impact of disaster has been collected for a number of events: Hurricanes Hugo (Guimaraes *et al.*, 1993), and Andrew (West and Lenze, 1994), the Northridge (Rose *et al.*, 1995 and Cochrane, 1996), and Kobe (Chang, 1996) earthquakes, and the Grand Forks flood (Chang, 1999). In general, the results seem to indicate that rebuilding expenditures provide a powerful stimulus to a disaster stricken region. Figure 3.3 showing the growth in Dade County sales before and after Hurricane Andrew provides a compelling example.

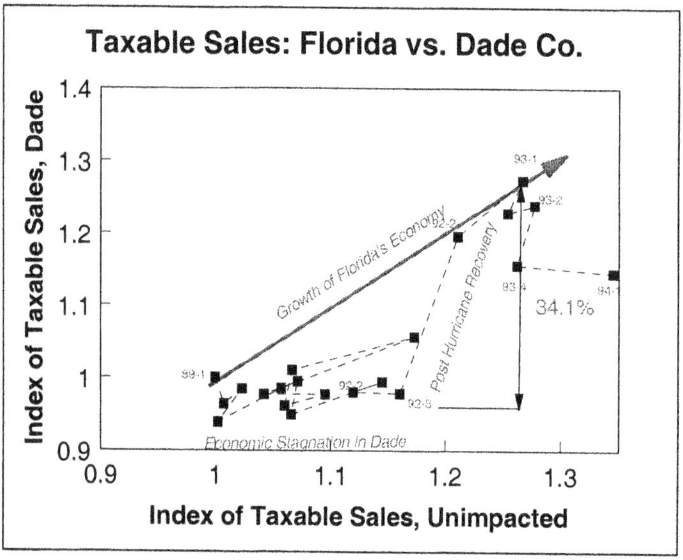

Figure 3.3 Taxable Sales: State of Florida versus Dade County

In this illustration, taxable sales collected in Dade County were used as a proxy for the change in gross regional product observed as a result of the damages caused by Hurricane Andrew. Taxable sales for Florida's unimpacted counties were used to normalize for the presence of ongoing economic trends. The plot suggests that the Dade County economy had weakened relative to that of the state prior to the storm. The post Andrew trend in taxable sales shows a dramatic reversal, amounting to a 34.1 percent rise from the pre event level. Note too that the stimulus weakens as rebuilding progresses, a pattern the HAZUS algorithm would have predicted. The same recovery pattern was detected in Charleston (after Hurricane Hugo[11]) and in Los Angeles County (after the Northridge Earthquake[12]).

Evidence to date also suggests that disasters are unlikely to strike a critical sector. The heavy damage suffered by Kobe's port facility (Hygo Ken Nanbu earthquake, 1995) and the disruption of semiconductor production in Taiwan (Chi-Chi earthquake, 1999) did not appear to produce significant forward or backwardly linked losses. Such losses were avoided in both cases for the same reason. There were sufficient inventories (in Taiwan) on hand and excess capacity in nearby port facilities (in Japan) to accommodate the disruption. Even in the case of the Northridge earthquake where freeways were blocked, commuters found alternative means of transportation, or discovered telecommuting.

3.11 Systemic Losses

Systemic loss is a financial variation of indirect loss. The most concise and clearest definition of systemic risk (loss) is found in Eisenbeis (p. 57):

> *Systemic risk is defined to be the transmission of problems at weak institutions to healthy institutions in ways that threaten the viability of many institutions or the system as a whole.*

Systemic losses could result from insufficient financing or insurance coverage, when such shortages disrupt fundamentally sound businesses. The key to this argument is a presupposition that businesses will curtail operations in *direct proportion* to the amount of insurance and bank financing available. Furthermore, it is argued that interlinked arrangements among insurers and banks could produce a contagion effect – those bankrupted by disaster could infect healthier firms, thus exacerbating the problem of insurance and credit availability.

The key question is: *Are financial failures contagious and will such failures cause losses to producers, service providers and local governments*? If so, the loss sustained from disasters may be larger than heretofore imagined. And, if it turns out that systemic risks do induce sizable losses outside of the insurance/financial sectors of the economy, then the argument for government involvement is strengthened. If not, then insurer/bank insolvencies are simply a matter best resolved by affected companies and industry associations.

---

[11] Guimaraes *et al.*, 1993.

[12] Cochrane, 1996.

3.11.1 Systemic Loss and Insurance

Much has been written about the fragility of the industry and the potential for large systemic losses resulting from a sudden reduction in surplus. The following is a synopsis of the industry's position.

A series of catastrophic natural disasters will deplete the insurance industry's surplus. Surviving firms will be forced to ration coverage by: 1) restricting availability and 2) raising premiums. Without coverage, firms not suffering direct damage will be forced to shut down and substantial national indirect economic loss will result.

What is the underlying reason for this position? Theoretically, a company's book of business is proportionate to its surplus. So, surplus serves the same function as required reserves in commercial banking. As claims are processed and paid, surplus diminishes, thereby reducing the amount of coverage the company can "safely" handle.

There are two problems with the rationing story. First, it is not clear that recent natural disasters caused firms to ration. It is difficult to argue that insurance was rationed after two of the nation's costliest natural disasters, hurricane Andrew and the Northridge earthquake. Claims about insurance rationing after the World Trade Center attack are addressed in Section 12. Second, the author is aware of no serious systematic study to estimate the indirect losses stemming from insurance rationing. The Congressional Budget Office (1994) concluded it didn't know what the indirect losses might be.

Has the supply of insurance been restricted by recent events? — Unfortunately, empirical tests of insurer behavior are hampered by lack of data; measures of coverage are simply unavailable. So other means have been employed to determine how loss of surplus impacts coverage. Industry profitability is commonly utilized as a proxy. The reasoning is as follows. The demand for insurance is price inelastic, meaning that, when faced with a shortage, buyers are willing to pay a disproportionately high premium in order to obtain coverage. Hence, when supply is restricted, the price paid for coverage rises and companies look more profitable, particularly those companies that paid no claims but benefited from price rationing. The resultant increase in profit should be observable in the equities market. According to the proponents of this technique, a rising stock price suggests the occurrence of insurance "rationing". From that it is *presumed* that financial indirect loss will ripple to other sectors of the economy.

Event studies of insurance company stock returns don't lend support to the contention that disasters constrict the national supply of insurance. In the case of Northridge, stock traders appeared to believe that insurance companies writing books of business in the disaster stricken area would prosper (Lamb and Kennedy, 1996). On page 9 they suggest a reason. "*Perhaps the gain in exposed firm stock values is due to an expected increase in demand for insurance in California (especially earthquake coverage) that will more than offset the potential earthquake losses.*" The Lamb and Kennedy findings call into question whether the risk insurers face is as great as the industry has publicized. The industry may argue that the reason why stock prices rose despite Northridge losses is due to their subsequent actions. That is, by curtailing exposure in California they limited future losses, thereby provoking a favorable market reaction. But, if this were the case, then why didn't stock values move in proportion to company exposure? They did not. Lamb and Kennedy found that exposure to loss was irrelevant to stock price movements. Zenith National Insurance with 88 percent of its direct written premiums in California moved the same as Allstate with only 13 percent.

The industry has also argued that back-to-back catastrophic losses such as Hurricane Andrew and the Northridge earthquake would cause unimpacted firms to ration insurance (due to depleted surplus) thereby creating higher prices and possibly increasing the profitability of some firms (those without substantial payouts). Again the Lamb and Kennedy (1996) and Lamb (1995) results do not support this conclusion. Firms with no exposure in California (Northridge) and in Florida (Andrew) show *no* excess returns or losses. In other words, their stocks were not bid up or down as a result of the events. Lamb's analysis of property and casualty stock price movements after Hurricane Andrew did detect a significant downward trend for exposed companies, something that should have been anticipated. According to Lamb (1995, p. 120), "*The market efficiently interpreted information generated by the hurricane and discriminated among property-liability insurers based on the existence and magnitude of insurance written.*"

These results are at odds with the conclusions reached by Doherty, Kleffner, and Posey (1992, p. 66) who concluded that a $50 billion loss of surplus would result in a reduction in the amount of insurance sold. The combined payouts for Andrew and Northridge are not as great as the amount contemplated in the Doherty *et al.* scenario, but they are a significant portion nonetheless. It is tempting to conclude that the Doherty *et al.* findings had indeed come to pass in Florida and California where insurers attempted to restrict coverage. However, it is more likely that this attempted withdrawal from both markets was motivated more by the political economics of regional insurance pricing and regulation than a national contraction in the supply.

Has the supply of insurance been restricted by recent events? The answer is *probably not.* Disasters, such as Andrew and Northridge have not drained industry reserves enough to limit the availability of other product lines. Prices have not changed appreciably for property and casualty coverage in general (as opposed to earthquake and wind). It appears that supply restrictions are concentrated in product lines where the industry perceives the risks to be *uninsurable.* The evidence available thus far does not support the contention that surplus shocks produce general rationing (aside from region specific adjustments motivated by factors other than supply considerations).

To what extent is the availability of insurance linked to real income losses? — Even if the above observation is ultimately proven wrong, the existence of systemic risk hinges on another factor, the link between (re)insurance availability and real economic dislocations. Numerous alternatives are available to firms and consumers, *e.g.*, to purchase alternative forms of insurance (such as catastrophe bonds); self-insure; seek coverage from the global insurance market; recapitalize via stock offerings; and merge with other financial institutions (such as banks or credit companies). At this stage, little is known about the adjustments firms and households make in the face of insurance rationing; the financial repercussions of rationing are equally murky. The Congressional Budget Office (1994), however, was not sanguine that a case for insurance related systemic losses could pass a common sense test. The report begins by suggesting that the lack of insurance may not prove to be a large problem for businesses. According to the CBO, "*the cost of risk is on average a small fraction of their costs*" (p. 29), and the cost of risk ranges from a high of 2.81 percent of revenue for the transportation services sector to a low of 0.132 for insurance.

The cost for all sectors averages about one half of one percent. The report concludes that "*the higher price of insurance may have little impact on overall economic activity in most cases*" (p. 30).

Will higher premiums cause indirect loss? — When it comes to consumers, the CBO report suggests that shortages of personal lines of insurance could induce households to save more, thereby altering the composition of spending in the economy (p. 32). The CBO construes such a shift as contributing to indirect loss. Is this an accurate portrayal? Basic economic theory predicts just such a shift. But, rather than labeling it an indirect loss, it would be characterized as an efficient price adjustment necessary to redirect scarce resources toward the rebuilding effort (and away from consumption). Therefore, there is nothing inefficient or costly about this shift in spending. High priced insurance produces the same result as high interest rates; it induces households and business to temporarily relinquish their claim on production, thereby permitting resources to be redirected to the rebuilding process.

How will mergers, acquisitions and more integrated global markets impact the picture? — Gaps in coverage, if they do materialize, may be filled by alternate financial arrangements. Additional capacity offered by a maturing foreign insurance industry could help make up for shortages. So too, a more integrated and concentrated domestic financial sector, is likely to weather surplus shocks more efficiently and smoothly than the current fragmented system. What is clear here is that predictions about the fallout from future disasters should not be grounded in outdated institutional arrangements.

Financial markets and sudden liquidations — As in the case of insurance, disaster induced systemic losses in the financial arena have few proponents. Systemic losses in banking, for example, are given much less credence than once thought. A conference on banking, financial markets and systemic risk (Kaufman, 1995) concluded, "*All things considered, systemic risk appears to be at a low level as of late 1994*" (p. 15). "*The potential for contagion and systemic risk in these instances is grossly exaggerated*" (Schwartz, 1995, p. 29). "*The research evidence seems to be fairly consistent that the relationship between financial panics and changes in real economic output runs from the real sector to the financial sector, rather than from the financial sector to the real sector*" (Eisenbeis, 1995, pp. 72-73). "*Financial crises ... are not the causal agents of recessions*" (Eisenbeis, 1995, p. 82).

What has been written about disasters and financial markets can be summarized succinctly. Capital markets are too vast to be disturbed by natural disasters. The municipal bond market which had been thought to be sensitive to sudden (disaster induced) liquidations seems to be moved instead by 1) interest earned on alternative investments and 2) current and proposed tax laws. Most post event stock price movements simply mirror the real losses corporations sustain. As such, stock price movements are a measure of direct loss, not of indirect loss (Cochrane *et al.*, 1992).

3.12 Economic Fallout from the World Trade Center Attack

Much of what has been developed thus far may be called into question in light of the first wave of economic assessments conducted shortly after the attack[13]. Loss estimates ranged from several hundred billion dollars to $16 billion, the former reflecting a reduction in stockholder wealth (attributed to the attack) and the latter incorporating the stimulative effect of reconstruction spending. Many of these early assessments were flawed. Most projected sizable losses and attributed the City's deteriorating post-attack economy to the attack itself. Only one study took into consideration the positive effect of rebuilding (financed predominantly by external sources). Later studies conducted by insurance and real estate interests broadened the range of losses to include the systemic effects resulting from a shortage of terrorism insurance coverage [14]

There are a number of issues that can be raised regarding these studies:

Baselines for estimating loss are inconsistent and misleading. — Many of the studies indicated that their baseline estimates of City income had disentangled the ongoing deterioration of the city's economy from the effects of the attack. They then accounted for the loss of post attack financial service personnel as a cost of the attack. This seemed appropriate at the time these reports were prepared (only months after the attack). In retrospect, and in light of new data, it appears that these baselines overstated the general health of the New York economy. A simple plot of net income earned by Merrill Lynch[15] from commissions, investments, and underwriting fees indicates that its economy was deteriorating prior to the disaster and continued to do so long after the event. See figure 3.4.

The trends show that employment is tied directly to the company's net income. It is only logical to address a follow-up question. To what extent was Merrill Lynch's net revenue (commissions, underwriting fees, etc.) impacted by the attack, as opposed to the bursting of a stock bubble, an event that would in all likelihood have occurred independent of the attack? It now appears that the stock price escalation observed during the 1995-2000 time period was due to irrational exuberance and accounting irregularities. The market's collapse since then can be arguably called a correction, fueled by accounting scandal. It appears that Merrill Lynch, much like other brokerage firms, are simply bringing their expenses (which means employment) in line with a shrinking revenue base. If so, it is highly likely that the baseline estimates of income reviewed by the GAO were overly optimistic[16], and as a result the income losses attributable to the attack are too high.

[13] New York City Office of the Comptroller, New York Governor and State Division of the Budget, New York City Partnership and Chamber of Commerce, Fiscal Policy Institute, New York State Finance Committee, Milken Institute, New York Assembly Ways and Means Committee

[14] Economic Perspectives on Terrorism Insurance (JEC Study -- May 2002).

[15] Merrill Lynch was chosen because of its size and the fact that it had relocated some of its operations out of the City, a decision which some argued contributed to the City's economic woes.

[16] Bram, Orr, and Rappaport (2002) employed time-series techniques to assess how the New York economy would have performed had the attack not occurred. An autoregressive technique was utilized linking regional employment to national employment. The trend in national employment was then used to derive a baseline estimate of the City's economic performance. Even this careful attempt to disentangle economic deterioration unrelated to the attack from the attack itself led to an overly optimistic employment picture. The fact remains that New York's economy is heavily influenced by the financial services sector, a sector

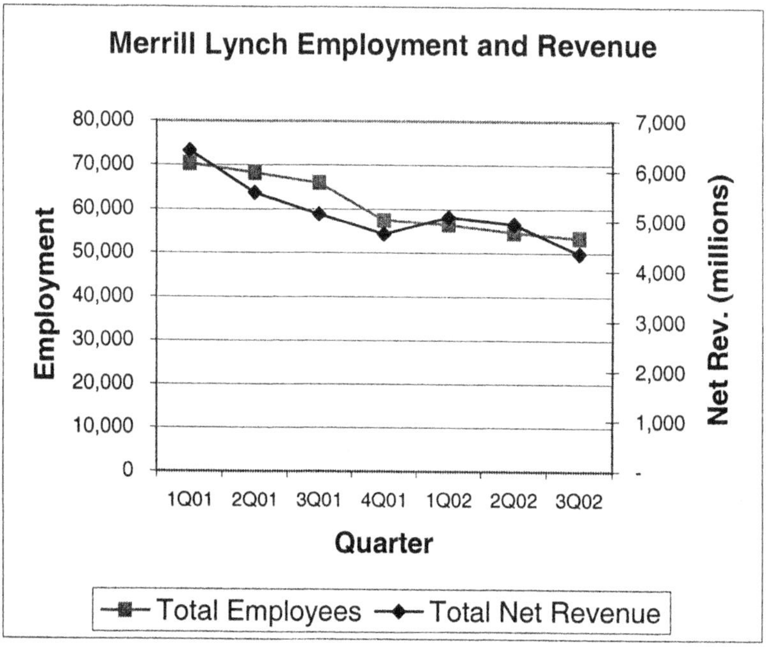

Source: Merrill Lynch quarterly reports

Figure 3.4 Merrill Lynch Employment and Revenue

Failure to account for the reconstruction stimulus — Most of the studies failed to account for the additional spending, and the associated multiplier effects, that would occur as a result of reconstruction (GAO, 2002).

Failure to distinguish between losses sustained from the attack vs. the cost of adjusting to the terrorist threat — The former is the loss sustained as a result of the disaster; the latter is an ongoing expense that is incurred both in New York and elsewhere. This may seem like a trivial distinction, especially to the victims, but it is an important distinction to those trying to assess disaster losses. Lumping everything into the disaster category would paint a misleading picture of the extent of indirect loss to be expected after a large urban catastrophe.

The existence of post attack systemic loss is still and open question. — Shortly after the World Trade Center disaster, banking, insurance, and real estate interests began lobbying for passage of legislation to provide federal terrorism insurance. They argued that without terrorism insurance the following would occur: 1) terrorism insurance will be unavailable and unaffordable; 2) new commercial construction will be hampered (curtailed); and real estate transactions postponed. These assertions are problematic for several reasons. First, terrorism insurance was already available on a standalone basis. Second, the claim regarding the cancellation of commercial

directly responsible for 33 percent of the City's income. Had the autoregressive technique concentrated on this sector, it would have resulted in a more realistic measure of baseline incomes.

construction projects is anecdotal and fails to pinpoint the cause. Vacancy rates in New York have risen and rents have fallen since the September 11 attacks, factors not conducive to the creation of additional commercial office and retail space. Whether the slowdown in building is traceable to the availability of insurance or a broader and more chronic slowdown in the City's economy remains an open question.

3.13 Conclusions

The Chapter provides a foundation for conceptualizing and measuring indirect loss and argues that systemic loss is a special case of indirect loss. It points out the pitfalls inherent in economic impact studies, and the difficulty of developing a simple rule of thumb tying indirect to direct loss. It suggests that the indirect economic loss module of HAZUS provides a *consistent* means of evaluating loss, a quality that appears to be lacking in a number of economic impact studies conducted in the wake of the World Trade Center attacks. At a minimum, the algorithm would have helped maintain a credible baseline, and would have accounted for injections of outside aid.

The Chapter questions the existence of disaster induced systemic loss. Did the insurance "crises" after the World Trade Center attack, Hurricane Andrew, and the Northridge earthquake lead to real economic dislocations? Anecdotal evidence aside, a careful economic analysis of this issue has yet to be conducted. The insurance industry's vulnerability to the most recent string of catastrophes is debatable. The links between insurance and macroeconomic performance have not been credibly demonstrated. And, since financial institutions are in the throws of a institutional revolution (including mergers, acquisitions and global integration), it seems unlikely that the newly emerging insurance sector will behave as in the past; what resources the "new" financial sector can draw upon, and how vulnerable it might be is open to debate.

References

Boisvert, R. 1992. "Indirect Losses from a Catastrophic Earthquake and Local, Regional, and National Interest," in *Indirect Economic Consequences of a Catastrophic Earthquake*. Washington, DC; National Earthquake Hazards Reduction Program, FEMA.

Bram, J., A Haughwout, and J. Orr. 2002. "Has September 11 Affected New York City's Growth Potential?" *FRBNY Economic Policy Review, November*.

Bram, J., J. Orr, and C. Rappaport. 2002. "Measuring the Effects of the September 11 Attack on New York City," *FRBNY Economic Policy Review, November*.

Brookshire, D. and M. McKee. 1992. "Indirect Loss Measurement and Computable Equilibrium Models," in *Indirect Economic Consequences of a Catastrophic Earthquake*. Washington, DC; National Earthquake Hazards Reduction Program, FEMA.

Chang, S.E. 1996. "Regional Economic Impact of the January 17, 1995 Kobe, Japan Earthquake," paper presented at the 43rd North American RSAI Meetings, Washington, DC, November 14-17, 1996.

Chang, S.E. 1999. "The Red River Flood of 1997: An Assessment of Damage and Long Term Economic Recovery," EQE International.

Cochrane, H. 1974. "Predicting the Economic Impact of Earthquakes," in H.C. Cochrane *et al.* eds. *Social Science Perspectives on the Coming San Francisco Earthquake*. National Hazards Research Paper, No. 25, NHRAIC, Boulder, CO; U. of Colorado.

Cochrane, H. 1996. "Is Northridge A Model for Future Events? What are the Economic Consequences With and Without Insurance and Federal Assistance?" Paper presented at EERI/FEMA Conference Analyzing Economic Impacts and Recovery, Pasadena, CA.

Cochrane, H. 1997. "Indirect Economic Losses," in *Development of Standardized Earthquake Loss Estimation Methodology Vol. II*. Menlo Park, CA; Risk Management Solutions, Inc.

Cochrane, H., M.P. Laub, and J.R. Barth. 1992. "The Effect of Earthquakes on Banking and Financial Markets, and Computable Equilibrium Models," in *Indirect Economic Consequences of a Catastrophic Earthquake*. Washington, DC; National Earthquake Hazards Reduction Program, FEMA

Congressional Budget Office. 1994. "The Economic Impact of a Solvency Crisis in the Insurance Industry," April, Washington, DC.

Cummins, J.D. and P.M. Danzon. 1991. "Price Shocks and Capital Flows in Liability Insurance," in J.D. Cummins, S.E. Harrington, and R.W. Klein eds. *Cycles and Crises in Property/Casualty Insurance: Causes and Implications for Public Policy*. Kansas City, MO; National Association of Insurance Commissioners.

Doherty, N., A. Kleffner, and L. Posey. 1992. *Insurance Surplus: Its function, Its Accumulation, and Its Depletion*. Boston, MA; The Earthquake Project.

Eisenbeis, R.A. 1995. "Systemic Risk: Bank Deposits and Credit," in George Kaufman ed. *Research in Financial Services Private and Public Policy, Banking, Financial Markets and Systemic Risk, Vol. 7*. Greenwich; JAI Press, 55-86.

Government Accounting Office. 2002. "Review of the Studies of the Economic Impact of the September 11, 2001Terrorist Attacks on the World Trade Center."

Gron, A. 1989. "Capacity Constraints and Cycles in Property-Casualty Insurance Markets," working paper, University of Chicago.

Guimaraes, P., F.L. Helfner, and D.P. Woodward. 1993. "Wealth and Income Effects of Natural Disasters: An Economic Analysis of Hurricane Hugo," *Review of Regional Studies*, 23 (2): 97-114.

Kaufman, G. 1995. "Comment on Systemic Risk," in G. Kaufman ed. *Research in Financial Services Private and Public Policy, Banking, Financial Markets and Systemic Risk, Vol. 7*. Greenwich; JAI Press.

Lamb, R P. and W.F. Kennedy. 1996. "Insurer Stock Prices and Market Efficiency Around the Los Angeles Earthquake," submitted to the Journal of Insurance Issues, June 1996.

Lamb, R. 1995. "An Exposure-Based Analysis of Property-Liability Insurer Stock Values Around Hurricane Andrew," *Journal of Risk and Insurance*, 62 (1): 111-123.

RMS Inc. 1996. "Development of a Standardized Earthquake Loss Estimation Methodology," draft technical manual prepared for the National Institute of Building Sciences, Washington, DC.

Rose, A., J. Benavides, S.E. Chang, P. Szczesniak, and D. Lim. 1995. "The Regional Economic Impact of an Earthquake: Direct and Indirect Effects of Electricity Lifeline Disruptions," paper presented at the 42nd North American Meeting of the Regional Science Association Meeting International, Cincinnati, November 10, 1995.

Schwartz, A.J. 1995. "Systemic Risk and the Macroeconomy," in G. Kaufman ed. *Research in Financial Services Private and Public Policy, Banking, Financial Markets and Systemic Risk, Vol. 7*, Greenwich; JAI Press, 10-30.

West, C.T. and D.G. Lenze. 1994. "Modeling the Regional Economic Impact of Natural Disaster and Recovery: A General Framework and Applications to Hurricane Andrew," *International Regional Science Review*, 17: 121-150.

4 Has September 11 Affected New York City's Growth Potential?*

Jason Bram, Andrew Haughwout, and James Orr

Federal Reserve Bank of New York, USA
e-mail: Jason.Bram@ny.frb.org, Andrew.Haughwout@ny.frb.org, James.Orr@ny.frb.org

4.1 Introduction

The terrorist attack on the World Trade Center claimed close to 3,000 lives and caused billions of dollars in property damage. In the aftermath, New York City's economy contracted briefly but sharply: many businesses were forced to shut down, mostly temporarily, and tens of thousands of workers were either dislocated for a short time or lost their jobs (Bram, Orr, and Rapaport, 2002). The economic implications for the city, however, clearly extend beyond the first few months following the attack.

In this article, we analyze the effects of September 11 on the longer run prospects for the New York City economy. We find that, on the one hand, several downside risks to the city's growth outlook have arisen. In a worst-case scenario, the concentration of the attack on Lower Manhattan has raised the possibility that financial firms might relocate outside the city, which could generate a cumulative downward spiral of job and income growth. On the other hand, the city's industrial structure and its quality-of-life amenities—namely, an industrial mix weighted toward high-growth sectors and an environment desirable to workers and firms—suggest favorable economic prospects. However, whether or not these prospects translate into actual growth going forward depends to a large degree on the city's policy response to the economic pressures arising from the attack. The key elements of this response will be the ability to avoid budgetary decisions that reduce the long-run attractiveness of a New York City location and, through rebuilding, to maintain the productivity of the city's capital stock.

We begin by describing recent earnings and land price trends in the city. To help interpret these data, we apply a model that emphasizes the importance of local property markets as an indicator of trends in a mature urban economy. The New York experience is then discussed in relation to the model, and the city's economic strengths and weaknesses both before and after the attack are evaluated. We also identify several challenges that local policymakers will face if the city is to recover fully. Finally, our study presents evidence on land prices after September 11 indicating that a strong demand for New York City property still exists.

---

* The original version of this chapter appears in Jason Bram, Andrew Haughwout, and James Orr (2002) "Has September 11 Affected New York City's Growth Potential?" *Economic Policy Review*, 8 (2), pp. 81-96: Federal Reserve Bank of New York. The authors thank two anonymous referees, as well as Nathan Barczi for valuable research assistance. The views expressed are those of the authors and do not necessarily reflect the position of the Federal Reserve Bank of New York or the Federal Reserve System.

4.2 Growth Trends in New York City

Three important patterns characterize the trends in the New York City economy over the past three decades: steady but cyclical employment, rising real earnings, and appreciating land prices.

Economic growth at the national level is usually measured as the average annual rate of expansion of real (inflation-adjusted) GDP over some period. At the city level, however, an official output measure is not available and thus growth is often measured as the trend rate of growth in jobs and/or real income. Although in many localities these measures move together, the measures in New York City present different perspectives on the process of economic growth. Average total employment in the city essentially has been unchanged for three decades (figure 4.1). Even with the rapid expansion of jobs in the mid-to-late 1990s, only in 1999 did total employment exceed its previous cyclical peak, in 1989. Indeed, the all-time peak level of total employment was reached in 1969. In contrast, real earnings of city workers have trended upward since 1980 at an average annual rate of about 3.5 percent.

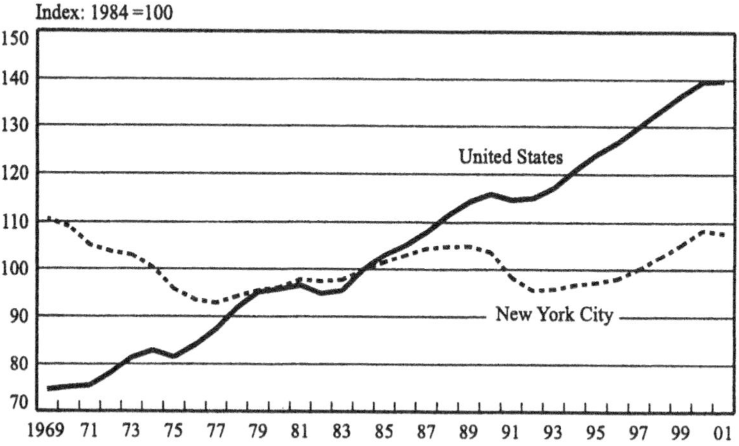

Source: U.S. Department of Labor, Bureau of Labor Statistics.

Figure 4.1 Total Employment in New York City and the United States

New York's pattern of steady employment and rising real earnings is attributable to occupational and industrial restructuring. Average earnings in the city have risen because of accelerating productivity growth in existing jobs and a shift toward higher paying jobs. During the 1980s, the expansion of jobs in the high-paying FIRE (finance, insurance, and real estate) sector helped raise average real earnings citywide. Many of these job gains were temporarily reversed in the recession of the early 1990s. However, the city's recovery since the mid-1990s, in conjunction with the job expansion in some relatively high-paying service sectors, again boosted real earnings.

Over this same period, earnings per job increased nationwide, but New York's earnings per job rose even more rapidly (figure 4.2). Thus, the city increased its earnings relative to the rest of the country by enhancing the productivity of existing employment and capturing a rising share of high-productivity jobs.

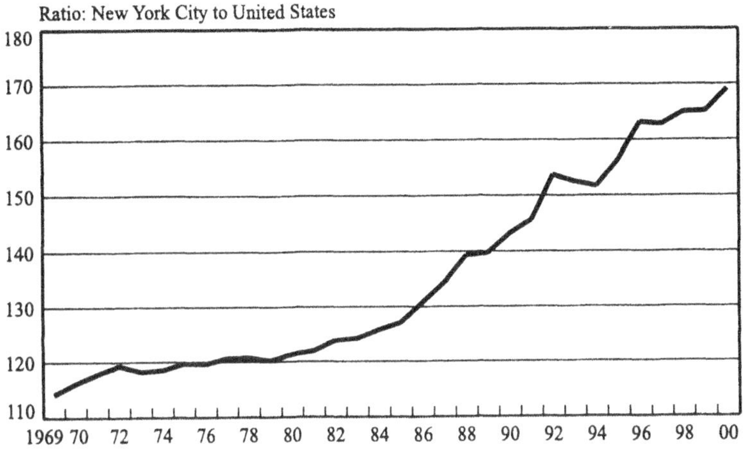

Sources: U.S. Department of Commerce, Bureau of Economic Analysis (earnings); U.S. Department of Labor, Bureau of Labor Statistics (total employment).

Note: Earnings are nominal.

Figure 4.2 Relative Earnings per Worker in New York City

Furthermore, the price of housing in New York has also risen relative to the nation since 1976 (figure 4.3). Here, we use the repeat-sales price index, which controls for the quality of the structure, as a measure of price change. Constant-quality housing price changes provide a relatively clean measure of the attractiveness of the bundle of local traits available to residents of these homes. These data indicate that the price of residing in New York has climbed relative to the rest of the nation—with a particularly sharp rise in the second half of the 1990s.

4.3 Modeling Growth in a Mature City

To analyze the effects of the terrorist attack on the New York City economy, we adopt the model of urban economies developed by Roback (1982) and refined in Blomquist, Berger, and Hoehn (1988), Gyourko and Tracy (1991), and Haughwout (2002). In this framework, metropolitan areas are viewed as small, open economies to which labor and capital are elastically supplied (see the appendix for more details on the model). Since each city is just one of many places where firms and households may choose to locate, it must offer competitive levels of profit for firms and utility for households.

The value of a city to firms and households determines their "bids" in the city's local land and labor markets. Thus, if a location has fixed characteristics that are extremely productive, firms will offer high bids for sites and high wages to attract workers there. Locations may vary in their attractiveness to firms for a variety of reasons, ranging from access to valuable existing infrastructure stocks to proximity to markets or sources of raw materials.

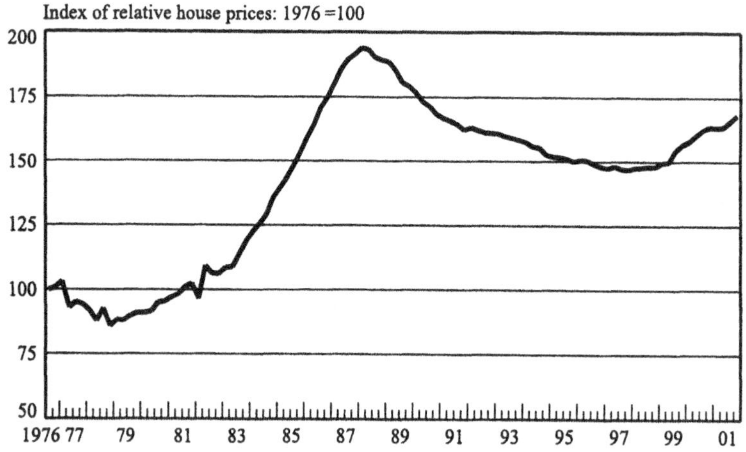

Sources: Office of Federal Housing Enterprise Oversight; Federal Reserve Bank of New York calculations.

Notes: The index is based on the ratio of the repeat-sales pricemeasure for existing single-family homes in the New York City metroarea to that of the United States overall; the index is designed to control for changes in the mix of homes sold. Data for the city itself are not available, so we present the index for the New York City primary metropolitan statistical area (PMSA), which consists of the city's five boroughs (the Bronx, Brooklyn, Manhattan, Queens, and Staten Island) and Westchester, Putnam, and Rockland counties in New York State. Approximately 80 percent of the population of the New York PMSA lives within the city boundaries.

Figure 4.3 New York City Area House Prices Relative to U.S. Average

Given this set of productive amenities, firms will relocate if wages and/or land prices are too high for them to make equilibrium profits. If costs are so low that incumbent firms can make excess profits, new firms will enter, bidding up local prices. A similar logic applies to households. In a mixed economy that provides sites for households as well as firms, local land and labor prices are determined by the satisfaction of firm and household equilibrium conditions. The local price equilibrium is thus attained when no firm or household wishes to relocate.

When a city's productive or residential environment changes over time, such changes will be reflected in local prices. Improvement in a city's appeal to firms and/or households will lead to a rise in relative land prices over time[1]. When a city's

[1] By "relative," we mean in comparison to other locations within the broad capital and labor market. In practice, the nation is a sensible benchmark, as national labor supply is relatively inelastic. Capital may be

attractiveness is at its highest to firms (for instance, if taxes paid by firms fall but nothing else changes), wages will tend to rise. When households receive the lion's share of the benefits of an amenity change (perhaps a costless reduction in the rate of violent crimes), relative wages would be expected to fall.

Wage and land price data for New York City indicate that the trends prior to September 11 were favorable, as both wages and relative land prices had been increasing over long and short horizons (figures 4.2 and 4.3). Land price increases are a sign of growing attractiveness to firms and/or households, while wage increases indicate that businesses increasingly value a New York location relative to households. Interpretation of the wage increases observed in the city, however, is also confounded by the change in the population's characteristics over time, as we shall discuss.

Prices of housing and land, both of which are durable assets, reflect not just current conditions, but also market participant views of a location's future. The demand for a property is determined by the present value of the stream of net benefits expected to accrue to the property over its lifetime. When an individual purchases a piece of real property, he or she must evaluate future conditions in the area where the property is located. If conditions are expected to deteriorate next year, or in five years, then purchase prices this year will be reduced, as the stream of returns into the future falls. Relative property prices in any period thus indicate, in part, expected future conditions in that location. Note that this predictive power of property prices does not extend to either property rentals or wages—both of which reflect current conditions or, more precisely, conditions expected to pertain to the duration of the contract.

Our discussion provides some insight into the sources of the apparent land market strength in New York City. Among the factors explored in some detail in the academic literature are changes in local fiscal policies, such as a fall in local taxes without compensating service decreases or an increase in local infrastructure provision. On the household side, safety from crime, the quality of the public educational systems, and taxes are all considered important factors determining local quality of life and thus local land and labor prices (see Blomquist, Berger, and Hoehn, 1988; and, Gyourko and Tracy, 1991). Measured by several of these factors, conditions in New York have improved over the past thirty years, with especially sharp improvements occurring in the past decade.

4.3.1 Applying the Model to New York City

Here, we extend the model presented above to account for the various types of firms and households that are locating in a modern economy.

The model dramatically simplifies the structure of mature real-world economies, wherein many kinds of firms and households coexist, particularly in urban areas. As a practical matter, the way that firms and households sort themselves over space is relevant as an indicator of both evaluations of locations and the prospects of particular areas. Thus, the fact that New York's relative (constant-quality) home values have

more elastically supplied at the national level, but in this context we take the national supply of capital as given.

increased significantly versus those of the nation indicates that the city has become more attractive to households and/or firms over the past few decades[2].

Still, the benefits of a New York City location are clearly not the same to all actors in the economy. New York's industrial and demographic structures differ from those of the nation because some groups are willing to pay more than others for the city's particular amenities. Most obviously, New York's long domination of financial services employment provides significant incentives for firms in that industry to locate there[3]. But this same feature is less attractive to producers in, say, the automotive industry. Examples of the city's critical household amenities include relatively easy access to high-skill service employment opportunities and a diverse set of consumption opportunities such as theaters, museums, and other cultural offerings. However, the city's public schools, by most criteria, fall short of their suburban counterparts on several crucial dimensions[4].

The fact that New York's appeal as measured by its relative land prices is near an all-time high seems to be contradicted by population and employment figures that have fallen relative to the nation. However, New York's situation is more complex than simple figures demonstrate: the city has substituted high-paying, high-productivity jobs in a few industries for low-skill, low-paying jobs in others. On the household side, corresponding changes in the composition of the local population have also tended to support a strong housing market, as we shall explain.

Clearly, the city's future depends on the growth prospects of the types of firms and households that have revealed a preference for New York. If, for example, the nation's financial services industry were expected to decline significantly, the city's concentration of these jobs would become a liability rather than an asset. Likewise, if concentrations of high-skill individuals or immigrants were detrimental to employment or population growth, the city's prospects would be dimmed.

New York's future—and the effect of events such as September 11—will therefore depend on the answers to two important questions:

- Can New York continue to provide the amenities valued most by those industries vital to its economy?
- Will the industries that benefit most from these amenities continue to thrive at the national level?

Our analysis suggests that the answers point to favorable prospects for New York City, although the future will not be without policy challenges. Our evidence indicates that New York is increasingly specialized in the production of skill-intensive services and in the provision of residences for high-skill workers and immigrants. These factors have supported rising land prices and wages in the city over the past three decades[5].

---

[2] This conclusion is also based on the assumption that higher prices of constant-quality housing reflect growth in demand for the land on which a house sits or for durable capital in place there, and that firms and households are competitors in the land market.

[3] See Harrigan and Martin (2002) and Quigley (1998) for discussions of the local benefits of industry concentrations.

[4] For example, for evidence on city schools' math test scores, see New York City Department of Education (2002).

[5] Because of improvements in the relative skill level of New York City employers and residents over time, we must be cautious in interpreting any income data. If the local skill-adjusted wage is higher than it is elsewhere, it can be interpreted as evidence that firms are willing to pay relatively more to workers here, implying that these workers are more productive when they work in New York.

4.4 The City's Growth Fundamentals

According to the model we employ, the desirability of a city springs from its productive and residential amenities. We begin our discussion of New York City's growth fundamentals by highlighting two key improvements over the past two decades that have made the city attractive to firms and workers: an improved fiscal position and a reduced crime rate. These improvements reflect both explicit policy choices and a healthier local economy. Next, we examine the associated changes in the city's industry and population mixes by reviewing the city's industrial structure and considering the medium-term prospects for these industries. Trends in the financial services sector are emphasized, because its fortunes exert a strong influence on the city's overall performance. We find that the local economy tends to have a relatively high concentration of industries expected to lead the nation in growth over the next decade. Finally, by profiling the changes in the local labor force, we find a more advanced educational profile in the city than in the nation as a whole—an advantage that has increased over the past ten years.

4.4.1 New York's Productive and Residential Environments

Both firms and households have benefited from the dramatic improvements in the city's financial condition since 1975, when New York experienced its most severe fiscal crisis since the Great Depression. The city has substantially reduced its reliance on debt finance, lowered property taxes, and enhanced its financial reporting since then (Haughwout, 1997). Although the overall tax burden remains high, the share of the local tax dollar used to service short-term debt has been sharply reduced. Improvements in the city's public transportation system, particularly during the 1990s, also have presumably benefited firms and households.

Public safety is another key aspect of a location's attractiveness. In that regard, New York's crime rate tells a compelling story (figure 4.4). The rate, which actually rose during the economic boom of the 1980s, began a steady and steep decline in 1991 (amidst one of the city's worst recessions) that continued through the end of the decade. Although the 1990s saw a downward trend in crime nationwide and the city's improving economy clearly helped to lower the crime rate, New York's decline in crime was even more pronounced. Moreover, neighborhood-specific data from New York City suggest that the sharpest declines in crime over this period tended to occur in the poorest areas and highest crime areas. The U.S. Census Bureau's periodically conducted New York City Housing and Vacancy Survey (HVS)—which includes a resident-reported measure of "neighborhood quality"— indicates that city dwellers perceived significant improvements in crime in their own areas, with the sharpest improvements noted in the poorest neighborhoods.

Although these improvements in the business and residential environments have helped to enhance relative land and labor prices, New York's portfolio of amenities is not without certain liabilities. Perhaps the most important of these are the city's relatively high taxes on residents and businesses, and public schools that fall short of suburban competitors. Nonetheless, the evidence suggests that these liabilities have been outweighed by improvements in recent decades.

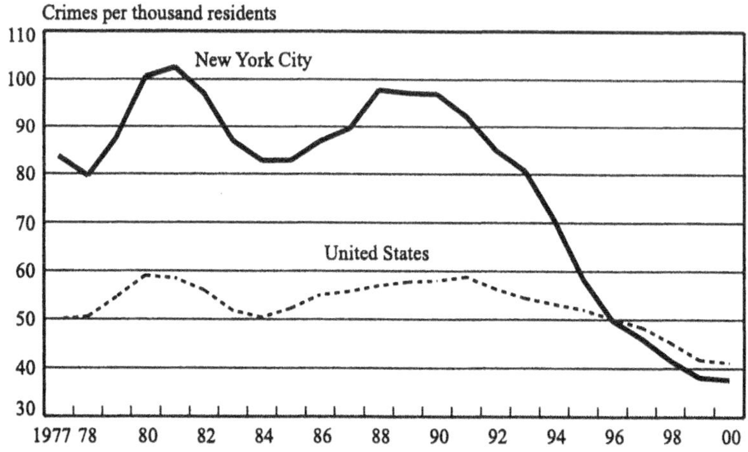

Sources: Federal Bureau of Investigation; New York State Division of Criminal Justice Services.

Figure 4.4 Annual Crime Rate of New York City and the United States

4.4.2 Industry Mix

A city's industry mix indicates which sectors benefit most from its particular package of amenities and helps determine medium-term growth prospects. New York has long had a significant concentration of employment in the financial services industry. Firms in this sector appear to strongly value proximity to one another, implying that New York's main advantage in this sector has promoted its growth over the past several decades[6].

In addition, areas with a high concentration of growth industries tend to benefit from trends in the broader economy. Accordingly, we examine New York City's current industry mix, which has resulted in part from the aforementioned changes in the city's environment. We find that the current mix is strongly weighted toward growth sectors. By comparing the performance of these industries in the city with their national counterparts, we find that most local industries have tended to lag moderately in terms of income growth but fairly substantially in terms of job growth. This finding is consistent with the model of New York City as a mature economy, with less potential for expansion than most parts of the nation—if for no other reason than land and space constraints.

In recent years, industries with a high concentration in New York City have tended to register relatively brisk job growth at the national level and are projected to continue doing so over the next decade. In particular, the securities, business services, motion picture, legal, social, educational, and management services industries have all registered solid job and earnings growth at the national level, and their attraction to New York City has contributed to strong overall gains for the city. Conversely, the industries with the weakest trends in employment and earnings—almost all of which

[6] See Harrigan and Martin (2002).

are in the manufacturing sector—are, for the most part, currently underrepresented in New York City. One notable exception is apparel manufacturing, which, though far less important than it was a few decades ago, is still one of the city's key industries.

Here, we illustrate the overall net effects of these industry dynamics by separating out the marginal effects of *local factors* and *industry mix* on overall income and job growth. Specifically, we decompose the differential between local and national growth into two components. *Local-factor effects* represent the performance of local industries compared with their national counterparts; these effects represent the portion of the overall growth differential not explained by local industry mix. *Industry-mix effects* represent what the differential would be if local job growth matched national job growth exactly within each industry; these effects represent the extent to which growth is enhanced or diminished by virtue of the local area's particular industry mix.

We conduct this exercise by first looking at changes in earnings over the 1995-2000 period to gauge the dynamics of the city's recent economic boom. Figure 4.5 presents local relative to national earnings growth for most two-digit Standard Industrial Classification industries, with selected key industries highlighted. As we see, some industries (those above the line) grew faster locally than they did nationally, while others (those below the line) grew more slowly. On balance, though, New York City's industries lagged their national counterparts in growth by an estimated 0.6 percentage point. However, because some of the nation's fastest-growing industries were overrepresented locally, New York City benefited disproportionately from its industry mix, as aggregate earnings growth was boosted by an estimated 2.3 percentage points (see table 4.1).

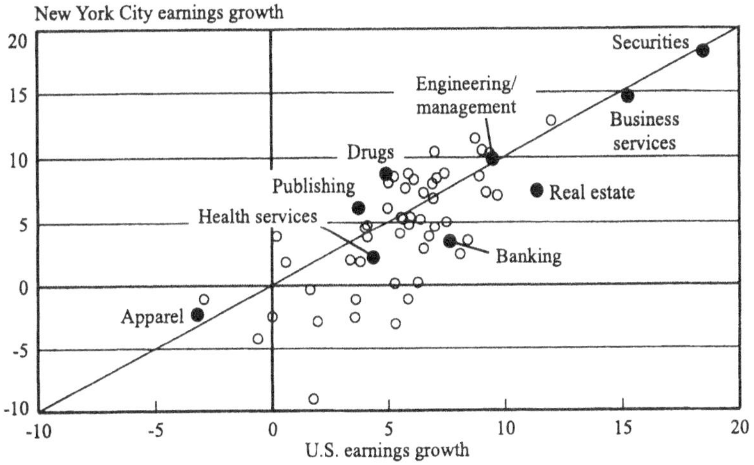

Sources: New York State Department of Labor; U.S. Department of Labor, Bureau of Labor Statistics.

Notes: Industries above the diagonal line registered faster earnings growth in New York City than they did nationally; those below the line experienced below-average growth in New York City. The dark circles represent selected two-digit Standard Industrial Classification industries that are important to the New York City economy.

Figure 4.5 Earnings Growth by Industry: United States versus New York City, 1995-2000

Table 4.1 Average Annual Growth in Earnings and Employment (percent)

Category	Total Earnings 1995-2000	Employment 1995-2000	Earnings per Worker 1995-2000	Employment 2000-2010
National growth	7.1	2.5	4.5	1.7
Local-factor effect	-0.6	-0.5	-0.1	—
Industry-mix effect	+2.3	+0.5	+1.7	+0.3
Local growth (New York City)	8.8	2.5	6.1	—

Source: U.S. Department of Labor, Bureau of Labor Statistics

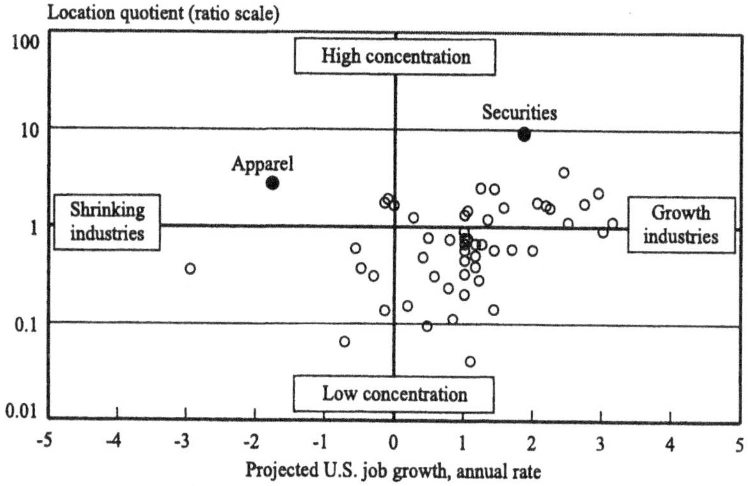

Sources: New York State Department of Labor; U.S. Department of Labor, Bureau of Labor Statistics; Federal Reserve Bank of New York calculations.

Notes: A location quotient is a measure of local industry concentration. It is defined here as an industry's share of employment in New York City divided by its national share. For example, the securities industry accounts for 4.9 percent of jobs in New York City, but only 0.57 percent of jobs nationwide, so its location quotient is 4.90/0.57=8.6. Thus, this industry's share of New York City employment is 8.6 times the U.S. average. The dark circles represent the chief two-digit Standard Industrial Classification industries in New York City.

Figure 4.6 Projected U.S. Job Growth in 2000-2010 versus New York City Concentration

In terms of employment growth between 1995 and 2000, these effects were somewhat less pronounced, but they generally worked in the same direction. Here, positive industry-mix effects fully offset negative local-factor effects, leaving overall local job growth on a par with that of the nation. Over this five-year period, the local-factor effect in earnings per job was very near zero. That is, all of the city's relative growth in earnings per job was attributable to industry mix.

To gauge the likely impact of industry mix going forward, we performed a similar analysis using Bureau of Labor Statistics *projected* job growth rates by industry. The projected rates are for 2000-2010; comparable rates would presumably apply for

2002-12. Although we cannot predict local-factor effects, based on this information, industry-mix effects should continue to contribute positively, albeit modestly, to local job growth, as demonstrated in the table. As figure 4.6 illustrates, industries with above-average expected job growth tend to be more concentrated in the city than those with subpar growth or job losses. Still, to the extent that local industries continue to lag their national counterparts in job growth, it is unclear whether local job growth would keep pace with the national average.

Because New York City has been moving toward increasingly high-value-added (and high-salary) industries and businesses over time, a comparison of local employment trends with corresponding national trends tends to understate the city's true relative performance. In fact, a study by McCarthy and Steindel (1997) points to the metropolitan region's persistently strong income growth as an indication of healthy demand for its goods and services. The authors argue that if income is holding up fairly well, then spending on regionally produced goods and services is high and business in the region may be more robust than the employment data suggest. As depicted in figure 4.2, average earnings per worker have consistently grown more strongly in New York City than nationwide; between 1995 and 2000, this growth was due entirely to *industry mix*. If these long-standing trends persist, New York City should continue to fare better in terms of income growth than in job growth.

Looking forward, we note that to the extent that local-factor effects on earnings and employment remain neutral or small—a condition that depends on New York City remaining an attractive site for business locations—the city may be expected to continue its long-standing pattern of rising earnings and cyclical but steady employment. Were city conditions to deteriorate, however, the positive industry-mix effect could be insufficient to offset large, negative local-factor effects in employment and even earnings per job. The role of the terrorist attack in this scenario is discussed later.

4.4.3 The Financial Sector

New York City has a favorable and diverse industry mix—at least within the service sectors—yet financial services stands out as a key sector driving the local economy. The securities industry alone accounted for 5 percent of city employment in 2000 and an estimated 20 percent of city earnings. These shares were higher than they were at any point in history and roughly eight times the respective U.S. figures. Not surprisingly, then, trends in the financial sector dominate the local economic landscape. In fact, a recent study (Bram and Orr, 1999) finds that cycles in the financial sector—particularly the securities industry—tend to presage cycles in the broader local economy.

Accordingly, the steep contraction in Wall Street employment over the past year—driven by the national recession, particular weakness in the financial markets, and the September 11 attack—is a matter of serious concern and appears to pose the greatest threat to the city's medium-term growth prospects. The fact that securities employment has also fallen, albeit somewhat less sharply, in the rest of the country suggests that these job losses reflect a combination of industry- and city-specific factors. Financial services will be an important sector to monitor when gauging the overall health not only of Lower Manhattan, but also of the broader regional economy. Over the longer term, the Internet and general advances in information and communications technology are likely to continue to have important effects on the broad structure of the financial services industry (see Orr and Rosen, 2000, for a more complete discussion).

4.4.4 Labor Force Composition

New York City's population growth accelerated over the past decade. According to the decennial U.S. census, the city's population grew 9.4 percent during the 1990s—the strongest growth since the 1920s. Moreover, adjacent counties across the metropolitan area also saw increasingly brisk gains. Because New York City's economy tends to be a major hub of information-based industries, the metropolitan area's labor force, not surprisingly, has a much higher proportion of college graduates than the nation overall. Much of the area's highly educated population resides in the suburbs, but the city itself has attracted a very well-educated workforce. In the city proper, an estimated 30.5 percent of adults are college graduates—a share well above the national average of 25.2 percent. More important, though, is the *change* in educational attainment: in 1991, just 22.1 percent of New York City adults were college graduates, only slightly more than the national figure of 21.4 percent[7]. This indicates that New York is growing in attractiveness as a residential location for well-educated Americans. Because high concentrations of well-educated residents are beneficial to urban growth (Rauch 1993), this component of the city's population base is a positive factor for its future prosperity. In addition, the city is home to many immigrants, presumably because of its historical role as an immigration gateway. Although the effect of large stocks of immigrants on local labor markets is unclear, recent research suggests that these individuals tend to support a vigorous housing market (Saiz, 2002). Thus, a reduction in immigration to the United States would disproportionately affect New York, if only in terms of population growth.

In sum, our empirical examination of New York City's changing industry and population mix shows the city to be attractive to those industries that benefit from its productive amenities and to a relatively highly skilled and highly educated labor force. These results are consistent with the model of a mature urban economy as well as with the rising land prices and real wages observed in the city over the past two decades.

Although the city is currently well positioned for growth over the medium term, the aftermath of September 11 could disturb this equilibrium. For example, a change in the medium-term industrial outlook is possible, but it does not appear likely at this time. A more likely factor that could affect this equilibrium is the city's policy response to the economic pressures arising from the attack. The task now faced by New York City is to restore and maintain the productive amenities and attractive environment for both firms and workers.

[7] Data refer to all persons twenty-five years of age and older. City data are from the U.S. Census Bureau (1991, 1999); national data are from the U.S. Census Bureau (2002). In 1991, the city's college-educated share of the population was 22.1 percent, while the national share was 21.3 percent in 1990 (U.S. Census Bureau 1991, 2002).

4.5 Policy Challenges after September 11

Although the city's industrial and population structures on September 10 were favorable for its continued prosperity, the loss of life and property on September 11 represents a journey into uncharted waters. Accordingly, we now explore certain effects of the attack in more detail and examine whether the specific types of damage to the city's economy might jeopardize its future. In the language of the model described earlier, the attack most likely induced shifts in the household and firm curves. But how large were these shifts, and what role can policy play in restoring the city to its pre-attack equilibrium? To answer these questions, we consider the future policy implications and challenges for the city in terms of its finances, its infrastructure, and the subsequent federal aid package.

4.5.1 Closing the City's Budget Deficit

The September 11 attack caused a significant short-run fiscal drain on the city. Although it is difficult to isolate the attack's effect on New York City's tax revenue, city agencies estimate that the attack will have reduced revenue by 3 percent to 7 percent in fiscal year 2001-02 and by 2 percent to 6 percent in fiscal year 2002-03[8]. Although the losses represent a sizable shock to the city's revenue stream, they are considerably less severe than the losses faced by the city during its 1975 fiscal crisis. In 1975, the city had to refinance a net cash deficit of more than 25 percent of total tax revenue. The city revenue costs of the September 11 attack are much more in line with the reversal in the city and national economies in fiscal year 1990. That year, revenues from the city's four major taxes (personal income, sales, property, and general corporation) fell approximately 0.5 percent after having grown by more than 3.7 percent the previous year—a swing of 4.2 percent in revenue growth rates (Haughwout, 1997). Thus, the tax revenue shortfall related to September 11 is of a magnitude similar to the effects of the city's 1990 recession.

Revenue shortfalls of this magnitude require concerted action to avoid ending the fiscal year with a deficit. In New York, deficit spending led to the severe fiscal crisis of 1975; since then, the city has been under a strict balanced budget regime. In 1990, faced with a shortfall of similar magnitude, the city raised personal income tax rates sharply to bring the budget into balance in fiscal year 1991. Today, the issue facing city officials is how to react to the current shortfall, which has been compounded by the effects of the recession and declines in equity markets. The historical pattern in New York has been to raise taxes to close budget gaps, and this is indeed the quickest way to generate cash. However, tax increases—particularly at this point in time—may be costly in the long run.

Recent research into the relationship between tax rates, bases, and revenues in four U.S. cities, including New York, indicates that these cities are very near the top of their local "revenue hills" (Haughwout *et al.*, 2000). This means that increases in tax rates can reduce the base over time in such a way as to generate very little additional revenue in the long run. Instead, they may distort the local economies, leading to

[8] The lower figures in the range (for both years) are from the New York City Office of the Comptroller (2001); the higher figures are May 2002 estimates from the New York City Office of Management and Budget, as reported by the U.S. General Accounting Office (2002).

various forms of tax avoidance, especially the loss of economic activity to other jurisdictions.

Therefore, although tax increases could help close the city's current budget gap in the short run, they might well come at a substantial long-run cost, as higher taxes would make the city a less attractive location for mobile firms and households. Haughwout *et al.* assert that the deleterious effects of city tax increases can extend beyond the tax base by also reducing the size of the real economy and the job base. Such a reduction in the job base, unlike the temporary displacement of jobs from the September 11 attack itself, could last for the duration of the tax increases. Of course, temporary tax increases might have smaller and less durable effects.

The first policy challenge for the city, then, is to close its current budget gaps at the least possible cost to its long-term vitality. The alternative to tax increases, of course, is expenditure reductions, themselves costly. In addition, it may be viable for New York City to obtain federal and state aid to help it through this difficult period.

4.5.2 Rebuilding Damaged and Destroyed Infrastructure

Compounding the fiscal problems facing the city and its residents is the fact that a significant part of the damage to Lower Manhattan was sustained by public facilities, notably the Port Authority Trans-Hudson (PATH) train station in the World Trade Center. In addition, some streets and city subway lines were heavily damaged. These facilities allow Lower Manhattan to function as an employment center for the city and the region. Without ready access to Lower Manhattan, one of the city's growth engines—the financial services and related industries concentrated around Wall Street—is cut off from the rest of the region.

More generally, evidence from the academic literature indicates that the productivity of city public capital (that is, its value to firms) is modest, but positive. Eberts (1990) and Haughwout (2002), for example, estimate output elasticities of around 0.04, meaning that a loss of 1 percent of a city's public capital stock is associated with a 0.04 percent decline in city productivity. Interestingly, household costs are estimated to be somewhat higher, yielding a total land price elasticity of about 0.12. These figures suggest that infrastructure stocks exert a significant influence on a city's attractiveness.

Initial reports indicate that New York's public facilities experienced approximately $1.4 billion in damage from the September 11 attack (Bram, Orr, and Rapaport, 2002). This amount is approximately 1.5 percent of the city's preexisting infrastructure stock (Haughwout and Inman, 1996), implying a 0.18 percent decline in city land values as a result of lost infrastructure, assuming that the losses are permanent[9]. The losses, however, are unlikely to be permanent.

The importance of these facilities to New York City can be illustrated by way of a rough calculation of the value of the World Trade Center's PATH station. With the destruction of that facility on September 11, some 20,000 New Jersey–Lower Manhattan commuters were forced to find alternate means of transportation to work. For some, this required a shift to more expensive ferries crossing the Hudson River from Hoboken, New Jersey. But for the substantial number who had commuted from the PATH station in Newark, New Jersey, the quickest option was to take a commuter

[9] Haughwout and Inman's data are for 1992 and should thus be viewed as illustrative of the order of magnitude of the September 11 attack.

rail to Pennsylvania Station in New York City and then a city subway to Lower Manhattan. New Jersey Transit, the state's commuter line, estimates that daily ridership to Pennsylvania Station increased by 15,000 passengers following September 11, despite the immediate loss of jobs experienced (New Jersey Transit, 2001).

For these riders, a twenty-two-minute ride from Newark to Lower Manhattan was replaced by a twenty-minute train ride to midtown plus a twenty-to-thirty-minute subway ride downtown, with a similar lengthening of the homeward commute. The additional time cost is thus approximately one hour per commuter per day. If we assume that 15,000 workers who value their commuting time at $25 per hour (half their hourly wage) are spending an additional hour per day commuting, we obtain a daily cost of $375,000, implying that the first year's loss of the World Trade Center PATH station cost nearly $100 million in lost time[10]. Thus, simply replacing the station would prevent this annual cost from becoming a perpetual loss, with a present value of nearly $2 billion[11].

These calculations are intended to give an order-of-magnitude estimate of the value of Lower Manhattan's infrastructure systems to the city and the region. In the short run, New Jersey commuters and their employers are paying the costs of this lost infrastructure, as workers must either spend less time at work or less time with their families and friends. Although it is less certain exactly who will gain the most, rebuilding the PATH station will clearly benefit the metropolitan area's economy as a whole. Moreover, any additional improvements to Lower Manhattan's transportation linkages should further enhance the regional economy's potential.

As of this writing, the Port Authority of New York and New Jersey has publicly committed to rebuilding the PATH station, and the city has made substantial progress in repairing other damaged elements of the downtown infrastructure. Ensuring accessibility to established job centers in the region is thus a crucial policy challenge facing the city; doing so should reverse the initial negative effect of September 11 significantly.

4.5.3 Using Federal Compensation Effectively

Since the attack on the World Trade Center, much has been done to foster the recovery of New York City, and Lower Manhattan in particular. An outpouring of support nationwide has sought to provide spiritual and financial aid to the victims and to the city as a whole. Several blue-ribbon committees have been formed to assess the damage and monitor the progress made in rebuilding. To date, the federal government has committed $21.7 billion to the rebuilding effort. These funds will be allocated to various efforts to sustain and rebuild New York, ranging from subsidies to residential properties, to the accelerated depreciation of business plants and equipment, to the repair and reconstruction of local streets (New York City Independent Budget Office, 2002).

It is therefore important to address the difficult question of the degree to which federal aid will help in the city's recovery. The city is awaiting a final accounting of

[10] The value-of-time estimate as half the gross hourly wage is standard in the urban transportation literature; see Small (1992).

[11] These calculations exclude the additional out-of-pocket cost to New Jersey commuters, as these amounts are transfers to regional transportation authorities. Lost time, however, is gained by no one. Present-value calculations assume 3 percent real discount and 2 percent depreciation rates.

the unreimbursed losses before it can know with certainty what proportion of the losses the federal aid will offset. Nevertheless, a recent agreement among federal, state, and city officials should greatly increase the flexibility in the allocation of federal aid for rebuilding. Specifically, roughly $9 billion has been set aside by the Federal Emergency Management Agency (FEMA) and earmarked for specific functions related to site cleanup, including debris removal, restoration of damaged public facilities and equipment, and police and fire department overtime. According to the New York City Independent Budget Office (2002), this overall funding allocation is likely to be far greater than the unreimbursed cleanup losses, including the cost of restoring damaged facilities and equipment, which are now estimated to be more on the order of $3 billion. Under the recent agreement, part of the remaining $6 billion of FEMA assistance can now be redirected to help rebuild and improve the entire transportation infrastructure in Lower Manhattan, including such projects as a new PATH terminal and a new transit center that would connect to the terminal and link the downtown subway lines. The use of FEMA funds for transportation projects that go beyond the repair and restoration of damaged facilities is unprecedented.

These issues point to the third challenge facing city policymakers: to apply federal funds in the most efficient way possible to maximize their return. Overall, the federal aid package seems likely to cover the rebuilding costs of the physical facilities lost on September 11. There may even be sufficient funds, if applied judiciously, for Lower Manhattan to transform itself into an even more appealing location than it was on September 10. The proposed transportation improvements, for example, could make Lower Manhattan a more desirable place for businesses, while subsidies to residential development could enhance the neighborhood's various amenities.

4.6 Land Prices Suggest a Positive Economic Outlook

In conjunction with the model, we can gain insight into the city's longer term economic outlook by examining land prices[12]. Although these prices cannot be observed directly, the selling prices of houses, apartments, and commercial properties can serve as a rough proxy.

In the days and weeks immediately following the terrorist attack, there was widespread concern that large numbers of people would no longer want to live in New York City. As it turns out, following a brief but sharp dip in September and October, housing markets in New York City's nearby suburbs, the outer boroughs, and even Manhattan are reported to have rebounded strongly (Bram, Orr, and Rapaport, 2002). As shown in figure 4.3, the selling prices of existing homes throughout the metropolitan area have continued to rise, not only in absolute terms, but also relative to the national average. These results suggest that the terrorist attack thus far has had little if any negative effect on land prices, even in areas close to the World Trade Center.

Manhattan's office market, however, has been more mixed. The rental market has clearly been weak: at midyear 2002, office vacancy rates were reported to be sharply

[12] Bram, Orr, and Rapaport (2002) discuss current economic indicators since the attack in order to assess short-term prospects.

higher than they were a year earlier and rents were reported to be sharply lower. However, the buyer's market for office property continued to be characterized as strong, again suggesting underlying strength in land prices.

Thus, conditions in New York's important property markets currently appear to reflect continued strong demand. To the extent that business and household evaluations of a location's prospects are reflected in bids, New York City's land markets suggest a very positive indicator for the city. Naturally, market participants are operating with very limited information about the future. Yet it appears that, from what they know, firms and households still view New York as an attractive location. Going forward, the strength of the city's property markets will be an important barometer of its economy.

4.7 Conclusions and Directions for Future Research

Economic growth in New York City for most of the past three decades has been characterized by a transformation of employment from relatively low-paying jobs to increasingly higher paying ones—that is, overall employment levels have remained roughly constant but real earnings levels have risen both in absolute terms and relative to the nation.

In light of the destruction that occurred on September 11, several downside risks to the city's economic growth prospects have arisen. The worst-case scenario includes the possibility that financial firms located in Lower Manhattan will leave the city, which could generate a cumulative downward spiral of job and income growth.

In assessing this and other potential risks to New York, we examined the city's economic position with respect to several key growth fundamentals prior to September 11. We found that New York's industrial mix, which is weighted toward high-growth sectors, along with the city's attractiveness to workers and firms, point to favorable prospects for growth. However, whether these conditions translate into actual growth going forward depends to a large degree on the city's policy response to the economic pressures arising from the terrorist attack. The main elements of this response will be the ability to manage relatively large budget deficits and rebuild New York City's damaged and destroyed infrastructure—while maintaining the productivity of the capital stock. Addressing the latter issue will require a vision of Lower Manhattan's role in the overall city economy as well as an efficient allocation of the anticipated federal aid.

The models examined here suggest that New York City maintains significant strength in spite of the terrorist attack. Going forward, future research in our view should proceed on four interrelated fronts:

- *Researchers could explore the short-run impact of natural (and man-made) disasters on the price-rental ratio and the ultimate consequences for local growth.* To the extent that disasters destroy infrastructure and cause short-run fiscal stress in an identifiable area, the model predicts that they should depress current *rentals*. At the same time, the market's view of the long-run viability of the location ought to be reflected in property (that is, land) *prices*. Has the long-run signal sent by property markets ultimately been borne out in locations that have had time to rebuild following a disaster?

- *Researchers could concentrate on the effects of disasters on industrial and population structures.* Our positive assessment of New York's resiliency is based in large measure on its favorable industrial structure and its well-educated population. Evidence from other disasters has demonstrated that the overall *size* of a city may, at least temporarily, decline in terms of both jobs and population. Disasters' effects on industrial and population *composition* are less well understood, but may be important in assessing the effects of future disasters. Do some kinds of disasters disproportionately affect particular industrial sectors? Do they induce certain elements of the population to move elsewhere?
- *Researchers could focus on understanding the role of federal, state, and local government policies in affecting expectations about disaster recovery and rebuilding.* In the New York case, expectations of significant federal aid in response to the September 11 attack may have played a crucial role in supporting demand for city locations. But the effect of these expectations is difficult to distinguish from the underlying strength of the city as a place to live and work. Does demand for favorable locations depend on favorable pre-existing conditions, federal aid, or both? How important are state and local responses to disasters?
- *Future research might also focus on defining a metric that can quantify the immediate effect of a disaster.* Measuring the scale of a disaster is never easy, especially when it is of an unprecedented nature. Such metrics might be based on the relationship between long run outcomes and general measures of a disaster's immediate impact.

Disasters, whether man-made or natural, are a fact of life in modern economies. How and why cities respond as they do will likely remain an important policy topic, and one that ought to engage significant research in coming years.

References

Blomquist, G., M. Berger, and J. Hoehn. 1988. "New Estimates of Quality of Life in Urban Areas," *American Economic Review*, 78 (1): 89-107.

Bram, J. and J. Orr. 1999. "Can New York City Bank on Wall Street?" Federal Reserve Bank of New York, *Current Issues in Economics and Finance*, 5 (11, July).

Bram, J., J. Orr, and C. Rapaport. 2002. "Measuring the Effects of the September 11 Attack on New York City," Federal Reserve Bank of New York, *Economic Policy Review*, 8 (2): 5-20.

Eberts, R. 1990. "Cross-Sectional Analysis of Public Infrastructure and Regional Productivity Growth," Federal Reserve Bank of Cleveland Working Paper, No. 9004.

Gyourko, J. and J. Tracy. 1991. "The Structure of Local Public Finance and the Quality of Life," *Journal of Political Economy*, 99 (4): 774-806.

Harrigan, J. and P. Martin. 2002. "Terrorism and the Resilience of Cities," Federal Reserve Bank of New York, *Economic Policy Review*, 8 (2): 97-116.

Haughwout, A. 1997. "After the Fall: An Examination of Fiscal Policy in New York City since 1974," in Proceedings of the Eighty-Ninth Annual Conference on Taxation, Washington, DC; National Tax Association: 96-105.

Haughwout, A. 2002. "Public Infrastructure Investments, Productivity, and Welfare in Fixed Geographic Areas," *Journal of Public Economics*, 83 (3): 405-28.

Haughwout, A. and R. Inman. 1996. "State and Local Assets and Liabilities, 1972-1992," Unpublished paper, Federal Reserve Bank of New York.

Haughwout, A., R. Inman, S. Craig, and T. Luce. 2000. "Local Revenue Hills: A General Equilibrium Specification with Evidence from Four U.S. Cities," NBER Working Paper, No. 7603, March.
McCarthy, J. and C. Steindel. 1997. "National and Regional Factors in the Metropolitan Economy," Federal Reserve Bank of New York, *Economic Policy Review*, 3 (1): 5-19.
New Jersey Transit. 2001. "Call to Action: Addendum." December. Addendum to "NJ Transit's Call to Action: An Investment for the Future." September 10.
New York City Department of Education. 2002. "1999-2002 State and City Mathematics Test Results, Grades 3, 4, 5, 6, 7, and 8." Available at <http://www.nycenet.edu/daa/2002math/default.asp>.
New York City Independent Budget Office. 2002. "World Trade Center Aid: Too Much for Some Needs, Not Enough for Others?" *Newsfax*, No. 100, May 28.
New York City Office of the Comptroller. 2001. "The Impact of the September 11 Attack on New York City's Economy and Revenues." October 4.
New York City Partnership and Chamber of Commerce. 2001. "Economic Impact Analysis of the September 11th Attack on New York City." November.
Orr, J. and R. Rosen. 2000. "The Financial Services Sector in London and New York," in *The London-New York Study*, London, UK; The Corporation of London: 11-24.
Quigley, J. 1998. "Urban Diversity and Economic Growth," *Journal of Economic Perspectives*, 12 (2): 127-38.
Rauch, J. 1993. "Productivity Gains from Geographic Concentration of Human Capital: Evidence from Cities," *Journal of Urban Economics*, 34 (November): 380-400.
Roback, J. 1982. "Wages, Rents, and the Quality of Life," *Journal of Political Economy*, 90 (6): 1257-78.
Saiz, A. 2002. "Immigration and Housing Rents in American Cities." Unpublished paper, Harvard University.
Small, K. 1992. *Urban Transportation Economics*. New York, NY; Harwood Academic.
U.S. Census Bureau. 1991 and 1999. "New York City Housing and Vacancy Survey." Washington, DC; Available at <http://www.census.gov/hhes/www/nychvs.html>.
U.S. Census Bureau. 2002. "Table 215." *2001 Statistical Abstract*, Washington, DC.
U.S. General Accounting Office. 2002. "Review of the Estimates for the Impact of the September 11, 2001, Terrorist Attacks on New York Tax Revenues." GAO-02-882R. July 26; Available at <http://www.gao.gov>.

Appendix

A Model of Local Prices in Mature Urban Economies

Free mobility means that firms and households must pay, albeit indirectly, for attractive local characteristics. Firms and households pay for local amenities through higher land prices and wages. Thus, when a place offers household amenities like an attractive climate or unique consumption opportunities, households will be willing to pay to locate near these amenities, bidding more for land there. Yet, since wages represent an income source to households, they will also be willing to accept *lower* wages to locate in places with features they value. If a place is a "bargain" to households in the sense that its land prices are low and its wages are high relative to its attractive amenities, households will move in, bidding land prices up and wages down. Firms, however, bid more for attractive sites in both land and labor markets.

Figure 4.A.1 presents equilibrium land and labor bids by a competitive firm (labeled $\Pi = \Pi_0$) and a representative household (labeled $V = V_0$) for a place with a given set of production and consumption amenities (A_0). Note that the firm curve (an isoprofit curve in the price space) is downward-sloping, indicating that as wages in a jurisdiction rise, land rents must fall if firms are to remain profitable. The household indifference curve in the price space is labeled V, and it slopes upward: higher wages are associated with higher land prices. With higher incomes, households must pay higher land rents or they will earn excess utility—leading to new households bidding in local land and labor markets. The point where both the firm and household equilibrium curves intersect is the local price equilibrium (W_0, R_0).

Figure 4.A.2 depicts the effect of a change in local amenities (to $A_1 < A_0$) on this equilibrium. When a city becomes less attractive to firms, the $\Pi = \Pi_0$ curve shifts downward. Changes that reduce the attractiveness of a location to households will shift the $V = V_0$ function downward. For given wage levels, firms and households are willing to pay less for land in unproductive places. Note that a reduction in attractiveness has an unambiguously negative effect on land prices, but that the effect on wages will depend on which curve (household V or firm Π) shifts more. When firms are disproportionately hurt by a change, the fall in equilibrium land rents will be accompanied by a *reduction* in wages. A relatively large impact on households would lead to *increases* in wages, as house-holds demand higher compensation to induce them to reside and work in an unattractive, low-amenity place (Haughwout (2002) formally derives these comparative statistics and provides an example of the use of wage effects in the determination of the incidence of policy benefits).

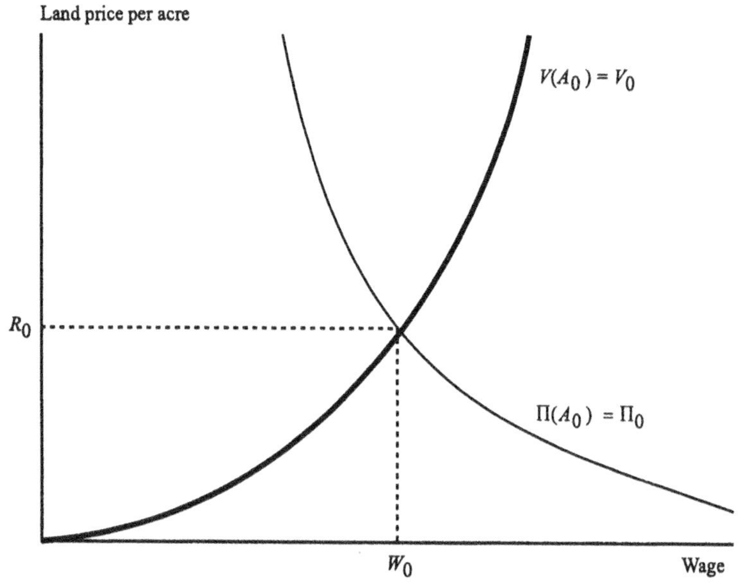

Figure 4.A.1 Wage and Land Rent Equilibrium

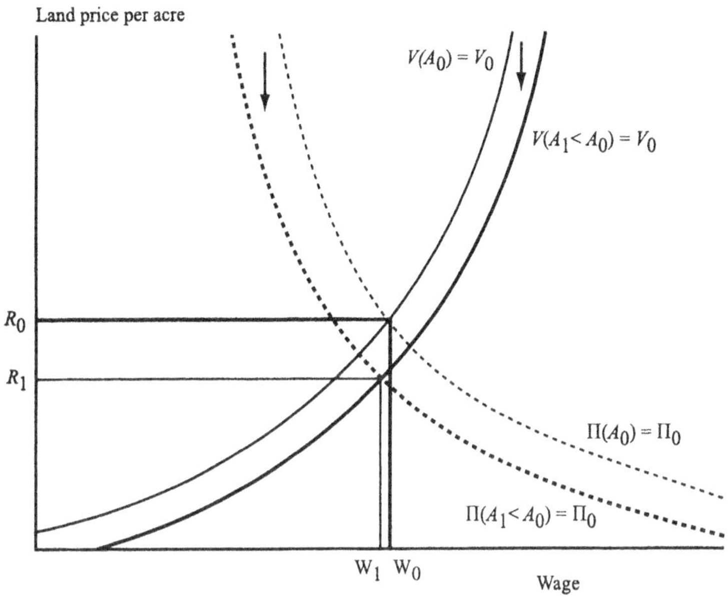

Figure 4.A.2 Wage and Land Equilibrium with Fewer Amenities

Part II: Economic Models

5 Measuring Economic Impacts of Disasters: Interregional Input-Output Analysis Using Sequential Interindustry Model*

Yasuhide Okuyama,[1] Geoffrey J. D. Hewings,[2] and Michael Sonis[3]

[1] Regional Research Institute, West Virginia University, USA
e-mail: yaokuyama@mail.wvu.edu
[2] Regional Economics Applications Laboratory, University of Illinois, USA
e-mail: hewings@uiuc.edu
[3] Regional Economics Applications Laboratory, University of Illinois, USA and Bar Ilan University, Israel
e-mail: sonism@mail.biu.ac.il

5.1 Introduction

The damages and losses by disasters, such as earthquakes, floods, tornadoes, and other major natural disasters, or man-made disasters, have significant and intense impacts on a region's economy. In addition, the impacts from the damages will spread over time, and will bring serious economic effects to other regions in a long run. Furthermore, the impacts of disasters are very complex, including not only the negative effects from damages and losses, but also the positive economic effects from the recovery and reconstruction activities. Most economic models and techniques cannot confront these significant changes in a relatively short time period, since they assume incremental, and/or predictable changes in systems over time. And, the unexpected nature of these events, especially in the case of earthquakes, creates a further complication of measuring the indirect impacts (Okuyama *et al.*, 2002). At the same time, most available data for the direct damages and losses and of the recovery processes are engineering oriented, *i.e.*, physical damages and disruption of lifelines and their repair and restoration, and the dimension and unit of these data are quite different from the economic counterpart—very detailed and short time span in engineering data while aggregated and longer time span in economic models. Consequently, these differences pose great challenges in order to model economic impacts of disasters.

While the economic modeling issues for measuring such disruptions and the impacts are more complex (for an excellent summary, see West and Lenze, 1994), the questions of the potential economic effects of a disaster have been studies and discussed in various aspects (for example, Cochrane, 1974; National Science of Academy, 1978; Chang, 1983; Ellson *et al.*, 1984; and Guimaraes *et al.*, 1993, among others). Input-output analysis has been employed in many studies to measure and

* This research was supported by the National Science Foundation, Award CMS-0099724. The authors are grateful to the encouragements and comments from Eliahu Romanoff and Stephen Levine, and to the comments given to the previous versions in various conferences.

evaluate the economic impacts of disasters, mainly because of the ability to reflect the structure of regional economy in great detail (for example, Cochrane, 1974, 1995, and 1999; Wilson, 1982; Kawashima *et al.*, 1991; Boisverst, 1992; Gordon and Richardson, 1996; Cole, 1997; Rose *et al.*, 1997; Rose and Benavides, 1998; Okuyama *et al.*, 1999a, among others). Whereas they provide useful information regarding the economic impacts and consequences and about the resource allocation strategies to minimize the losses and impacts, many of these studies have failed to investigate the dynamic nature of impact path over space and time, due mainly to the difficulties to obtain such data and due partly to the static nature of input-output framework.

This is an inherent problem for the impact analysis of disasters; as West and Lenz (1994) pointed out, the sophisticated regional impact models requiring precise numerical input have to be reconciled with imperfect measurements of the damages and losses. Moreover, measuring the economic effects of a disaster poses a great challenge for modeling the event *per se* and its consequences—damages and losses occur across various geographical areas and in a relatively short period of time, while the economic effects spread over a larger region (and, sometimes, to other regions, too) and, in some cases, may last for a relatively long period of time. In addition, it presents complexities to link a physical/engineering nature of damages and losses with an economic model, in which the degree of details is widely different from direct damage data. Several attempts have been made to integrate the engineering data (sometime models) with an economic model. The most common way to connect between the damages and losses from a disaster and its economic consequences is to directly input the engineering data into an economic model (for example, Rose *et al.*, 1997 and Rose and Benavides, 1998, for modeling the economic impacts of electricity lifeline disruption in Memphis area due to a hypothetical catastrophic earthquake). This approach is simple and straightforward in terms of modeling strategy and requires less modification to the original models; however, it has no dynamic interactions (feedbacks) between engineering and economic models and it may exhibit some problems, such as sensitivity differences between engineering scenario and economic results.

More engineering oriented approaches (or, integrative approaches combining engineering model and economic model) to measure the economic effects of a disaster have been implemented in recent years (for example, Chang *et al.*, 1999; Okuyama *et al.*, 1999b; and Cho *et al.*, 2001). While the connection between engineering and economics components becomes closer and enables feedback between them, these models become either highly aggregated and complex (Okuyama *et al.* and Cho *et al.*) or too simple and broad to provide the details of the economic consequences (Chang *et al.*). This scheme also has a common difficulty in data requirement and availability. More comprehensive approach to measure overall impacts of a disaster can be found in HAZUS software developed by National Institute of Building Science. Using the Geographic Information System (GIS) as the user interface, HAZUS includes a wide range of related modules, from geological and direct physical damage modules to lifeline-utility system and indirect economic losses modules, in order to estimate the effects of a natural hazard in a comprehensive fashion. While the strategy and structure of the indirect economic losses module in HAZUS (see Cochrane, 1997 and 1999 for the details) are similar to Rose *et al.* and Rose and Benavides' models above in the sense that these models employ input-output framework with supply/capacity constraints, it is tailored to deal with smaller geographical areas (county-level) and to include time dimension for more detailed analysis of the recovery process. However, the links between the indirect economic losses module and other engineering modules, such as lifeline-utility systems module, are less clear, and the time dimension implemented in the indirect economic loss module is rather ad hoc.[1]

In order for attempting to measure the economic impacts and consequences of a disaster over space and time, Sequential Interindustry Model (SIM) introduced by Levine and Romanoff (1989), Romanoff (1984), and Romanoff and Levine (1977, 1981, 1986, 1990a, 1990b, 1991, and 1993) is employed in this chapter to investigate the dynamic process of the impact paths and recovery process of a disaster while maintaining the simplicity of input-output framework. The SIM framework aims to modify the static framework of standard input-output table into a dynamic formulation, incorporating with production chronology. This framework of SIM is particularly useful to simulate the dynamic process of impact propagation and recovery process after a disaster in short and long runs. Moreover, the SIM provides an opportunity to connect the macroeconomic nature of input-output framework with the microeconomic process of production.

In the next section, a brief discussion of dynamic modeling based on the input-output framework and the analytical framework of SIM are presented and discussed. Section 3 discusses the advantages and issues for adopting the SIM framework to modeling economic impacts of a disaster. In Section 4, the SIM framework is applied to measure the interregional economic impacts of the Kobe Earthquake in Japan. Finally, Section 5 summarizes and concludes this chapter, and addresses some future research needs.

5.2 Modeling Dynamic Process in Interindustry Production: Sequential Interindustry Model

Early interest in the dynamics of interindustry production within the framework of input-output analysis can be seen in Goodwin (1947) and Leontief (1951). These models were expanded by Dorfman, Samuelson, and Solow (1958) and Kuenne (1963) and further advanced by Morishima (1964); and, these lines of analysis extended their attention to the integration with linear programming, and/or to Computable General Equilibrium (CGE) modeling to make the model more operational with the specific objective function (linear programming) and to incorporate with more theoretical underpinnings. As a model becomes more sophisticated, however, the level of aggregation tends to be less detailed and the size of the model becomes larger, especially in the case of CGE models.

As another line of the effort to make the static input-output framework a dynamic model, a dynamic version of input-output model was first introduced by Leontief (1953)[2] and was later modified (Leontief, 1970). The dynamic input-output model aims to analyze and determine the structural and the technological changes of an

---

[1] In HAZUS99 Technical Manual (1999), the time dimension of the indirect economic losses module is described that for the first two months after an earthquake weekly time intervals are used; between two months and 24 months, the economy is evaluated on a monthly basis; from two years to 15 years, the economy is evaluated annually.

[2] In his first model in 1953, Leontief formulated investment as the rate of change in required capital stock using the time derivative of total input vector: $\mathbf{x} = \mathbf{A}\mathbf{x} + \mathbf{C}\dot{\mathbf{x}} + \mathbf{f}$.

economy (or economies) by including an intertemporal mechanism of capital accumulation. Leontief (1970) developed a discrete approximation of model using a system of difference equations with dated technical matrices reflecting structural change in an economy:

$$\mathbf{x}_t = \mathbf{A}_t\mathbf{x}_t + \mathbf{C}_{t+1}(\mathbf{x}_{t+1} - \mathbf{x}_t) + \mathbf{f}_t \quad (5.1)$$

where $\mathbf{C}_{t+1}(\mathbf{x}_{t+1} - \mathbf{x}_t)$ represents the investment requirements in addition to productive stock during t and $t+1$ in order to expand their capacity output from $\mathbf{x}_t$ to $\mathbf{x}_{t+1}$. Leontief's dynamic input-output model has a clear advantage to include the changes in flows and stocks (as a form of capital formulation), and can be suited to incorporate with damages to capital stock (production facilities) and flows (backward and forward linkages) from a disaster. While the mathematical properties of this dynamic model have been studied by many (for example, Zaghini, 1971; Schinnar, 1978; de Mesnard, 1992; and Guangzhen, 1993, among others), however, the model has been used in few empirical works due to various problems involved: first, the implementation of the dynamic model requires the assembly of capital requirement matrices that distinguish between replacement and expansion of the capital; and second, the model could produce implausible results due to its structure.[3]

Using a different approach to introduce a dynamic structure in the static input-output framework, a group of lagged input-output models with distributed activities were proposed (for example, ten Raa, 1986; and, Cole, 1988 and 1989). Concerning the time used in production process and taking into account the time of labor market adjustment, among other things, these lagged input-output models aim to capture the process of impact (or growth) from a production expansion within the input-output framework. The distributed activity model developed by ten Raa (1986) is based on the formulation of the Leontief's dynamic input-output model, solving some of the drawbacks in that framework (singular capital structures, unbalanced growth, and different time profiles of investment) while preserving the formal structure and simplicity of the original Leontief dynamic model. Ten Raa's model is highly theoretical and aggregated, and few practical applications based on his formulation have been implemented. On the other hand, Cole's models (Cole, 1988 and 1989) are highly operational and are based on the empirical examples of western New York and

[3] Leontief (1970) implemented his dynamic model using 1947 and 1958 US data, and it revealed the two major inherent drawbacks of the model, which could produce implausible results. Leontief solved the model employing the backward-looking way--determine the final impacts first, and then solve the model for the requirements in previous years. This backward-looking solution is stable, yet unrealistic, since it assumes that the economy has a perfect foresight of the future. Although the forward-looking solution has been studied [Szyld (1985), Steenge (1990a), Heesterman (1990), and Steenge (1990b)], it has been found that a set of non-negative solutions for $\mathbf{x}_t$ exists only if the initial conditions lie on the "balanced growth path". This drawback comes from the assumption of full capacity utilization: the entire physical productive capacity will be utilized. Another difficulty to derive the solution of the Leontief dynamic model is the singularity of the capital matrix, $\mathbf{C}$. As Duchin and Szyld (1985) pointed out, most theoretical works have been carried out based on the assumption that the $\mathbf{C}$ matrix is invertible, whereas the $\mathbf{C}$ matrix may be invariably singular, with rows of zeros corresponding to the sectors not producing durable goods. In order to overcome these problems, Duchin and Szyld (1985) proposed the new formulation of the dynamic input-output model, and this formulation was used in Leontief and Duchin (1986) study. More recently, Campisi and his colleagues developed a series of models based on the Duchin-Szyld formulation, with an extension to multiregional context, and applied to the Italian economy (Campisi *et al.*, 1990; Campisi, Nastasi, and Bella, 1992; and Campisi and Nastasi. 1993).

of Aruba. While the Cole's approach is highly practical for simulation analysis, some theoretical and structural issues of modeling framework need to be addressed.[4] As a similar approach but with more emphasis on production chronology, Romanoff and Levine introduced the Sequential Interindustry Model (SIM) in order to incorporate more engineering aspect of production process.

5.2.1 Sequential Interindustry Models

Levine and Romanoff (1989), Romanoff (1984), and Romanoff and Levine (1977, 1981, 1986, 1990a, 1990b, 1991, and 1993) introduced the Sequential Interindustry Model (SIM) in response to the need to analyze interindustry production in a dynamic economic environment, such as large construction projects where the effects on production and employment are transitory. Assuming for simplicity that time is divided into discrete intervals of equal duration, the SIM enhances the static input-output model to the dynamic one by supplementing the structure of production with a production chronology. In the SIM, production is not simultaneous as in the static input-output model, but rather occurs sequentially over a period of time (Romanoff and Levine, 1981). The interval of an industry production process is divided into two components: the production interval and the shipment interval with inputs and product inventories, as illustrated in figure 5.1. In order to create the dynamics of SIM, a distinction is made among three events in a production process: demand stimulus occur when goods are ordered; yield or supply happens when goods are delivered; and production yield occurs when goods are produced. In SIM, demand is not restricted to final demand but includes intermediate demand along the production sequences, as in the standard input-output framework. Final demand stimulus is the ultimate system input, while final yield or final supply is the net system output.

Determining the dynamics of interindustry production, two simplified production modes are proposed: *anticipatory production* mode and *responsive production* mode. The anticipatory production mode is typical in agriculture and many manufacturing industries, in which the production is made in anticipation of future orders. In this mode, producers' specifications result in ready-made standards products and in holding product inventory. On the other hand, responsive production takes place after the receipt of orders, responding to customers' specification by producing to meet the unique requirements, while product inventory is unlikely. This production mode is typical of some manufacturing, most of construction and ordnance industries, and most of services industries[5] (Romanoff and Levine, 1981, and 1986).

In the time-varying SIM, some specific time indices are defined (Romanoff and Levine, 1990). These are given in parentheses and pertain to production technique in use and to temporal events or intervals. The time indices are:

---

[4] Extensive discussions were made regarding the theoretical underpinning and the formulation of Cole's models in Jackson *et al.* (1997), Cole (1997), Jackson and Madden (1999), Cole (1999), and Oosterhaven (2000).

[5] Most of services industries can be considered as *just-in-time production* mode, in which the production takes place and the goods delivered, as the order is placed—less time is needed for production comparing to manufacture and construction industries. However, just-in-time production mode can be considered a special case of responsive production mode when ordering lead time and production interval are minimal, as shown in later part of this section.

t : time interval of input application;

σ : time interval of output or production completion;

h_{ij} : application period of an input from industry i used by industry j, referenced from the initial application interval t, to product completion σ;

h_j : production period of industry j, equivalent to the longest application period among h_{ij};

$t-\sigma$: input duration, indicating the period from an input application interval t to the time of output completion, σ, measured backwards in time from σ (equal or longer than h_j); and

ϕ_{ij} : transportation delay associated with the shipment from industry i to industry j, representing the components of the transportation delay matrix, $\mathbf{\Phi}$.

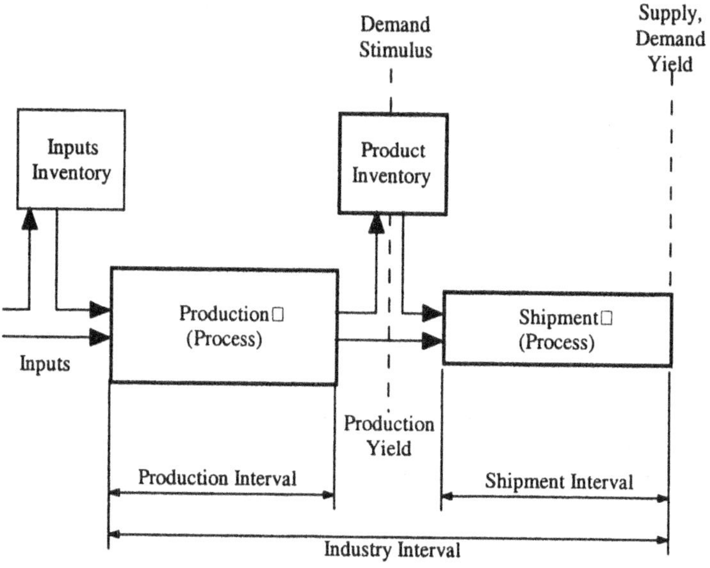

Figure 5.1 Production Intervals (after Romanoff and Levine, 1981)

5.2.2 Anticipatory Production Mode

Assuming just-in-time production on the input side (no input inventory), the input price from industry i to j is given by $p_i\left(t-\phi_{ij}\right)$, since the input price is determined at the time that it leaves supplying industry i. The quantity of input from i to j is defined as $q_{ij}\left(\mu_{i,ij};t,\sigma\right)$, processed by technology μ at time t in industry j using input from i in order to complete the product at σ. The value of transaction form i to j is:

$$x_{ij}\left(\mu_{t,ij};t,t-\phi_{ij},\sigma\right)=p_i\left(t-\phi_{ij}\right)\cdot q_{ij}\left(\mu_{t,ij};t,\sigma\right) \tag{5.2}$$

The total output of industry j, x_j, completed at σ is priced at the time of product completion for anticipatory production mode is:

$$x_j\left(\sigma,\sigma\right)=p_j\left(\sigma\right)\cdot q_j\left(\sigma\right) \tag{5.3}$$

where $q_j\left(\sigma\right)$ is the quantity of output produced by j at σ. Using (5.2) and (5.3), the time-phased technical coefficients can be derived as follows:

$$a_{ij}\left(\mu_{t,ij};t,t-\phi_{ij},\sigma,\sigma\right)=\frac{x_{ij}\left(\mu_{t,ij};t,t-\phi_{ij},\sigma\right)}{x_j\left(\sigma,\sigma\right)} \tag{5.4}$$

Romanoff and Levine (1990) claims that the coefficients represent the relative costing out of i for the production process of the schedule of j; thus, in equilibrium, the coefficients sum to the technical coefficient in the static input-output framework over time for each i.

Total intermediate output produced by industry i becomes:

$$w_i\left(t,t\right)=\sum_{\sigma}\sum_{j}\sum_{\mu}a_{ij}\left(\mu_{t,ij};t,t-\phi_{ij},\sigma,\sigma\right)\cdot x_j\left(\sigma,\sigma\right) \tag{5.5}$$

which is a generalized convolution indicating the dynamics of intermediate production. Then, the accounting identity of industry i will become as follows:

$$x_i\left(t,t\right)=\sum_{\sigma}\sum_{j}\sum_{\mu}a_{ij}\left(\mu_{t,ij};t,t-\phi_{ij},\sigma,\sigma\right)\cdot x_j\left(\sigma,\sigma\right)+u_i\left(t,t\right)+y_i\left(t,t\right) \tag{5.6}$$

where $u_i\left(t,t\right)$ is the output to product inventory (adjusting the differences between anticipated output level and demand level) and $y_i\left(t,t\right)$ is the final demand for industry i. Using matrix notation, (5.6) becomes:

$$\mathbf{x}\left(t,t\right)=\sum_{\sigma}\mathbf{A}\left(\mathbf{M};t,t-\phi_{(\cdot)},\sigma,\sigma\right)\cdot\mathbf{x}\left(\sigma,\sigma\right)+\mathbf{u}\left(t,t\right)+\mathbf{y}\left(t,t\right) \tag{5.7}$$

where $\mathbf{M}$ is the technology matrix, and $\phi_{(\cdot)}$ indicates the appropriate elements of the transportation delay matrix, $\mathbf{\Phi}$. This is a formally specified formulation of their simpler version of SIM (called Core SIM) presented in Romanoff (1984) and Romanoff and Levine (1981), with an exception of inventory. If the function of inventory is included in the Core SIM, it can be written for anticipatory production mode as follows:

$$\mathbf{x}_t=\mathbf{A}\mathbf{x}_\sigma+\mathbf{u}_t+\mathbf{y}_t \tag{5.8}$$

The major assumption here in terms of inventory is that, unlike final demand that is exogenous of the system, inventory is an endogenous function based on inventory policies of anticipatory producer. This inventory function is further discussed in the next section.

5.2.3 Responsive Production Mode

While producers' price, $p_i\left(t-\phi_{ij}\right)$, may be better suited for anticipatory producers, they may also be applicable to responsive producers (Romanoff and Levine, 1990). Therefore, equation (5.2) holds also for responsive production mode. What make responsive production mode different from anticipatory mode is in the time of output pricing, *i.e.*, for responsive producers, the price of output is determined when an order is issued. Hence, this ordering lead time, ε_j, is set ahead of beginning of the production interval, h_j from the production completion at σ. Then, the total output of industry j becomes:

$$x_j\left(\sigma,\sigma-h_j-\varepsilon_j\right)=p_j\left(\sigma-h_j-\varepsilon_j\right)q_j\left(\sigma\right) \tag{5.9}$$

Consequently:

$$a_{ij}\left(\mu_{t,ij};t,t-\phi_{ij},\sigma,\sigma-h_j-\varepsilon_j\right)=\frac{x_{ij}\left(\mu_{t,ij};t,t-\phi_{ij},\sigma\right)}{x_j\left(\sigma,\sigma-h_j-\varepsilon_j\right)} \tag{5.10}$$

The total output of responsive production mode is:

$$\begin{aligned} &x_i\left(t,t-h_j-\varepsilon_j\right)= \\ &\quad=\sum_\sigma\sum_j\sum_\mu a_{ij}\left(\mu_{t,ij};t,t-\phi_{ij},\sigma,\sigma-h_j-\varepsilon_j\right)\cdot x_j\left(\sigma,\sigma-h_j-\varepsilon_j\right)+ \\ &\quad+y_i\left(t,t-h_j-\varepsilon_j\right) \end{aligned} \tag{5.11}$$

Note that responsive production mode is, of its nature, without production inventory.[6] The corresponding matrix form is:

$$\begin{aligned} &\mathbf{x}\left(t,t-h_j-\varepsilon_j\right)= \\ &\quad=\sum_\sigma \mathbf{A}\left(\mathbf{M};t,t-\phi_{(\cdot)},\sigma,\sigma-h_j-\varepsilon_j\right)\cdot\mathbf{x}\left(\sigma,\sigma-h_j-\varepsilon_j\right)+\mathbf{y}\left(t,t-h_j-\varepsilon_j\right) \end{aligned} \tag{5.12}$$

This formulation can also be compared to a simpler version in Core SIM, and corresponding responsive production mode in Core SIM can be re-written as:

$$\mathbf{x}_t=\mathbf{A}\mathbf{x}_{\sigma\text{-}h\text{-}\varepsilon}+\mathbf{y}_t \tag{5.13}$$

5.2.4 Combined Anticipatory-Responsive Production Model

Since the input-output model is on interindustry framework, it is natural to assume that anticipatory production and responsive production industries are coupled with each other in the model. The anticipatory production mode in (5.7) and the responsive

[6] For responsive production mode, it is more likely to have input inventory; however, in this formulation (Romanoff and Levine, 1990), assuming just-in-time production for simplicity, input inventory is not considered in either mode. However, the production inventory of anticipatory industries can work as the input inventory for responsive industries, although it may not fully reflect the complexity of real world production process.

production mode in (5.12) are distinguished in a way that the anticipatory mode does not require ordering lead time ($h_j + \varepsilon_j = 0$), whereas the responsive mode does not include the production inventory ($\mathbf{u}(\bullet) = \mathbf{0}$). Therefore, the combined anticipatory-responsive model encompasses both properties as follows:

$$\mathbf{x}(t, t - h_j - \varepsilon_j) = \\ = \sum_\sigma \mathbf{A}(\mathbf{M}; t, t - \phi_{(.)}, \sigma, \sigma - h_j - \varepsilon_j) \cdot \mathbf{x}(\sigma, \sigma - h_j - \varepsilon_j) + \\ + \mathbf{u}(t,t) + \mathbf{y}(t, t - h_j - \varepsilon_j) \tag{5.14}$$

While this version of SIM combines the anticipatory mode and responsive production mode industries, each industry is classified into one of the production modes. In the real world, some industries may produce their goods using a combination of these modes (for example, computer manufacturing industry that produces both pre-configured (with anticipation) and customer-configured (by taking orders) computers). Although this type of details becomes important in an analysis of specific industries, with the aim of modeling in a regional context, it is beyond the scope of this chapter. Again, the model (5.14) can be re-written for corresponding to the Core SIM version as follows:

$$\mathbf{x}_t = \mathbf{A}\mathbf{x}_{\sigma\text{-h-}\varepsilon} + \mathbf{u}_t + \mathbf{y}_t \tag{5.15}$$

Base on this combined model in (5.15), the Leontief inverse can be obtained as follows:

$$\mathbf{x}_t = \boldsymbol{\beta}[\mathbf{A}(\bullet), \mathbf{u}_t, \mathbf{y}_t] \tag{5.16}$$

where $\boldsymbol{\beta}[\bullet]$ is a vector function, or calculation rule, representing a general extension of the Leontief inverse. The specification depends on the inventory function (Romanoff and Levine, 1990). In the static version, with $\mathbf{u}_t = \mathbf{0}$, model (5.16) becomes $\mathbf{x} = \boldsymbol{\beta}[\mathbf{A}, \mathbf{y}] = (\mathbf{I} - \mathbf{A})^{-1} \mathbf{y}$, the standard Leontief inverse.

5.3 Issues of SIM for Modeling Economic Impacts of Disasters

In this section, the applicability of Sequential Interindustry Model (SIM) to measure economic impacts of unscheduled events is discussed. In particular, the issues of modeling economic impacts of disasters, such as uncertainty, technology replacement, and inventory, are examined.

5.3.1 Uncertainty

One of the less-emphasized aspects in modeling economic impacts of a disaster is that the occurrence and the consequences of such event involve uncertainty. For example, the occurrence of the event is usually unpredictable, especially in the case of earthquakes, and this unpredictability creates a surprise effect to anticipatory production industries (Okuyama *et al.*, 2001). This surprise factor can be further analyzed with the inventory function in the SIM framework; for example, whether or not the increased product inventory, which is suddenly realized right after the event, can be offset by the influx of demand injections for recovery and reconstruction activities over time, or how the damages and/or capacity limitation of product inventory can affect the production level after the event.

Uncertainty arises after a disaster, because: first, the extent and range of direct damages are unknown right after the event; second, the trends of economic activities, especially the fluctuation of demand, become unclear in a short run; and, third, the influx of demand injections for recovery and reconstruction activities makes the long run forecast (anticipation) of economic growth difficult, particularly in the region. These uncertainties after a disaster can be treated and evaluated in the SIM framework in the way that some of the anticipatory production industries may not be able to anticipate the future intermediate demand stream and may decide the production level based on the current demand level. Because of the difficulty to anticipate the future intermediate demand, the production inventory need play an important role. On the other hand, while the recovery and reconstruction schedule will become available and these activities will occupy a large share of the economic activities in the region for a while after the event, the degree of uncertainty over time requires a careful treatment.

5.3.2 Technology Replacement

Damages and losses from a disaster may become significant to the older structures and facilities in the region. These older structures and building are more likely to be equipped with older machines and production lines. Or, even in newer buildings and facilities, older equipments may exist and be damaged due to the age (mechanical fatigue) of equipments. After the disaster and during the recovery stage, these damaged equipments and production facilities will be replaced with new ones with newer technology. Although these technology replacements may not take place all the facilities within a particular industry in the region, which is different from technological progress in a specific industry in the economic growth context, it may become significant if the area with a disaster includes the concentration/cluster of a particular industry (for example, Chemical Shoes Industry located in Nagata Ward, Kobe in the Kobe Earthquake).

Within the SIM framework presented in the previous section, this technology replacement is reflected in the time variant technology, $\mu_{t,ij}$, and thus, the technical coefficient also becomes time variant, $a_{ij}\left(\mu_{t,ij};t,t-\phi_{ij},\sigma,\sigma-h_j-\varepsilon_j\right)$ in the combined model presented in (5.14). Romanoff and Levine (1990) discussed the issue of technical change in the SIM framework, with a simple example of "*pure technical replacement*" using a logistic function as the transition process. While the logistic curve of technology transition is plausible under the normal circumstances of economic growth, in a disaster situation technology replacement may become rather arbitrary and exogenous of production schedule, depending mainly on the availability and schedule of replacement funds (insurance and/or savings). In the current version of SIM, this technology replacement can be handled with the manipulation of time variant technology, $\mu_{t,ij}$, and its sensitivity analysis in terms of replacement timing may lead to some policy implications for the recovery and reconstruction process.

5.3.3 Inventory

One of the criticisms against the SIM framework is its "perfect knowledge" assumption (Mules, 1983), in which each sector is aware of both the direct and indirect demands made upon it as a result of any given initial impact. Romanoff and Levine responded to include the inventory for the adjustment between anticipated demand and actual orders, and the product inventory, $\mathbf{u}(t,t)$, is a function of the different inventory strategy among anticipatory industries (Romanoff and Levine, 1990 and 1991). This inventory function in SIM is more realistic than the previous perfect knowledge assumption, and is particularly useful under a disaster situation. As indicated in the discussion of uncertainty, product inventory becomes a key to analyze the surprise aspect of the occurrence of an unexpected disaster and to investigate its adjustment role for the influx of demand injections in some sectors for recovery and reconstruction activities. In addition, some disasters may damage the inventory facilities; and, the analysis on the impacts of damaged inventory becomes possible with the SIM framework.

However, this function of product inventory needs to be treated carefully in terms of its modeling structure. While a product flow to the product inventory occurs when the anticipated demand and actual orders/shipments do not match,[7] the anticipated demand is used only for the production of intermediate goods; therefore, the following relationship should hold:

$$\mathbf{A}\mathbf{x}_{\sigma} + \mathbf{u}_{t} = \mathbf{A}\hat{\mathbf{x}}_{\sigma} \tag{5.17}$$

where $\mathbf{x}_{\sigma}$ is the actual orders/shipments at σ and $\hat{\mathbf{x}}_{\sigma}$ is the anticipated demand for σ at t. An inherent problem of input-output framework arises in this equality, similar to the one that the original Leontief dynamic model, (5.1), faced: since input-output framework is a flow (transaction) model, it does not include the concept of stock—accumulation of flows. And, inventory is a stock—accumulation of flows to the inventory including carry-over from the previous period. The product inventory, $\mathbf{u}_{t}$, is the flow to the product inventory at t; therefore, when actual order/shipment level, $\mathbf{x}_{\sigma}$, is larger than the anticipated demand level, $\hat{\mathbf{x}}_{\sigma}$, some of the elements in $\mathbf{u}_{t}$ may become negative value. This creates the problem in a flow model (while the equality of the formulation holds, there is no indication of actual inventory level), whereas this can be dealt with in the stock model—the inventory level becomes lower due to the greater orders than anticipated/produced. In this regard, the SIM framework needs to be further modified to incorporate with inventory stock.

[7] This is a narrow definition of product inventory. Some industries always keep a particular level of inventory as a part of the production strategy.

5.3.4 Compatibility with Engineering Models

Originally, SIM was developed by supplementing the structure of production with production chronology (Romanoff and Levine, 1981). This inclusion of production chronology in the static input-output framework made possible to analyze lagged impacts, such as labor market expansions and man power issues (Romanoff, 1984). Production process, especially for most of manufacturing industries, is well represented in this SIM framework (with anticipatory production mode), and the framework was used in evaluating the impacts of large construction projects for its lagged impacts (with responsive production mode; Romanoff and Levine, 1990a). This close representation of production scheduling made SIM coupled with engineering project scheduling methods, such as Critical Path Method (CPM) in order to analyze the dynamic economic impacts of a large, complex, and lengthy production, construction, and related macro-engineering undertakings (Levine and Romanoff, 1989). This feature of SIM is particularly useful and effective to model the dynamic process of recovery and reconstruction activities after a disaster. The version of SIM presented in the chapter is especially valuable for incorporating with various time schedule of production process (input application period, product completion period, order lead time, and transportation delay), production technology, and inventory, which may be different from before and/or may be fluctuating after a disaster.

While SIM can deal well with engineering information regarding production scheduling and transportation delay, other engineering losses, such as disruption and/or fluctuation of lifelines, can be handled as capacity constraints. In the standard input-output framework, capacity constraints can be imposed as supply constraint (shock) by rebalancing transaction matrix as in Cochrane (1997, and 1999) or by deriving final demand change from potential output reductions as in Rose *et al.* (1997) and Rose and Benavides (1998). Romanoff and Levine (1986) studied the capacity limitations with the use of inventory in the SIM framework; however, the formal representation of capacity limitation in their modified SIM was not explicit. Another characteristic of lifeline losses is that the duration of the outages might be quite short, especially for power outage (a few weeks for full recovery of the entire system at longest), and it can be shorter than the production duration in the SIM framework for most industries. Although some industries have a very short production period, the major assumptions[8] of and the nature of SIM make difficult to incorporate with such short and fluctuating changes. In this sense, SIM may need to be modified further to integrate with the damage data sensitive to a very short time period.

5.4 The Great Hanshin Earthquake: Empirical Application

At 5:46 a.m., on January 17, 1995, the worst disaster in postwar Japan struck the second largest region of Japan--the Kinki region. The City of Kobe and surrounding

[8] Romanoff and Levine (1981) stated that two aspects of production interval are essential to the formulation of SIM: 1) the industry production interval is the same for all industries and remains constant in time; and, 2) all industry production intervals are synchronized. Although the first assumption of fixed production intervals are relaxed in their later model (Romanoff and Levine, 1986; the formal representation is not clear, though), the second assumption remains essential all versions of SIM.

municipalities experienced massive destruction of houses, buildings, roads, rails, and infrastructure. The direct damages from the Great Hanshin Earthquake were estimated at about 10 trillion yen (100 billion dollars) according to the Hyogo Prefecture Government, equivalent to about 2.1% of Japan's GDP (Gross Domestic Product) and 11% of Kinki's GRP (Gross Regional Product). These direct damages were concentrated in the destruction of buildings (including houses and production facilities), of transportation facilities (port, roads, and rails), and utilities (water, sewage, gas, and electricity). Although the damaged geographical area[9] is only 4% of Kinki, it includes 15% of Kinki's population. These direct damages, inevitably, may have significant effects not only on the Kinki region but also on other regions. The loss of capital stocks, however, was 0.8% of Japan's total, while it was 10.5% in *the Great Kanto Earthquake* in 1923 (Yomiuri Newspaper, June 20, 1995).

Immediately after the event, various studies assessing the direct and indirect damages from the event were carried out by many institutions. The increase of final demand, especially in construction sector, for the recovery and reconstruction activities, furthermore, has also been estimated in various ways (see the summary and critique in Miyao, 1995). Utilizing these estimates of the damages and losses, the interregional effects were evaluated using the 1985 interregional input-output table published by the Ministry of International Trade and Industry's (MITI) (see Okuyama *et al.*, 1999 for the details of the assumptions and settings).

The previous study (Okuyama *et al.*, 1999), which employed a two-region (Kinki and Rest of Japan) interregional input-output table with the Miyazawa's (1976) extended framework, found that the estimated interregional impacts of the Great Hanshin Earthquake exceed the intraregional impacts, in terms of both income formation and output generation. This is particularly important to justify the public policies which allocate significant amount of the national government's budget for recovery and reconstruction activities in the damaged area. However, because of the static and macroeconomics nature of the input-output framework, the results cannot reflect the temporal impacts that may be severer right after the event and gradually decrease afterwards. These temporal indirect impacts may create demand-supply mismatch in some cases, and then create the further complication of economic impacts over time and space. The Sequential Interindustry Model (SIM) can extend the static input-output framework to include a dynamic process with a sequence of discrete time periods as discussed in the previous sections and, thus, can deal with the analysis of the dynamic process of recovery and reconstruction activities. In the following section, the SIM is employed to examine the temporal impacts of the Hanshin Earthquake and the sensitivity of recovery schedule. The assessment of direct damages and the base model utilizes the assumptions and the interregional input-output table with the Miyazawa's extended formulation in the previous study. In this section, the two-region interregional input-output model used in the previous model is modified to the SIM formulation with a sequence of discrete time periods (quarters). The emphasis of the SIM application here is, however, not for estimating the temporal impacts with the detailed specification for the SIM extension; rather, it intends to illustrate how the SIM framework is useful in evaluating the economic impacts of unscheduled events in an empirical setting, especially for the sensitivity analysis of different recovery schedules. Therefore, for simplification, the direct damages are linearly adjusted assuming the smooth recovery and reconstruction, and are distributed over quarters depending on

[9] Damaged area consists of the City of Kobe, Ashiya, Nishinomiya, Takarazuka, Akashi, Sumoto, and six towns on Awaji Island.

the period to be fully recovered by sector. It is also assumed that the final demands without the event are set to be equal across quarters so that the quarterly system does not have any seasonal fluctuations.

5.4.1 Production Mode Assignment

In order to apply the SIM, the sectors in the model need to be assigned as either anticipatory, or responsive. As indicated above, Romanoff and Levine (1986 and 1991) defined that anticipatory sectors are agriculture, mining, and most of manufacturing industries; responsive sectors are some manufacturing, construction, ordnance, and most of services sectors. However, this original classification is a little too broad: for example, agriculture and mining sectors may anticipate further in the future than most manufacturing sectors do, because of their production cycle. Furthermore, services sectors, such as Trade and Sales, and Finance among others, may adjust their production schedule relatively quickly within a quarter. Thus, the sectors are assigned in the following four modes (the full list of sectors can be found in appendix 5.A.1):

a) Anticipatory (long) sectors: anticipate four-quarter (one year) ahead:
   Agriculture and Mining
b) Anticipatory sectors: anticipate one quarter ahead:
   Food, Textile, Wood Products, Pulp and Papers, Chemicals, Petroleum, Plastic, Ceramic, Irons and Steel, Metal (non-Iron), Metal (others), Machinery (general), Machinery (electric), Machinery (transportation), Machinery (precision), and Other Manufactures
c) Responsive sector: response to one quarter ago[10]:
   Construction
d) Just-in-time sectors: adjust production within a quarter:
   Utilities, Trade and Sales, Finance, Transportation Services, Services, and Others

These sector assignments can be further specified, perhaps with more disaggregated sector classification, for more detailed estimation of the temporal impacts; however, in this chapter, a simpler assignment is used for examining the usefulness of the SIM framework.

5.4.2 Temporal Changes in Gross Output

The temporal impacts from the event were derived using the SIM version of the two-region input-output model with the damage assessment used in the previous study (Okuyama *et al.*, 1999; see the summary of the settings in appendix 5.A.2). Based on

[10] In this case study, Construction sector is assumed to process their production over several phases—demolition, cleaning up the land; structure, exterior; interior, installation of equipment, etc. While an actual building, for example, can requires a few years to be constructed from the initial order, in this case study, the buildings and structures are constructed over several stages. In this sense, the response time of one quarter can be justified, and also the scenarios below with flat and decreasing demand schedules for recovery and reconstruction can be useful.

the model (5.15),[11] the quarterly outputs by sector were calculated. Unlike a static model that uses the Leontief inverse matrix to derive the quarterly outputs for a particular quarter, the SIM model here requires the calculation based on a time series of final demand schedule over infinite quarters with time-phased technical coefficient matrices. Consequently, the quarterly outputs are derived by summing up to the 28th quarter from the initial quarter so that the estimated quarterly output by sector may deviate only less than ± 1% from the ultimate value obtained from an infinite summation.

In order to illuminate the sensitivity of reconstruction demand injection over time, two cases are set to be investigated: Case 1 has a constant amount of demand injection over three years to highlight the recovery of damages over the periods; Case 2, on the other hand, has a more realistic trend of demand injection with linearly decreasing amount over three years. The gross outputs of each block are calculates over eight quarters (two years).

Case 1: Constant Demand Injection

The paths of regional gross output changes (deviations from the level without the event) over time are shown in figure 5.2. Gross outputs are increasing during the first four quarters, especially in the first three quarters. The Kinki-Kinki block has the trend of steep increase during the first three quarters; the next three quarters have a relatively flat trend; and in quarters 7 and 8, the gross outputs are decreasing slowly. These trends are more significant in the Rest of Japan (ROJ)-Kinki block than in the Kinki-Kinki block. The initial increases are flatter than the Kinki-Kinki's; however, after five quarters, the positive impacts, mainly from the demand injection in Kinki, are decreasing. The Kinki-ROJ and the ROJ-ROJ blocks have the similar trends: initial steep increases and flatter increase in the latter quarters. These trends are mainly resulted from the anticipatory sectors and from the just in time sectors: in the Kinki-Kinki and ROJ-Kinki blocks, the anticipatory sectors start to decrease as early as in quarter 5; on the other hand, the just in time sectors continue to increase--steep increases in the first year and flatter increases in the second year--across the blocks. The responsive sector--Construction sector--has the constant positive impacts in the Kinki-Kinki block, and very small impacts in the other blocks, originated from the constant demand injection in Kinki. These different paths among the modes result mainly from the difference of production chronology, while the interindustry relationship and the various direct damages and recovery periods among sectors may partially contribute to these differences. A typical example is the trends of the anticipatory (long) sectors, although these sectors do not have significant impacts in total. In the Kinki-Kinki and ROJ-Kinki blocks, the trends are significantly altered between quarter 4 and 5, because of their one-year-ahead anticipation, and because the demand injections last only for three years.

If the static quarterly model can be defined as $\mathbf{x}_t = (\mathbf{I} - \mathbf{A}_t)^{-1} \cdot \mathbf{y}_t$, the comparison between this static quarterly model and the sequential model may provide the characteristics of the sequential processes in the model. Figure 5.3 shows the deviations of the static quarterly model from the sequential model. The static quarterly model underestimates the gross outputs across the blocks during the first three quarters,

[11] A simplified version of model (5.15) is used without inventory function. The mathematical specification of this simplified version can be seen in appendix 5.A.3.

and overestimates them after quarter 4. The most significant underestimation is in the ROJ-ROJ block, and the largest overestimation is in the ROJ-Kinki block (the deviation increases significantly during the latter quarters). These deviations originate mainly from the anticipatory (long) and anticipatory sectors. Because these sectors anticipate the future intermediate demand, in the earlier quarters they produce more products based on the anticipated recovery than in the static quarterly model. During the latter quarters, these sectors take into account the fact that the demand injection will last only three years; consequently, they will decrease their production, gradually. In this context, the static quarterly model may provide misleading impacts from the event, especially the positive impacts from the demand injection.

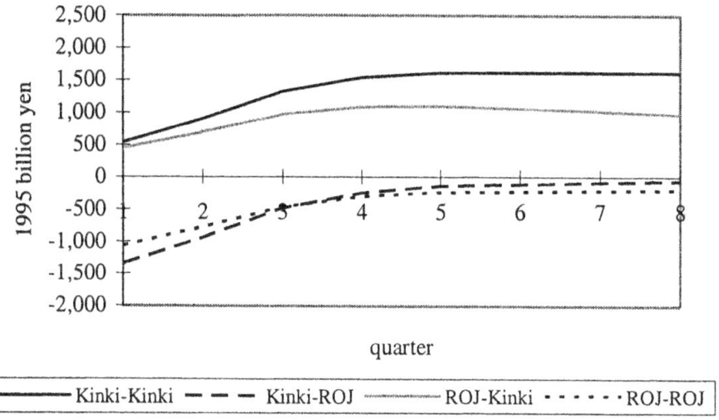

Figure 5.2 Changes in Gross Output over Time (Case 1)

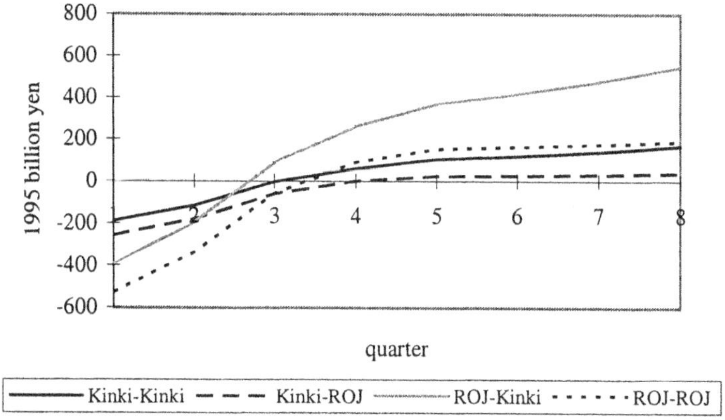

Figure 5.3 Differences between Sequential Model and Static Quarterly Model (Case 1)

Case 2: Decreasing Demand Injection

Recovery and reconstruction activities may start immediately after the event, and, as assumed in the previous study (Okuyama *et al.*, 1999), the most damaged sectors are restored and recovered in the relatively early quarters. On the other hand, in some sectors, especially Other Manufactures and Transportation Services sectors, their damages require as long as eight quarters (two years) for full recovery. Therefore, it is plausible to assume that the reconstruction demand injection in Kinki may be the largest at the first quarter, and then gradually decrease over time. The paths of regional gross output changes over time in Case 2 are shown in figure 5.4. In the Kinki-Kinki and ROJ-Kinki blocks, the gross outputs increase in the early quarters reflecting the recovery of production, but the increases are relatively smaller than in Case 1. After the third quarter, the gross outputs of these blocks start to decrease: the speeds of decline seem to be accelerated in the latter quarters, because of the decreased demand injection at those quarters. The paths of the Kinki-ROJ and ROJ-ROJ seem the similar trends in Case 1. The anticipatory and responsive sectors have relatively steep declines over quarters in the Kinki-Kinki and ROJ-Kinki blocks, and in the Kinki block, respectively. The gross outputs of the just-in-time sectors increase during the first four quarters (first year), and then start to decrease in the Kinki-Kinki and ROJ-Kinki blocks. Since the demand injection gradually decreases, the anticipatory (long) sectors do not have the sudden declines in the Kinki-Kinki and ROJ-Kinki blocks observed in Case 1, and they decreases relatively smoothly.

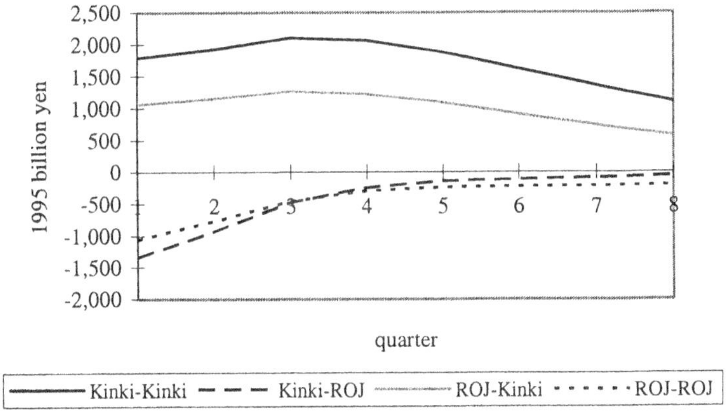

Figure 5.4 Changes in Gross Output over Time (Case 2)

The differences between the sequential model and the static quarterly model are shown in figure 5.5. The paths are quite different from those in Case 1 for the Kinki-Kinki and ROJ-Kinki block, and are quite similar for the Kinki-ROJ and ROJ-ROJ blocks. In the Kinki-Kinki and ROJ-Kinki blocks, the static quarterly model overestimates the gross outputs throughout the eight quarters; however, the trends of overestimation are first increasing to the fifth quarter; then, they decrease during the latter quarters. Since the demand injection is also decreasing over time, the differences between two models may narrow. As in Case 1, the ROJ-Kinki block has the largest difference between the models, because the sequential model can anticipate the decrease of demand injection, and may derive the smaller gross outputs than the static quarterly model does.

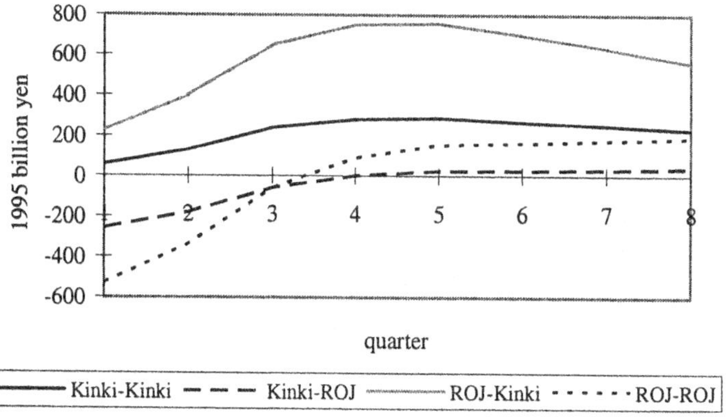

Figure 5.5 Differences between Sequential Model and Static Quarterly Model (Case 2)

5.5 Summary and Conclusions

In this chapter, the Sequential Interindustry Model (SIM) framework is examined and discussed in order for the use in modeling economic impacts of disasters. This modeling framework is found especially valuable to analyze the impacts of a disaster and its recovery and reconstruction processes over time. In addition, because of its temporal structure and the adoption of production chronology, SIM has the capabilities to handle many issues that other economic models may have difficulties, such as uncertainty, technology replacement, and inventory. The SIM framework also has great potentials to link with engineering data and models for the damages and losses, since its original intents aims very closely to the evaluation of engineering processes and projects. While some further theoretical sophistication are necessary, especially for the formal specification of inventory strategy and function, for the inclusion of uncertainty—relaxing the assumption of perfect knowledge, and for the adoption to the situations of demand/supply constraint, the SIM framework is flexible enough to handle a wide spectrum of situations and scenarios in disaster situations.

Another promising potential of the SIM framework is to analyze the process of recovery and reconstruction process in detail, while the assignment of production mode across the industries and the determination of different inventory management strategies among anticipatory industries may pose some difficulties in data requirements. Since SIM was originally developed for evaluating large construction projects (Romanoff and Levine, 1990), the structure of SIM is suited to examine the rapid and intense process of recovery and reconstruction after a disaster. With a determination of inventory function, SIM can simulate the case of supply constraints under the influx of recovery activities (shortage of construction material, etc.) and can evaluate the labor market adjustment process (a large increase in labor supply during the recovery and reconstruction and sudden decrease after the completions). The analysis of uncertainty due to a disaster is also an important issue in modeling the economic impacts that SIM can handle well, as studied in the previous study (Okuyama *et al.*, 2001).

By contrast with the static input-output model, the intertemporal process of economic impacts from a disaster can be examined. As discussed in Section 4, the static input-output model overestimated the impacts during most of the periods. In particular, the interregional block (ROJ-Kinki) has the largest deviations between the sequential model and the static model in both cases. This implies that the interregional spill-over effects from the demand injections in the region with disaster (Kinki) may become less significant than the static model predicted, as a result of the inclusion of production process, especially by anticipatory sectors. Moreover, the comparison between the constant injection and decreasing injection cases may reveal that the timing of demand injection has considerable effects on the impacts over time. In this chapter, however, the damage recovery processes in the coefficients and in the final demand and the reconstruction demand injection are not linked systematically.

While the sensitivity analysis using more detailed set-up and assignment of the production mode and disaggregated input-output model needs to be done for drawing any policy implications from the case study results, the following two points can be addressed. First, recovery and reconstruction activities after a disaster need to be planned and phased so that no significant supply constraints of intermediate goods to construction sector occur. As mentioned above, reconstruction activities, and usual construction activities, can be divided into several stages: demolition and clean-up the site; constructing the foundation; constructing the structure; exterior works; installation of the building equipments and lifelines; interior works; installation of furniture and other equipments, etc. Each stage requires different intermediate inputs, such as equipments, machines, materials, different type of labor (construction workers). It was the case in the Great Hanshin Earthquake that the demolition and cleaning-up activities for reconstructing damaged facilities were constrained due to the overwhelming demand. Hence, a policy toward smooth recovery requires prioritizing the reconstruction activities and scheduling to distribute them to different stages of construction phase in order not to create severe supply constraints of intermediate goods and primary inputs. Secondly, with the rich information on interindustry relationships embedded in interregional input-output table, temporal key sector analysis can be accomplished under a disaster situation for illustrating which sectors are more crucial for economy-wide recovery, in a particular stage of reconstruction. Although it may be difficult to concentrate on the recovery of particular sectors after a catastrophic disaster, this type of information can be utilized for creating retrofit priority to make the key sectors less vulnerable. Furthermore, this formulation may be used to derive the optimal resource allocation for recovery and reconstruction as a control problem, as Rose *et al.*, 1997, did with a static input-output table; however, it is beyond the scope of this chapter.

References

Boisverst, R. 1992. "Indirect Losses from a Catastrophic Earthquake and Local, Regional, and National Interest," in *Indirect Economic Consequences of a Catastrophic Earthquake*. Washington, DC; National Earthquake Hazards Reduction Program, Federal Emergency Management Agency: 207-265.

Campisi, D. and A. Nastasi. 1993. "Capital Usage and Output Growth in Multiregional Multisectoral Models: An Application to the Italian Case," *Regional Studies*, 27 (1): 13-27.

Campisi, D., A. La Bella, M. Gastaldi, B. Ludovici, and G. Schachter. 1990. "A Multi-Regional Multi-Sectoral Input-Output Model for Italy," in L. Anselin and M. Madden ed. *New Directions in Regional Analysis*. New York, NY; Belhaven Press: 187-210.

Campisi, D., A. Nastasi, and A. La Bella. 1992. "Balanced Growth and Stability of the Leontief Dynamic Model: An Analysis of the Italian Economy," *Environmental and Planning A*, 24 (4): 591-600.

Chang, S. 1983. "Disasters and Fiscal Policy: Hurricane Impact on Municipal Revenue," *Urban Affairs Quarterly*, 18 (4): 511-523.

Chang, S.E., W.D. Svekla, and M. Shinozuka. 2002. "Linking Infrastructure and Urban Economy: Simulation of Water Disruption Impacts in Earthquakes," *Environment and Planning B*, 29 (2): 281-301.

Cho, S., P. Gordon, J.E. Moore II, H.W. Richardson, M. Shinozuka, and S.E. Chang. 2001. "Integrating Transportation Network and Regional Economic Models to Estimate the Costs of a Large Urban Earthquake," *Journal of Regional Science*, 41 (1): 39-65.

Cochrane H.C. 1999. "Indirect Economic Losses," in *Earthquake Loss Estimation Methodology: HAZUS99 Technical Manual, Volume 3*. Washington DC; Federal Emergency Management Agency: 16-1 – 16-63 (http://www.fema.gov/hazus/dl_tech3.pdf).

Cochrane, H.C. 1974. "Predicting the Economic Impacts of Earthquakes," in H.C. Cochrane, J.E. Haas, M.J. Bowden, R.W. Kates eds. *Social Science Perspectives on the Coming San Francisco Earthquake*, Natural Hazards Research Paper No. 25, NHRAIC, Boulder, CO; University of Colorado.

Cochrane, H.C. 1995. "Forecasting the Economic Impact of a Midwest Earthquake," in B.G. Jones ed. *Economic Consequences of Earthquakes: Preparing for the Unexpected*. Buffalo, NY; National Center for Earthquake Engineering Research: 223-247.

Cole, S. 1988. "The Delayed Impacts of Plant Closures in a Reformulated Leontief Model," *Papers in Regional Science Association*, 65: 135-149.

Cole, S. 1989. "Expenditure Lags in Impacts Analysis," *Regional Studies*, 23 (2): 105-116.

Cole, S. 1997. "Closure in Cole's Reformulated Leontief Model: A Response to R.W. Jackson, M. Madden, and H.A. Bowma," *Papers in Regional Science*, 76 (1): 29-42.

Cole, S. 1997. "Decision Support for Calamity Preparedness: Socioeconomic and Interregional Impacts," in M. Shinozuka, A. Rose, and R.T. Eguchi eds. *Engineering and Socioeconomic Impacts of Earthquakes*. Buffalo, NY; Multidisciplinary Center for Earthquake Engineering Research: 125-153.

Cole, S. 1999. "The Phantom of the Matrix: Inverting the Case on Closure in Cole's Model," *Papers in Regional Science*, 78 (4): 429-436.

de Mesnard, L. 1992. "The Asynchronous Leontief Model," *Economic Systems Research*, 4 (1): 25-34.

Dorfman, R., P.A. Samuelson, and R.M. Solow. 1958. *Linear Programming and Economic Analysis*. New York, NY; McGraw Hill.

Duchin, F. and D.B. Szyld. 1985. "A Dynamic Input-Output Model with Assured Positive Output," *Metroeconomica*, 37 (3): 269-282.

Ellson, R., J.W. Milliman, and R.B. Roberts. 1984. "Measuring the Regional Economic Effects of Earthquakes and Earthquake Predictions," *Journal of Regional Science*, 24 (4): 559-579.

Forrrester, J.W. 1968. *Principles of Systems*. Cambridge, MA; Wright-Allen Press.

Goodwin, R.M. 1947. "Dynamical Coupling with Especial Reference to Markets Having Production Lags," *Econometrica*, 15: 181-204.

Gordon, P. and H. Richardson. 1996. "The Business Interruption Effects of the Northridge Earthquake," Lusk Center Research Institute, University of Southern California, Los Angeles, CA.

Guangzhen, S. 1993. "Economic Impact of the Structural Changes of Production Technology: A Robust Approach," *Economic Systems Research*, 5 (3): 235-242.

Guimaraes, P., F.L. Hefner, and D.P. Woodward. 1993. "Wealth and Income Effects of Natural Disasters: an Econometric Analysis of Hurricane Hugo," *Review of Regional Studies*, 23: 97-114.

Heesterman, A.R.G. 1990. "The Instability of the Dynamic Input-Output Model," *Economic Systems Research*, 2 (4): 351-355.

Jackson, R.W. and M. Madden. 1999. "Closing the Case on Closure in Cole's Model," *Papers in Regional Science*, 78 (4): 423-427.

Jackson, R.W., M. Madden, and H.A. Bowman. 1997. "Closure in Cole's Reformulated Leontief Model," *Papers in Regional Science*, 76 (1): 21-28.

Kawashima, K., H. Sugita, and T. Kanoh. 1991. "Estimation of Earthquake Induced Economic Damage." *Report of Public Works Research Institute,* 186: 1-57. (in Japanese)

Kuenne, R.E. 1963. *The Theory of General Economic Equilibrium*. Princeton, NJ; Princeton University Press.

Leontief, W. 1951. *The Structure of American Economy, 1919-1939. 2nd Edition*. New York, NY; Oxford University Press.

Leontief, W. 1953. "Dynamic Analysis," in W. Leontief, H.B. Chenery, P.G. Clark, J.S. Duesenberry, A.R. Ferguson, A.P. Grosse, R.N. Grosse, and M. Holzman ed. *Studies in the Structure of the American Economy: Theoretical and Empirical Explorations in Input-Output Analysis*. New York, NY; Oxford University Press.

Leontief, W. 1970. "The Dynamic Inverse," in A.P. Carter and A. Brody eds. *Contributions to Input-Output Analysis*. Amsterdam, The Netherlands; North-Holland Publishing: 17-46.

Leontief, W. and F. Duchin. 1986. *The Future Impact of Automation on Workers*. New York, NY; Oxford University Press.

Levine, S.H. and E. Romanoff. 1989. "Economic Impact Dynamics of Complex Engineering Project Scheduling," *IEEE Transactions Systems, Man and Cybernetics*, 19: 232-240.

Morishima, M. 1964. *Equilibrium, Stability and Growth*. Oxford, UK; Clarendon Press.

Mules, T.J. 1983. "Some Simulations with a Sequential Input-Output Model," *Papers of Regional Science Association*, 51: 197-204.

National Academy of Sciences. 1978. *A Program of Studies in the Socioeconomic Effects of Earthquake Predictions*. Washington, DC; National Academy of Sciences.

Okuyama, Y., G.J.D. Hewings, and M. Sonis. 2001. "Economic Impacts of Unscheduled Events: Sequential Interindustry Model (SIM) Approach," Paper presented in the North American Meetings of Regional Science Association International, Charleston, SC.

Okuyama, Y., G.J.D. Hewings, and M. Sonis. 1999a. "Economic Impacts of an Unscheduled, Disruptive Event: A Miyazawa Multiplier Analysis," in G.J.D. Hewings, M. Sonis, M. Madden, and Y. Kimura eds. *Understanding and Interpreting Economic Structure*. Berlin, Germany; Springer-Verlag: 113-144.

Okuyama, Y., G.J.D. Hewings, T.J. Kim, D.E. Boyce, H. Ham, and J. Sohn. 1999b. "Economic Impacts of an Earthquake in the New Madrid Seismic Zone: A Multiregional Analysis," in W.M. Elliott and P. McDonough eds. *Optimizing Post-Earthquake Lifeline System Reliability*. Technical Council on Lifeline Earthquake Engineering, Monograph No.16: 592-601.

Oosterhaven, J. 2000. "Lessens from the Debate on Cole's Model Closure," *Papers in Regional Science*, 79 (2): 233-242.

Romanoff, E. 1984. "Interindustry Analysis for Regional Growth and Development: The Dynamics of Manpower Issues," *Socio-Economic Planning Sciences,* 18 (5): 353-363.

Romanoff, E. and S.H. Levine. 1977. "Interregional Sequential Interindustry Modeling: A Preliminary Analysis of Regional Growth and Decline in a Two Region Case," *Northeast Regional Science Review*, 7: 87-101.

Romanoff, E. and S.H. Levine. 1981. "Anticipatory and Responsive Sequential Interindustry Models," *IEEE Transactions on Systems, Man, and Cybernetics,* 11 (3): 181-186.

Romanoff, E. and S.H. Levine. 1986. "Capacity Limitations, Inventory, and Time-Phased Production in the Sequential Interindustry Model," *Papers of the Regional Science Association,* 59; 73-91.

Romanoff, E. and S.H. Levine. 1990a. "Combined Regional Impact Dynamics of Several Construction Megaprojects," *Regional Science Review*, 17: 85-93.

Romanoff, E. and S.H. Levine. 1990b. "Technical Change in Production Processes of the Sequential Interindustry Model," *Metroeconomica,* 41 (1): 1-18.

Romanoff, E. and S.H. Levine. 1991. "Technical Change and Regional Development: Some Further Developments with the Sequential Interindustry Model," in D.E. Boyce, P. Nijkamp, and D. Shafer eds. *Regional Science: Retrospect and Prospect.* Berlin, Germany; Springer-Verlag: 251-275.

Romanoff, E. and S.H. Levine. 1993. "Information, Interindustry Dynamics, and the Service Industries," *Environmental and Planning A,* 25: 305-316.

Rose, A. and J. Benavides. 1998. "Regional Economic Impacts," in M. Shinozuka, A. Rose, and R.T. Eguchi eds. *Engineering and Socioeconomic Impacts of Earthquakes.* Buffalo, NY: Multidisciplinary Center for Earthquake Engineering Research: 95-123.

Rose, A., J. Benavides, S.E. Chang, P. Szczesniak, and D. Lim. 1997. "The Regional Economic Impact of an Earthquake: Direct and Indirect Effects of Electricity Lifeline Disruptions," *Journal of Regional Science,* 37 (3): 437-458.

Schinnar, A. P. 1978. "The Leontief Generalized Inverse," *The Quarterly Journal of Economics,* 92 (4): 641-52.

Steenge, A.E. 1990a. "On the Complete Instability of Empirically Implemented Dynamic Leontief Models," *Economic Systems Research,* 2 (1): 3-16.

Steenge, A.E. 1990b. "Instability Problems in the Dynamic Leontief Model: an Economic Explanation," *Economic Systems Research,* 2 (4): 357-361.

Szyld, D.B. 1985. "Conditions for the Existence of a Balanced Growth Solution for the Leontief Dynamic Input-Output Model," *Econometrica,* 53 (6): 1411-1419.

ten Raa, T. 1986. "Dynamic Input-Output Analysis with Distributed Activities," *Review of Economics and Statistics,* 68 (2): 300-310.

West, C.T. and D.G. Lenze. 1994. "Modeling the Regional Impact of Natural Disaster and Recovery: A General Framework and an Application to Hurricane Andrew," *International Regional Science Review,* 17 (2): 121-150.

Wilson, R. 1982. "Earthquake Vulnerability Analysis for Economic Impact Assessment," Washington DC; Information Resources Management Office, Federal Emergency Management Agency.

Zaghini, E. 1971. "Solow Prices and the Dual Stability Paradox in the Leontief Dynamic System," *Econometrica,* 39 (3): 625-632.

Appendix

5.A.1 Sectors in the Two-Region System (based on the 1985 MITI Interregional Input-Output Table)

1. Agriculture
2. Mining
3. Food
4. Textile
5. Wood Products
6. Pulp and Papers
7. Chemicals
8. Petroleum
9. Plastic
10. Ceramic
11. Iron and Steel
12. Metal (non-Iron)
13. Metal (others)
14. Machinery (general)
15. Machinery (electric)
16. Machinery (transportation)
17. Machinery (precision)
18. Other Manufactures
19. Construction
20. Utilities
21. Trade and Sales
22. Finance
23. Transportation Services
24. Services
25. Others

5.A.2 Summary of Settings in Okuyama *et al.*, 1999

Table 5.A.2.1 Changes in Direct Input Coefficients

Sector	Initial Damage Rate	Year to Recovery	Time-Adjusted Damage Rate	Damage Rate in Kinki	Interregionally Substitutable?
3. Food	-30%	0.5	-7.5%	-2.44%	yes
4. Textile	-30%	0.5	-7.5%	-0.14%	yes
11. Iron and Steel	-50%	0.7	-17.5%	-2.32%	yes
16. Machinery (transp.)	-25%	2	-18.75%	-2.33%	yes
18. Other Manufactures	-40%	1	-20%	-3.42%	yes
20. Utilities	-50%	0.3	-7.50%	-0.60%	no
21. Trade and Sales	-60%	1	-30%	-3.54%	yes
23. Transp. Services	-75%	2	-56.25%	-14.10%	no

Table 5.A.2.2 Changes in Final Demand

Sector	Initial Damage Rate	Year to Recovery	Time-Adjusted Damage Rate	Damage Rate in Kinki	Interregionally Substitutable?
3. Food	-30%	0.5	-7.5%	-0.49%	yes
4. Textile	-30%	0.5	-7.5%	-0.23%	yes
11. Iron and Steel	-50%	0.7	-17.5%	-1.64%	yes
16. Machinery (transp.)	-25%	2	-18.75%	-1.29%	yes
18. Other Manufactures	-40%	1	-20%	-2.13%	yes
20. Utilities	-50%	0.3	-7.50%	-0.63%	no
21. Trade and Sales	-60%	1	-30%	-1.92%	yes
23. Transp. Services	-75%	2	-56.25%	-3.63%	no

Changes in Consumption Coefficients

Using a linear time adjustment, the time-adjusted rate of consumption decrease is 7.5%. This decrease is also assumed to affect all sectors' consumption equally in the Kinki to Kinki block and in the rest of Japan to Kinki block.

Changes in Value-Added Coefficients

Assuming the unemployment due to the earthquake persists over the course of a year, it would result in a decrease in the value-added coefficient of 0.2%. It is also assumed that this decrease is spread over all sectors in the Kinki to Kinki block and in the rest of Japan to Kinki block.

5.A.3 Two-Region Interregional SIM Model

For the empirical case study on the Great Hanshin Earthquake, the combined anticipatory-responsive SIM model, (5.15), is modified to fit with the sector assignment and assumptions in the following manner. First, the model is simplified into the Core SIM version specified in Romanoff and Levine, 1981, using a simpler specification in time phase and production process. Secondly, the inventory function, $\mathbf{u}_t$, is removed from the model, since the lack of empirical data for inventory strategy of each sector and since it is assumed that demand and supply may be adjusted within a single period—one quarter (three months). Third, the sectors are assigned to four different production modes—anticipatory (long), anticipatory, responsive, and just-in-time modes—with the specified time lags. Then, the applicable model can be written as follows:

$$\mathbf{x}_t = \mathbf{A}_a\mathbf{x}_{t+1} + \mathbf{A}_{al}\mathbf{x}_{t+4} + \mathbf{A}_r\mathbf{x}_{t-1} + \mathbf{A}_j\mathbf{x}_t + \mathbf{y}_t \quad (5.A.1)$$

where $\mathbf{A}_a$ is the direct input coefficient matrix for the anticipatory sectors with one quarter anticipation; $\mathbf{A}_{al}$ is for the anticipatory (long) sectors with one year (four quarters) anticipation; $\mathbf{A}_r$ is for the responsive sectors with one quarter response period; and $\mathbf{A}_j$ is for the just-in-time sectors. The solution of this model becomes:

$$\mathbf{x}_t = \sum_{k=1}^{\infty}\mathbf{A}_a^k\mathbf{y}_{t+k} + \sum_{k=1}^{\infty}\mathbf{A}_{al}^k\mathbf{y}_{t+k+3} + \sum_{k=1}^{\infty}\mathbf{A}_r^k\mathbf{y}_{t-k} + \sum_{k=0}^{\infty}\mathbf{A}_j^k\mathbf{y}_t + \\ + \sum_{k=-\infty}^{\infty}\mathbf{G}_k\left(\mathbf{A}_a,\mathbf{A}_{al},\mathbf{A}_r,\mathbf{A}_j\right)\mathbf{y}_{t-k} \quad (5.A.2)$$

where $\mathbf{G}_k\left(\mathbf{A}_a,\mathbf{A}_{al},\mathbf{A}_r,\mathbf{A}_j\right)$ is a matrix function whose *ij* element contains the sum of synergetic path gains among different production modes from industry *i* to *j* with a total delay of *k*. Using this setting, a two-region interregional version of this simplified model can be specified as follows:

$$\begin{bmatrix} \mathbf{x}_{1,t} \\ \mathbf{x}_{2,t} \end{bmatrix} = \tilde{\mathbf{A}}_a \begin{bmatrix} \mathbf{x}_{1,t+1} \\ \mathbf{x}_{2,t+1} \end{bmatrix} + \tilde{\mathbf{A}}_{al} \begin{bmatrix} \mathbf{x}_{1,t+4} \\ \mathbf{x}_{2,t+4} \end{bmatrix} + \tilde{\mathbf{A}}_r \begin{bmatrix} \mathbf{x}_{1,t-1} \\ \mathbf{x}_{2,t-1} \end{bmatrix} + \tilde{\mathbf{A}}_j \begin{bmatrix} \mathbf{x}_{1,t} \\ \mathbf{x}_{2,t} \end{bmatrix} + \begin{bmatrix} \mathbf{y}_{1,t} \\ \mathbf{y}_{2,t} \end{bmatrix} \tag{5.A.3}$$

where $\mathbf{x}_{i,t}$ is the total output (input) vector of region i at time t, $\mathbf{y}_{i,t}$ is the final demand vector of region i ay time t. The direct input coefficient matrix for each production mode becomes also interregional specification: $\tilde{\mathbf{A}}_a = \begin{bmatrix} {}_{11}\mathbf{A}_a & {}_{12}\mathbf{A}_a \\ {}_{21}\mathbf{A}_a & {}_{22}\mathbf{A}_a \end{bmatrix}$; $\tilde{\mathbf{A}}_{al} = \begin{bmatrix} {}_{11}\mathbf{A}_{al} & {}_{12}\mathbf{A}_{al} \\ {}_{21}\mathbf{A}_{al} & {}_{22}\mathbf{A}_{al} \end{bmatrix}$; $\tilde{\mathbf{A}}_r = \begin{bmatrix} {}_{11}\mathbf{A}_r & 0 \\ 0 & {}_{22}\mathbf{A}_r \end{bmatrix}$; and $\tilde{\mathbf{A}}_j = \begin{bmatrix} {}_{11}\mathbf{A}_j & {}_{12}\mathbf{A}_j \\ {}_{21}\mathbf{A}_j & {}_{22}\mathbf{A}_j \end{bmatrix}$, where ${}_{mn}\mathbf{A}$ is the block matrix of direct input coefficients (of each production mode) for region m to n. Construction sector (responsive production mode) does not export nor import the final products across regions; thus, off-diagonal elements, interregional block matrices, are set to null. Similarly to the single region model above, the solution of this two-region model can be outlined with $\tilde{\mathbf{A}}_a$, $\tilde{\mathbf{A}}_{al}$, $\tilde{\mathbf{A}}_r$, $\tilde{\mathbf{A}}_j$, and $\tilde{\mathbf{G}}_k(\bullet)$, which not only includes synergetic interactions among production modes but also includes interregional interactions.

6 Geohazards in Social Systems: An Insurance Matrix Approach*

Sam Cole

Department of Urban and Regional Planning, University at Buffalo, USA
e-mail: samcole@ap.buffalo.edu

6.1 Introduction

The principle goal of the chapter is to explore ways of solving social accounting matrices (SAMs) that may be useful for estimating the impact of natural and other disasters and the effectiveness of recovery programs. To this end my chapter introduces a model to assess the way in which disasters affect social actors and propagate through society. It adopts findings from disaster research and anthropology, to embody social transactions in an insurance accounting matrix. This assumes that the pre-disaster secular and spatial structure of any economy is a reflection, at least in part, of the various uncertainties associated with regular production, transportation, financing, daily survival, and so on. These may result in formal expenditures - as with insurance or stocks, or informal effort - such as maintaining social networks. Each measures the opportunity cost against other income. With a major disturbance, the proportion of failed transactions is increased beyond the capability of these afforded precautions to cushion the impacts, resulting in a "disaster" or even a "catastrophe".

The method proposed here entails a number of novel adaptations of conventional social accounting matrices and methods of solution. This introduces the magnitude and time-scale of lifeline and secular failures, as well as specific mitigation, and reconstruction efforts so as to describe situations where damage from a hazard is temporally distributed across infrastructure and delivery systems. The approach accounts for the cross-impacts insurance failures between various businesses and household networks. In the chapter, after explaining the accounting structure and basic equations used in the solution, I shall provide a stylized empirical application to illustrate the plausibility and potential utility of the approach.

* This chapter develops ideas presented at the IV Latin American Symposium on Urban Geohazards, San Jose, Costa Rica, 9-13 September, 1996 and in Living with Uncertainty, Disasters, and Acts of God presented at the 93rd Annual Meeting of the Association of American Geographers in Fort Worth, Texas, April 1-5 1997, session on Spatial Analysis in Mathematical Models and Quantitative Methods Specialty Group/IGU Commission on Geographical Modeling. The initial work was supported by the National Center for Earthquake Engineering and Research with funding from the National Science Foundation.

6.2 Hazards and Their Socio-Economic Issues

In this chapter, a "disaster" is regarded as the unfortunate coincidence of several otherwise minor events, that together lead to a major event. Funtowitz and Ravetz (1994), for example, argue that industrial accidents are the coming together of several unusual events - someone doing the wrong thing, at the wrong time, on a machine with a design fault. The severity of a natural disaster - say a storm surge following a hurricane - is the result of the coincidence of seasonal, periodic, and transient and distributed natural and human events - seasonal currents and trade winds and frequency and path of hurricanes, the daily timing of tides, the time-of-day of the event, the prevailing building codes and their enforcement, the age of the infrastructure, the preparedness of emergency agencies, communities and individuals, and so on. The predictability of these events, or our ability to avert them, varies considerably (see, *e.g.*, Jackson, 1995, on the obstacles to forecasting very large earthquakes).

Economic actors – businesses, households, and government - adopt various strategies to minimize, or at least reduce, their total transaction costs including tactics to avoid the consequences of unexpected delays (*e.g.*, Olson, 1971; Landa, 1994). Even so, most actors typically can afford only to make selected preparations - that are limited by their financial, social, and other resources - for disasters of a certain magnitude, or for temporary breakdown (such as short-run backup water and electricity supply, or insurance and savings). Engineers use rule-of-thumb safety factors; building codes and public infrastructure are designed to cover up to a specified magnitude of earthquake or a "50 year flood". Producers and their customers "insure" against disruptions by maintaining buffer stocks, or by concentrating their business in nearby markets. In most cases there is a trade-off. For example, "just-in-time" production reduces inventory costs, but increases short-run vulnerability to interruptions, as witnessed by several recent labor disputes, and also requires that supplies are sourced locally (Allen, 1994).

In many cases, disasters have their most severe impacts on isolated localities and marginal urban communities (Ebert, 1982; Dynes and Tierney, 1994). For these people and places, the extreme severity of impacts often can be traced to poverty and inappropriate development, such as inferior infrastructure or housing. In some cases, victims are disadvantaged even further by deficient recovery programs (Cuny, 1983; Dyer, 1995). Several authors (Jones, 1981; Cuny, 1983; Aysan *et al.*, 1989) have insisted that disasters are an inextricable component of the process of development. In developing countries there are especially compounding links between disasters and development (see, *e.g.*, Ebert, 1982; Bender, 1990). Poor communities and households are generally, unable to make substantive contingencies, and tend to rely on supportive social networks or marginal activities (see, *e.g.*, Cohen, 1985; Bender, 1990; Cole, 1995; Smith, 1990; and Udry, 1993). Subject to repeated disaster, eventually they may become dependent on outside assistance because of their excessive vulnerability. For example, peasant farmers avoid the possible catastrophic risks associated with new production methods, by sticking to traditional techniques, or thwarting innovation. Thus, there are physical, spatial, economic, and social consequences of our protective behaviors.

6.3 Social Accounting Matrices

Empirically, a social accounting matrix (SAM) is a large table showing the annual transactions (income and expenditures) of economic actors in a society (see, *e.g.*, Stone, 1973; Pyatt and Roe, 1977). SAMs are an extension of the earlier input output models invented by the late Wassily Leontief and provide a consolidated quantitative representation of the flows and distribution of goods between businesses, households and public service, and regions, or even between neighboring micro-economies (Cole, 1994; Cole 1996).

A SAM is an especially useful way of representing the transactions in a region, before, during, and after a major disruption. During a period of disaster, some of the flows in the economic network described by the model are interrupted, with ramifications throughout the economy. Even when they are not impacted directly, people and businesses may be affected indirectly through damage to lifelines such as water supply or roads, or through the loss of livelihood or markets (*e.g.*, Ellson *et al.*, 1984; Boisvert, 1992). Calculations with a SAM allow the consequences of this damage for all types of participant in an area's economy, whether directly impacted or not, to be estimated. Moreover, because the SAM may incorporate a range of social, economic, and environmental variables, it is possible to provide more substantive indicators of the impacts on target populations and industries, such as changes in levels of output or household income.

In this chapter the SAM is modified to explicitly include expenditures and other activities to defend against "disasters". Underlying the new specification and solution is the idea that many disasters, whether human instigated, natural, or a combination of both, are a regular occurrence in the process of development and that some level of mitigation strategy is embodied in all structures and transactions. The disruption arising from a given event depends in good measure on how well prepared the society is. Under "normal" circumstances - characterized by small frequent hazards - some proportion of each transaction will be delayed unacceptably, unless producers, households, and other institutions can reduce their losses. With a major disturbance, the proportion of failed transactions is increased beyond the capability of these afforded precautions to cushion the impacts, resulting in a "disaster".

The aim of the chapter is to formalize this process within the frame of a modified SAM. For want of a better term, I shall call this an "insurance accounting matrix", but it might equally be called a "buffer matrix", or a "protection matrix," since it represents both formal and informal means for protection. This will allow the magnitude and time-scale of lifeline and sectoral failures, and mitigation, or reconstruction efforts, to be integrated into the solution of, so as to describe situations where damage from a hazard is spatially and temporally distributed across infrastructure and delivery systems, and impacts differentially on various businesses and households.

Overall, the pre-disaster economic, social, and spatial structure of any community is a reflection, at least in part, of the various uncertainties associated with regular production, transportation, financing, daily survival, and so on. These may result in formal expenditures - as with insurance or stocks, or informal - as with the effort of maintaining social networks insofar as it represents an opportunity cost against other income. In the present chapter, the costs of protection afforded for transactions are represented by an insurance matrix to be incorporated into the structure and solution of the social accounts.

The proposed solution of the matrix is rather different from that associated with traditional input-output models, since it accepts that economies are always in a disequilibrium state as a consequence of normal shocks and uncertainties, but will also occasionally suffer major calamities, beyond the normal response capabilities of the system. I shall use an adaptation of the delayed transaction method described in Cole (1988), where the lagged round-by-round calculation of the impacts are approximated by a revised Leontief inverse in order to calculate impacts at a given time horizon. In the present case, however, the systemic effects of transient production, shipment, and financial disruptions are incorporated explicitly into the round-by-round calculation.

In principal, the insurance matrix may be calibrated empirically against a simulated stream of events, so that the impacts of a major event take account of the normal defenses of a society, and then used to shed light on a number of questions. For example, what is the "optimal" level of insurance? Given constraints on possibilities for insurance, what is an appropriate social, economic, and spatial disposition of transactions? Or, how does under-insurance in some parts of a community affect well-insured parts, and vice versa?

6.4 The Insurance Accounting Matrix

The first step is to prepare the insurance accounting matrix from social accounts. For clarity, I shall use a stylized set of accounts for a small region, with a "dualistic" economic structure, that is characteristic of many developing countries. Table 6.1 shows the base social accounts. The individual entries in the social accounts represent the annual amounts, such as the intermediate purchases between sectors, wages paid by each sector to households, exports of goods, and financial flows. The column entries are the expenditures by each sector. For production sectors this includes factor and commodity expenditures (intermediate demand and imports) and for households consumption of goods and services, taxes, and so on. Adding the transactions along a row gives the total income for each account.

Activities are aggregated into traditional and modern production sectors, rich and poor households, government, and capital. Here, these transactions can be derived from more detailed social accounts, from case studies, and so on. The aggregation here is designed to highlight the critical structure of a dualistic developing society, such as developed-backward, modern-traditional, urban-rural, rich-poor, and so on. The literature on development shows that these attributes are highly correlated - economies with divergent technologies typically have an uneven income distribution (see, *e.g.*, Clarke, 1984; Ryan, 1991; Cole and Miles, 1984). External flows comprise imports and exports, receipts and repatriation of capital and development aid. The traditional and modern sectors use different technologies and produce non-substitutable commodities. Poor households receive wage income, and consume both commodities, and pay taxes. Rich households also receive investment income and save, and spend some income directly outside the region. In table 6.1 the sizes of the traditional (*e.g.*, rural or informal) sector and the modern sector are 50 and 80 units (*e.g.*, million pesetas or chungas) respectively while the income of poor and modern households is 20 and 40 units. This distribution shows the relative ownership of wage and investment income. All amounts are in constant units and prices are fixed.

The social accounts also include the new "insurance" activity. This is the total amount of resources expended by actors - formally and informally - to protect their transactions against the deleterious consequences of hazards. The row entries (presently blank) will become the regular (annual) expenditures on insurance activities but the amounts will have the same meaning as those for other production sectors noted above.

Table 6.1 Stylized Base Social Accounts

BASE SOCIAL ACCOUNTS (ANNUALIZED)									Base Prices=1
ACTORS	Traditional	Modern	Poor	Rich	Govt.	Capital	Protection	Trade	Total
Traditional	10	5	10	5	5			15	50
Modern	5	10	5	15	5	5		35	80
Poor	10	5			5				20
Rich	5	10			10	15			40
Government	5	10	5	10				5	35
Capital	5	15		5	10			5	40
Protection									0
Trade	10	25		5		20			60
Total	50	80	20	40	35	40	0	60	

Implicitly, table 6.1 includes informal transactions, since these are quite large for rich and poor communities in all countries, and have a significant protective role, although they are obviously difficult to measure (Pyatt and Roe, 1977; Adelman, *et al.*, 1988; Cole 1994). As remarked earlier, much insurance is informal, especially in developing countries or marginal societies (Humphrey and Hugh-Jones, 1985). For present purposes, it is assumed that all activities, including insurance related activities, can be measured and converted to monetary units. Note, however, that in dualistic economies many activities, and especially insurance activities, are non-monetary exchange-based transactions. For example, families, tribes, and other small groups have reciprocal obligations (Landa, 1994). These may be paid for through day-to-day help with farming, home construction etc. (payment for the insurance policy) but can be called upon in times of urgent need (payment from the insurance policy). Depending on circumstances these arrangements are capped or limited in potential disaster (Newberry, 1993; Cole, 1994).

The separate expenditures on insurance for each transaction may be recorded in the "insurance matrix" using the same structure as the social accounts, as shown in table 6.2. This table also shows the amount of "buffer" for each activity measured as the number of months for which the insurance strategy provides protection under normal conditions. Following the literature cited earlier, it is evident that these periods also reflect the dualistic character of the community, with traditional activities and poor households being less well protected than the modern sector or richer households. The total cost of "insuring" each activity is shown in the bottom row of this table.

Table 6.2 Insurance Matrix and Buffers

	Traditional	Modern	Poor	Rich	Govt.	Capital	Protection	Trade	
BUFFER	0.4	1.5	0.4	2.0	1.0	1.0	1.0	1.0	months

INSURANCE MATRIX (ANNUALIZED)								Base Prices=1	
ACTORS	Traditional	Modern	Poor	Rich	Govt.	Capital	Protection	Trade	Total
Traditional	0.3	0.6	0.3	0.8	0.4	0.0		1.3	3.8
Modern	0.2	1.3	0.2	2.5	0.4	0.4		2.9	7.8
Poor	0.3	0.6	0.0	0.0	0.4	0.0		0.0	1.4
Rich	0.2	1.3	0.0	0.0	0.8	1.3		0.0	3.5
Government	0.2	1.3	0.2	1.7	0.0	0.0		0.4	3.7
Capital	0.2	1.9	0.0	0.8	0.8	0.0		0.4	4.1
Protection									
Trade	0.3	3.1	0.0	0.8	0.0	1.7			6.0
Total	1.7	10.0	0.7	6.7	2.9	3.3		5.0	30.3

Table 6.3 Insurance Accounting Matrix (Stylized Social Accounts with Insurance Items)

REVISED SOCIAL ACCOUNTS (ANNUALIZED)								Base Prices=1	
ACTORS	Traditional	Modern	Poor	Rich	Govt.	Capital	Protection	Trade	Total
Traditional	9.7	4.4	9.7	4.2	4.6	0.0	3.8	13.8	50.0
Modern	4.8	8.8	4.8	12.5	4.6	4.6	7.8	32.1	80.0
Poor	9.7	4.4	0.0	0.0	4.6	0.0	1.4	0.0	20.0
Rich	4.8	8.8	0.0	0.0	9.2	13.8	3.5	0.0	40.0
Government	4.8	8.8	4.8	8.3	0.0	0.0	3.7	4.6	35.0
Capital	4.8	13.1	0.0	4.2	9.2	0.0	4.1	4.6	40.0
Protection	1.7	10.0	0.7	6.7	2.9	3.3	0.0	5.0	30.3
Trade	9.7	21.9	0.0	4.2	0.0	18.3	6.0	0.0	60.0
Total	50.0	80.0	20.0	40.0	35.0	40.0	30.3	60.0	

These last amounts are inserted into the "insurance" row, and the protection cost for each transaction subtracted from the corresponding entries in the original social accounts, and balanced via the insurance expenditures columns. This gives the new protected social accounts shown in the last table in table 6.3. The final row in this table shows the percentage loss, arising from a given (random) set of hazards, over the total time period that is to be calculated using the model. Since no excess hazards have been introduced these entries are zero.

The interpretation of these revised social accounts is subtly different from usual social accounting matrices and other input output tables. Conventional tables record the actual level of transactions, and total income, and expenditures of economic actors. The new table measures potential output (*i.e.*, the capacity of the economic system) that is never reached in reality and only through contrived experiments with the model. In principle, the income and expenditure accounts of the new matrix need not be balanced, since the capacities of actual production activities are only approximately matched. This point will be clarified below.

6.5 Overall Approach to Solution

The method of solution developed for the model rests on the general propositions, noted earlier, that: i) disasters of all kinds, whether human instigated, natural, or a combination of both, are a regular occurrence in the process of development; ii) the pre-disaster secular and spatial structure of any economy reflects the uncertainties associated with regular production, transportation, financing, and so on; and iii) during a major disturbance, the proportion of failed transactions is increased beyond the capability of these afforded precautions to cushion the impacts, resulting in a reduction in flows and outputs in the economy. This requires an elaboration of the lagged matrix inversion solutions to the transaction matrix (Cole, 1988).

As with any input-output table, every entry in the social accounts is both an expenditure and an income, and thus the table also describes a process whereby an increase in the income of any actor, leads to an increase in the income of other actors, and so on. The outcome of this process may be calculated using a round-by-round process (summing a convergent series), or via a Leontief inverse method (the solution to a set of simultaneous equations). With either method, given their steady-state equilibrium assumptions, the levels of output and income, and all other transactions, should be sustained at a constant level by the specified exogenous levels of exports and development assistance in absence of excessive fluctuations (*i.e.*, above the insurance level provided).

With solutions where a dynamic growth mechanism is assumed, such as a round-by-round calculation with distributed transaction lags, output should build up asymptotically to its equilibrium level, at a rate that depends on the lags used in the model. With the method of Cole (1988), the distributed lags in the round-by-round process are approximated by introducing a characteristic lag for each activity that is incorporated into the Leontief inverse, which, then, is solved for a specified time horizon (rather than as a Type I or Type II closure). This greatly simplifies the calculation since, in reality, lags arise at all stages in the economic process (ordering, production, shipment, and payments), and are distributed in both a statistical and a stochastic fashion (Cole, 1993).

In order to make the fluctuations in the flows explicit, I preserve the explicit round-by-round character of the calculation, and introduce an alternative approximation. There are a number of ways to solve this model. I shall adopt a fairly straightforward procedure that assumes that demand for the output of each activity may be accumulated in an escrow (or outstanding orders) account, unless delays become excessive, in which case, orders are canceled, production is written off, and income is reduced. This differs from the textbook round-by-round calculation (see, *e.g.*, Miller and Blair, 1985), since not all new and indirect income from a given transaction is recycled in the current time period, but is held over to future periods.

The procedure used to introduce exogenous events, which includes regular trade, hazards, and recovery measures, has been explained elsewhere (for example, Cole *et al.*, 1993). Essentially, this involves using an "event matrix" to overlay the transactions of the social accounts. In the present example, this is simplified to a steady stream of exogenous income from traditional and modern sector exports, and from development aid, as shown in the social accounts. This stream may be modified to reflect deliberate changes (in trade levels), or fluctuations from exogenous hazards. A suitably haphazard pattern for modeling natural hazards may be simulated by concatenating a series of random functions. I do not pretend that this offers a forecast; rather, this is merely a plausible pattern for the magnitude and frequency of events in regions of varying size.

Hazards are represented by additional delays in the transactions, over and above the lags used in the characteristic lags, and are applied to domestic activities (*i.e.*, transactions within the region disrupted by local events) and external activities (*e.g.*, imports and exports inhibited by disruptions in accessing the region). When the delays are much less than the period afforded by the normal transaction, there will be little or no impact (depending on the precise assumptions used). When delays are greater transactions are unacceptably delayed and they become "failed transactions" unless they are protected by additional insurance. This in turn leads to significant reductions in current activity that will undermine the average (*e.g.*, annual) level of activity, if lost production cannot be recovered. The deleterious effects may spill over from poorly protected activities to the entire economy.

6.6 Model Equations and Round-by-Round Computation Procedure

The empirical procedures and equations for the model implementation are now explained by following through a single step in the round-by-round computation.

The base social accounts matrix is defined as $A_{ij}(t)$ with a corresponding set of transaction coefficients $a_{ij}(t)$ (column entries divided by column totals) as explained for table 6.1. This includes all current domestic and import matrix coefficients (*i.e.*, with the insurance row and column blank). The advantage of this definition is that all coefficients may be treated in the same manner when carrying out the calculation and when a given account is to be excluded for sensitivity analysis or other reasons the characteristic transaction lag may be extended indefinitely to cancel any return flows through that account (see Cole, 1988). The coefficients matrix, in general, will be time dependent, because of technical change, price variations, or other variations in the level of insurance. For calculations below varying prices and techniques are not included, but the general formulation is retained.

The insurance buffer for the transaction matrix $A_{ij}(t)$ is defined as $B_{ij}(t)$. This corresponds to table 6.2 and includes all formal and informal, local and non-local, insurance expenditures as explained above. This then leads to the revised transaction matrix as illustrated by table 6.3 with $A'_{ij}(t) = A_{ij}(t) - B_{ij}(t)$, with a corresponding set of transaction coefficients $a'_{ij}(t)$ (again, column entries divided by column totals). The coefficients corresponding to table 6.3 are shown in table 6.4. The coefficients in the row labeled protection are the insurance coefficients, so $b_j(t)$ is the proportion of total expenditures used for insurance by sector *j*.

Table 6.4 Coefficients Matrix (Stylized Social Accounts with Insurance Items)

COEFFICIENTS MATRIX								
	Traditional	Modern	Poor	Rich	Govt.	Capital	Protection	Trade
Traditional	0.19	0.05	0.48	0.10	0.13	0.00	0.13	0.23
Modern	0.10	0.11	0.24	0.31	0.13	0.11	0.26	0.53
Poor	0.19	0.05	0.00	0.00	0.13	0.00	0.05	0.00
Rich	0.10	0.11	0.00	0.00	0.26	0.34	0.12	0.00
Government	0.10	0.11	0.24	0.21	0.00	0.00	0.12	0.08
Capital	0.10	0.16	0.00	0.10	0.26	0.00	0.14	0.08
Protection	0.03	0.13	0.03	0.17	0.08	0.08	0.00	0.08
Trade	0.19	0.27	0.00	0.10	0.00	0.46	0.20	0.00
Total	1.00	1.00	1.00	1.00	1.00	1.00	1.00	1.00

The hazard induced delay $h_j(t)$ is introduced as an exogenous variation. This may be different for each sector (in principle, such variation may be applied to any coefficient.). This means, for example, that a transaction is reduced in magnitude by an amount $h_j(t)$ unless the transaction is insured. This could apply to any or all of the external transactions - exports, household or business repatriation, transfers, development assistance, or other income earned outside the region. This might occur because of a failure in transportation systems due to an earthquake or mudslide for example, currency failures, or destruction of production capacity locally because of a flooding or civil disturbance, or invasion by an occupying army. As explained above the hazard is introduced as a delay in deliveries and so long as this delay is shorter than the buffer provided by insurance, production will not be significantly affected. However, a proportion of delays, generally those arising from a concatenation of events, will be excessive. For illustrative purposes here the same variation $h(t)$ is applied to the exogenous demand for output from all accounts.

For illustrative purposes below, the same pattern of disturbances will be applied to all sectors. The hazard induced delays, $h(t)$, are formulated as the product of several random processes *i.e.*, $h(t) = H \cdot rand1(t) \cdot rand2(t) \cdot rand3(t)$, where $rand1(t)$, etc. are independent random processes and *H* is a parameter that sets the typical size of the hazards relative to the output of the sector or economy. The key point about this formulation is that it leads to an irregular sequence with occasional spikes, quite characteristic of much real-world perturbations. For the experiments below, a time series of variations representing a fixed concatenation of events is generated and then applied to different configurations of a society, such as one with an under-insured versus an over-insured economy.

The main assumption in the above formulation is that output must be delivered in an acceptable time period. With manufacturers, for example, if too much demand is outstanding then, given the constraints on production, customers will simply cancel orders, and in turn, firms will reduce their own expenditures. What is "acceptable" obviously depends on many considerations, but may be thought of as the result of an ongoing negotiation of each actor with every other, whereby each attempts to maximize their overall well being in the light of opportunities and uncertainties (including hazards). Levels of stock carried by firms, the disposition of their suppliers and customers, levels of insurance and investment in sound construction, or time invested in building social and business networks, are aspects of this.

The computation is carried out as a time-step simulation with a time increment (step length), Δ. In a given time t the outstanding demand for sector j is $T_j(t)$. This demand is comprised of two parts, new orders received in the current time period and any outstanding orders from previous time periods. Thus, in any time period a proportion of orders will be "in the works", *i.e.*, ordered but not delivered (as explained further below). The downstream effect of the demand $T_j(t)$ in each period is calculated through the coefficients matrix, $a'_{ij}(t)$ introducing an exponential lag function $m_j(t) = \left(1-\exp\left(-\Delta/l_j\right)\right)$, where l_j are the characteristic lags associated with each sector. The utility of this formulation is that it allows the transactions to be distributed over time and related to physical and behavioral characteristics of each sector $b_j(t)$, such as the level of stocks as described in the lagged input-output formulation of Cole (1988).

Thus, the indirect demand or downstream intermediate demand for sector j is $I_j(t) = \sum_i T_j(t) a'_{ij}(t) m_j(t)$ and the unfulfilled portion of current demand for sector j is $U_j(t) = T_j(t)\left(1-m_j(t)\right)$.

The marginal demand insured in time period Δ will be related to Δ, the level of insurance, $b_j(t)$, and the size of the hazard, $h_j(t)$. For illustrative purposes in the calculations below the level of protection, $P_j(t)$, is taken to be a decaying function of the level of insurance relative to the characteristic lag, *i.e.*, $P_j(t) = \exp\left(-\left(l_j + h_j(t)\right)/\left(l_j + b_j(t)\right)\right)$. In general, the formulation should account for design characteristics of the system (engineering, recovery possibilities, vulnerability of materials and so on).

Orders that cannot be satisfied in the current time period $D_j(t) = P_j(t)U_j(t)$ are recycled. In effect the orders are held over in escrow until the next period, and added to any new order $N_j(t+1)$. Thus, there is a transient demand at the next time step $T_j(t+1) = N_j(t+1) + D_j(t)$ that is the sum of total outstanding orders. This amount is then injected into the model to begin the next time step iteration and so provides the new levels of outstanding orders for each activity. For computation, initial levels of demand are set at the steady-state levels of total income.

6.7 Illustration of Typical Results

6.7.1 Example One: Non-Hazard Growth

The first computation will check the effect of making expenditures alone, in absence of any hazards by comparing the long-run equilibrium of the model with and without insurance payments.

The insurance strategies of each actor are an expense deducted from other expenditures. In absence of any hazards, assuming that premiums are still paid, they should be expected to reduce the overall performance of each actor and the economy as a whole. A comparison of figures 6.1 and 6.2 shows that is the case. Figure 6.1 shows the period-to-period growth of income for each account when there are no hazards and no insurance paid. Note that with the round-by-round solution there are initial adjustments (because of the artificial balancing of the base matrix). Then, the income of each actor grows with each time-step and without disturbances would eventually reach the long-run equilibrium level (full-capacity) shown in the social accounts in table 6.1. For example, after 15 periods the level of non-insurance expenditures of the modern sector has reached about 75 units and is close to the long-run amount. The rate of growth depends on the relative lags in the model. Note here that the levels of output are driven upwards to the long-run equilibrium by the exogenous demand for traditional and modern sector commodities (*i.e.*, exports) and also exogenous payments to households (*e.g.*, development aid and foreign investment). This is different from the investment-driven process described by the dynamic Leontief model.

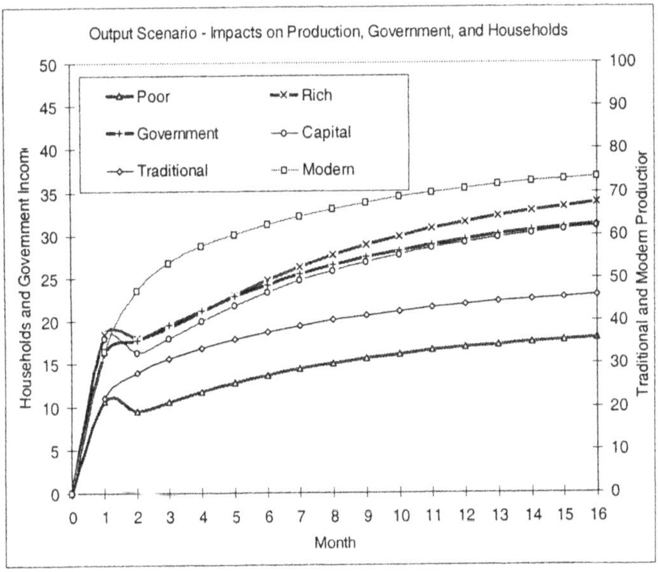

Figure 6.1 Hazard-Free Scenario without Insurance

Figure 6.2 shows the period-to-period growth of income in the model when there are no hazards and insurance is paid. The trajectories for all actors are towards lower long-run levels of "productive" output. For example, long-run output for the modern sector is now about 60 units. Of course, this sub-optimal situation would not continue very long in the real world, since, in absence of constant unpredictable fluctuations, participants would reduce stocks, premiums, etc. and reallocate the resources for current and future consumption.

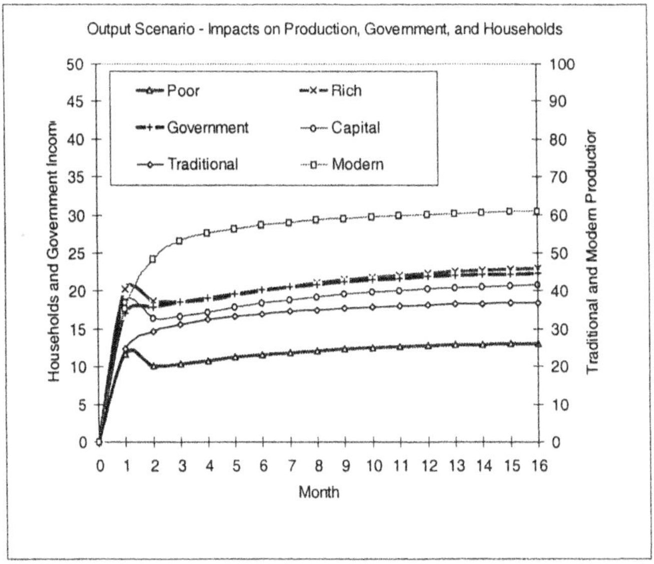

Figure 6.2 Hazard-Free Scenario with Insurance

6.7.2 Example Two: Growth with Concatenated Hazards

The second computation will illustrate the model trajectory when there are concatenated random hazards as described earlier.

The nominal levels of exogenous demand for this scenario are set at the steady-state level given in table 6.1 in order to focus on the consequences of the random hazards. The exogenously generated hazards, in this case representing interruptions in deliveries and levels of exogenous demand for all activities, for a selected model experiment are shown in figure 6.3.

The levels of income for each activity when insurance premiums are not paid are shown in figure 6.4. The effect of the exogenous fluctuations is to reduce levels of output for all sectors. Each individual major disruption causes a marked depression in income. More significantly, because the economy is unprotected, the repeated loss of inputs results in a long-run slow down in growth and the trend is well below that in figure 6.1. This behavior obviously depends on the precise pattern of the exogenous shocks but is quite characteristic of this type of scenario. Indeed, the combination of insurable (but uninsured) shocks and more severe shocks eventually causes output to decline all together. Thus, a series of disasters becomes a potentially unrecoverable catastrophe.

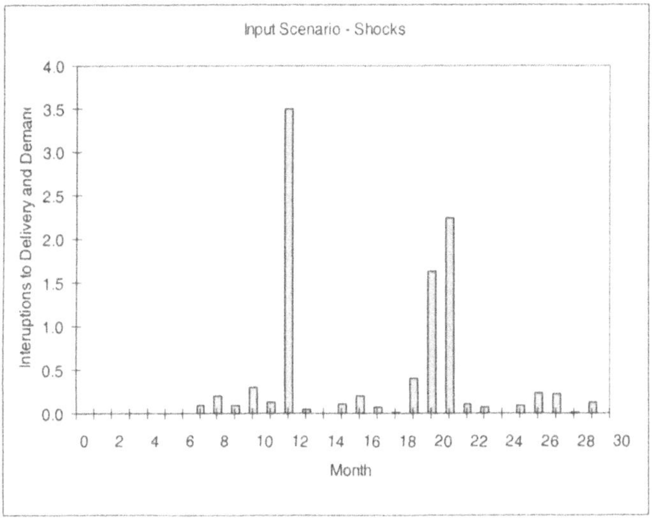

Figure 6.3 Concatenated Hazards - Exogenous Input to Model

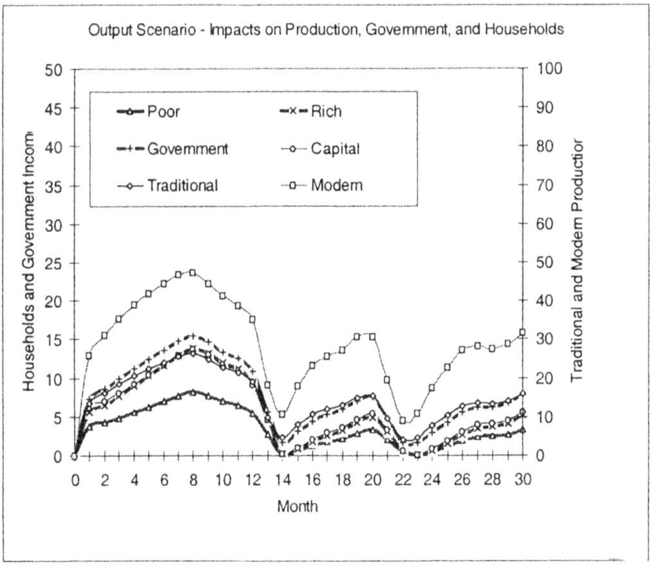

Figure 6.4 Concatenated Hazards with No Insurance Scenario

Figure 6.5 shows the results with identical shocks as the above experiment but with insurance payments reinstated. In this case small disruptions are cushioned by the insurance measures but each individual major disruption causes a marked depression in income. However, in the early stages of growth, the depressing effect of insurance costs (*i.e.*, as between figures 6.1 and 6.2) is offset by the benefits of insurance (*i.e.*, fewer losses arising from the exogenous fluctuations). Although uninsured (or more precisely under-insured) disasters still cause a major recession, recovery is aided by protection of smaller shocks in following time periods. Moreover, with insurance, the long-run trend is for output and incomes to rise, although at a lower level than with figure 6.2. The key issue here is whether delays are within the contingency preparations of the various actors so that production is not discontinued or temporarily postponed. Overall, it is nonetheless clear, that relatively under-insured sectors, notably the traditional and poor households are more severely impacted.

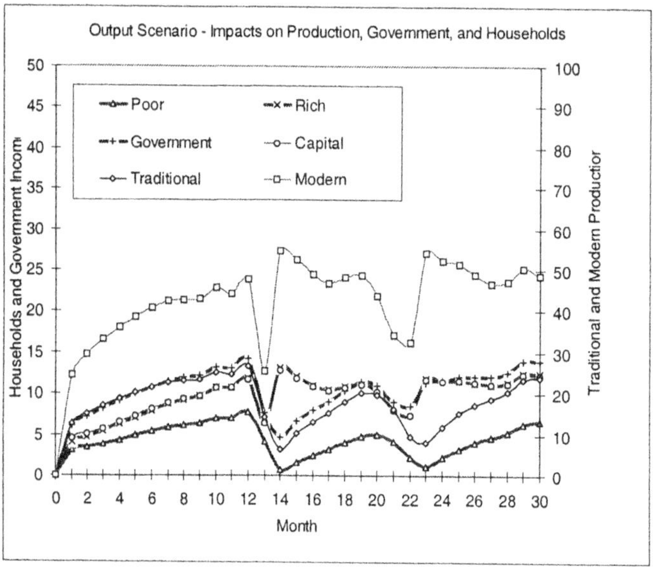

Figure 6.5 Concatenated Hazards with Insurance Scenario

6.8 Conclusions

The above formulation of insurance strategies against hazards within a transactions matrix in this chapter is only one of several within the overall framework explained in the first sections. It thus illustrates an approach and the kind of results that might be expected rather than provides a definitive formulation. This said, while the above results are tentative they are reasonably plausible and suggest that the insurance matrix approach and the method of solution could be useful for more realistic or detailed situations. Thus, the approach might be used to demonstrate how contingencies and protection in one sector or in one segment of society affects the vulnerability of another. Equally, the method might be adapted to examine questions such as what is an optimal level of insurance, or the level of insurance may also be allowed to respond to the severity of recent events.

References

Adelman, I., S. Taylor, and S. Vogel. 1988. "Life in a Mexican Village: A SAM Perspective," *Journal of Development Studies*, 25 (1): 5-24.

Allen, P. 1994. "Coherence, Chaos, and Complexity in a Social Context," *Futures*, 26 (6): 583-597.

Axelrod, R. 1985. *The Evolution of Cooperation*. New York, NY; Basic Books.

Aysan Y., A. Coburn, and R. Spence. 1989. "Turning Short-term Recovery into Long-Term Protection," Mimeo, Reconstruction after Earthquakes, An International Agenda to Achieve Safer Settlements in the 1990's, National Center for Earthquake Engineering Research, Buffalo, NY.

Bender, S. 1990. *Environment and Disasters: Tying Natural Disaster Protection to Development*. Washington, DC; Organization of American States/ World Bank.

Berke, P. and T. Beatley. 1992. *Planning for Earthquakes: Risk, Politics, and Policy*. Johns Hopkins.

Boisvert R. 1992. "Direct and Indirect Economic Losses from Lifeline Damage", in *Indirect Consequences of a Catastrophic Earthquake*. National Earthquakes Hazards Reduction Program (NEHRP), FEMA, July 1992.

Clarke, C. 1984. "Pluralism and Plural Societies: Caribbean Perspectives," in C. Clarke, D. Ley, and C. Peach eds. *Geography and Ethnic Pluralism*. London, UK; Unwin Hyman.

Cohen, A. 1985. *The Symbolic Construction of Community*. London, UK; Routeledge.

Cole, S. 1988. "The Delayed Impacts of Plant Closures in a Reformulated Leontief Model," *Proceedings of the Regional Science Association*, 65: 135-149.

Cole, S. 1994. "A Community Accounting Matrix for Buffalo's Eastside Neighborhood," *Economic Development Quarterly*, 8 (2): 107-126.

Cole, S. 1995. "Lifelines and Livelihood: A Social Accounting Matrix Approach to Calamity Preparedness," *Journal of Contingencies and Crisis Management*, 3 (4).

Cole, S. 2001. "In the Spirit of Miyazawa: Multipliers and the Metropolis," in G.J.D. Hewings, M. Sonis, M. Madden, and Y. Kimura eds. *Understanding and Interpreting Economic Structure*. Berlin, Germany; Springer Verlag, 263-287.

Cole, S. and I. Miles. 1984. *Worlds Apart: Technology, Distribution, and North-South Relations*. Brighton, NJ; Rowman & Littlefield.

Cole, S., E. Pantoja, and V. Razak. 1993. A Social Accounting Matrix Approach to Disaster Preparedness and Recovery Planning, National Center for Earthquake Engineering Research, NCEER 920002, Buffalo, NY.

Cuny, F. 1983. *Disasters and Development*. Oxford University Press.

Dyer, C. 1995. "An Analysis of the Variability of Institutional and Cultural Reaction to the Impact of Hurricane Andrew," mimeo, American Anthropological Association, Washington, November.

Dynes, R. and K. Tierney. 1994. *Disasters, Collective Behavior, and Social Organization*. Newark, DE; University of Delaware Press.

Ebert, C. 1982. "Consequences of Disasters for Developing Nations," *Impact of Science on Society*, 32 (1): 93-100.

Ellson, R., J. Milliman, and R. Roberts. 1984, "Measuring the Regional Effects of Earthquakes and Earthquake Predictions," *Journal of Regional Science*, 24: 561-579.

Funtowitz, S. and J. Ravetz. 1994. "Emergent Complex Systems," *Futures*, 26 (6).

Humphrey, C. and S. Hugh-Jones. 1985. *Barter, Exchange, and Value: An Anthropological Approach*. Cambridge University Press.

Jackson, D. 1995. "Forecasting Very Large Earthquakes," mimeo, Department of Geophysics, University of California at Los Angeles.

Jones, B. 1981. Planning for Reconstruction of Earthquake Stricken Communities, PRC-US Joint Workshop, Beijing, China.

Landa, J. 1994. *Trust Ethnicity and Identity. Beyond the New Institutional Economics of Ethnic Trading Networks, Contract Law, and Gift-Exchange*. Michigan University Press.

Miller, R.E. and P.D. Blair. 1985. *Input-Output Analysis: Foundations and Extensions*. Englwood Cliffs, NJ; Prentice-Hall, Inc.

Newberry, D. 1993. "Implications of Imperfect Risk Markets for Agricultural Taxation," in K. Hoff, A. Braveman, and J. Stiglitz eds. *The Economics of Rural Organization: Theory, Practice, and Policy*. World Bank.

Olson M. 1971. *The Logic of Collective Action*. Cambridge, MA; Harvard University Press.

Pyatt, G. and A. Roe. 1977. *Social Accounting Matrices for Development Planning with Special Reference to Sri Lanka*. Washington, DC; World Bank.

Ryan, S. 1991 *Social and Occupational Stratification in Contemporary Trinidad and Tobago*. St. Augustine; University of the West Indies.

Smith, M. 1990. *Perspectives on the Informal Economy*. New York, NY; University Press of America.

Stone, R. 1973. "A System of Social Matrices," *Review of Income and Wealth*, 19 (2): 143-166.

Udry, C. 1993. "Credit Markets in Northern Nigeria: Credit as Insurance in a Rural Economy," in K. Hoff, A. Braverman, and J. Stiglitz eds. *The Economics of Rural Organization*. New York, NY; Oxford University Press.

7 Computable General Equilibrium Modeling of Electric Utility Lifeline Losses from Earthquakes*

Adam Rose[1] and Gauri-Shankar Guha[2]

[1] Department of Geography, The Pennsylvania State University, USA
e-mail: azr1@psu.edu
[2] Department of Economics and Decision Science, Arkansas State University, USA
e-mail: gguha@astate.edu

7.1 Introduction

Losses from earthquakes are usually associated with building and other property damage. However, many businesses are forced to shut down, even if physically unscathed, when suppliers of lifeline services or other inputs are disrupted, or if their employees are unable to reach the workplace. Likewise, businesses may be forced to curtail operations if orders for their products are canceled by their customers, or if they are unable to deliver their products to market. Moreover, these impacts pertain not only to immediate suppliers and customers, but to successive rounds of upstream or downstream links. The totality of these impacts is some multiple of the businesses directly impacted, hence the typical application of some form of "multiplier" analysis.

Natural hazard impact analysis has been dominated by input-output models (see, *e.g.*, Cochrane *et al.*, 1997; Gordon *et al.*, 1998) or their mathematical programming extensions (see, *e.g.*, Cole, 1995; Rose *et al.*, 1997). Regional econometric models have gained some favor (see, *e.g.*, Ellson *et al.*, 1984; Guimaraes *et al.*, 1993), but several analysts have pointed out the advantage of computable general equilibrium (CGE) models (see, *e.g.*, Boisvert, 1992; Brookshire and McKee, 1992). With very few exceptions (see Rose *et al.*, 1999), however, applications of CGE modeling to hazards thus far have been limited to simple stylized examples. Still, these models have great potential to overcome major deficiencies of their precursors, such as linearity, lack of behavioral content, absence of quantity-price interactions, and neglect of resource constraints.[1]

---

* The research contained in this chapter was funded by a grant from the Multidisciplinary Center for Earthquake Engineering Research. The authors wish to thank Gbadebo Oladosu, Chunsheng Shang, and Shu-Yi Liao for their advice on the computer simulations. An earlier version of this chapter was presented at the American Society of Civil Engineers 5th U.S. Conference on Lifeline Earthquake Engineering, Seattle, WA, August 1999, and at the North American Regional Science Association International Meetings, Montreal, Canada, November, 1999. The authors thank the discussants of these conferences and an anonymous reviewer for helpful comments on this chapter.

[1] There is a need for accurate estimates of benefits associated with natural hazard mitigation/adaptation. Benefits in this context represent the avoided damages and not just directly but economy-wide. Mitigation

The purpose of this chapter is to design and apply a computable general equilibrium model to analyzing the regional economic impacts of an earthquake. Rather than cover all aspects of earthquake damage, however, we have chosen to focus on one key area—electric utility lifelines. This helps us avoid diffusing our efforts in favor of increased depth of analysis. Lifelines by their very nature are at the core of earthquake impacts, directly affecting all sectors and setting in motion a full array of general equilibrium effects. We will show how major characteristics of earthquake losses and adjustments are consistent with basic features of CGE models during most stages of the aftermath. We construct a model of Shelby County (Memphis) Tennessee, the metropolitan area nearest the earthquake zone where the largest earthquakes in the history of the U.S. previously took place. We then perform a number of simulations to further refine the model, specifically to test the sensitivity of the model to parameter changes. We show how various adjustments in these features can mimic the various stages of earthquake recovery and reconstruction in terms of resiliency, of hazard response, and the adjustment to equilibrium of the economy. Finally, we discuss some important policy implications relating to how the CGE model can yield insight into the optimal allocation of electricity services, including the superiority of its pricing results over the more standard computation of non-interruptible service premia.

7.2 Economic Context of Natural Hazards

Most regional economies have become increasingly vulnerable to natural hazards as they grow and become more specialized and more interdependent. Some economies become more self-sufficient through diversification of their industry base and hence reduce trading channels with the outside world. Others enhance their trade with the outside but then become more vulnerable to interruptions to transportation links. Growth creates more productive capacity at risk. Specialization leads to an economic system becoming more reliant on inputs with fewer substitutes, which, if cut off, can bring many sectors to a standstill. Greater interdependence causes these direct losses to ripple through the economy to a greater extent via many rounds of cost increases and lost purchases and sales. Greater self-sufficiency means an economy is more vulnerable to a disaster within its borders because of reduced trade with the outside world.

In modeling the effects of natural hazards, analysts must first set the stage by identifying key characteristics, which provide the basis for specifying assumptions and causal relationships of an analytical model. The characteristics for two different contexts (short run vs. long run) are enumerated in table 7.1. For example, characteristics 1 through 3 have implications for the extent and aggregation of capital asset variables in the model. They also identify the conduit through which impacts manifest themselves, or, more practically, indicate which variables are affected by the

---

costs likewise must incorporate not only direct expenditures on individual strategies such as structural reinforcement and land-use planning, but also the economy-wide burdens being imposed. The model to be presented below is capable of broader estimation of both sides of the ledger.

event. The capital asset variables raise an important distinction. The terms *short run* and *long run* pertain to the standard economic distinction between the period when some inputs (usually capital) are fixed and the period when all inputs are variable. In relation to earthquakes, the former refers to what is usually termed the period of "recovery," while the latter pertains to the period of reconstruction.

Other characteristics can be grouped into those related to the spatial dimension, the potential for recovery, the role of markets, the role of government, behavioral responses to uncertainty, and nature of equilibrium. In the following section we note how CGE modeling can incorporate them.

Table 7.1 Key Considerations in Modeling the Economic Impacts of Earthquakes

Short Run
1. capital stock is reduced immediately 2. loss is concentrated in man-made capital 3. damage manifests itself thru capacity reduction 4. disequilibrium is pervasive 5. losses are usually regionally isolated 6. production is curtailed 7. some prices may rise 8. new input combinations are used 9. imports are a major stop gap 10. use of savings provides an economic boost 11. government recovery aid provides an economic boost 12. insurance payments provide an economic boost 13. recovery may require some central planning 14. modest uncertainty exists 15. decision-making is myopic because of immediate needs
Long Run
1. rebuilding capital stock takes time 2. rebuilding often includes mitigation measures 3. damage stops when capacity is rebuilt or institutions are rearranged 4. equilibrium is re-established but may differ from the original 5. other regions lose if aid is not repaid 6. some or all of production can be recaptured 7. prices are likely to return to previous levels 8. some input combination changes may persist 9. imports return to pre-disaster norm or revised pattern 10. use of savings in S-R is a L-R drain 11. government reconstruction aid provides an economic boost 12. insurance options will decline if losses increase 13. reconstruction is helped by planning (including incentive based) 14. uncertainty declines 15. decision-making is somewhat myopic because of infrequency of events

7.3 Computable General Equilibrium Analysis

Computable General Equilibrium (CGE) analysis represents the state-of-the-art in regional economic modeling, especially for policy analysis. It represents multi-market simulation models based on the simultaneous optimizing behavior of individual consumers and firms, subject to economic account balances and resource constraints (see, *e.g.*, Shoven and Whalley, 1992). The CGE formulation incorporates many of the best features of other popular model forms, but without many of their limitations.

The basic CGE model represents an excellent framework for analyzing natural hazard impacts and policy responses. CGE models can be finely disaggregated to better distinguish the various degrees of vulnerability to hazards across sectors. The production functions are inclusive of all inputs, not just primary factors as in the case of many other economic models, which facilitates identification of materials shortages. At the same time, CGE models allow for the possibility of input substitution, which mimics real world responses beyond the very short run in minimizing hazard impacts. They also allow for the substitution of imported goods for regionally produced goods. CGE models are non-linear in form, thereby more closely reflecting real world conditions, such as economies of scale and non-linear damage functions. CGE models are more capable of analyzing disjoint change than are model forms based on time series data and which therefore simply extrapolate the past. They can also more readily accommodate engineering data or data based on informed judgment.

Another set of CGE model advantages pertains to the important role of prices and markets. Related to this is explicit consideration of behavioral response, including not just simple optimization but also instances of bounded rationality. This applies not only to consumption but also to mitigation and savings behavior. Most other macroeconomic models are based on a sectoral unit of analysis and therefore are not able to analyze individual businesses. Basic CGE models use a production function of a representative firm in each sector, which can be further disaggregated according to firm size, profit rate, or other considerations. Studies of disaster impacts typically have found that many of the firms that go out of business would have done so in the near future even if the disaster had not occurred (Tierney, 1997). However, a disaster imposes a cost through premature business closings, and a CGE model disaggregated by business profitability can incorporate this concept of the marginal business.

CGE models also lend themselves readily to socioeconomic analysis. The household component is typically disaggregated by income bracket and this feature together with its inherent sectoral and possible labor skill level disaggregation facilitates the analysis of impacts on various stakeholders.

CGE models have some distinct advantages in modeling recovery from natural hazards and thus the full extent and duration of losses. Dynamic versions can include interest rate and timing variables. Even static versions can be used to analyze the shift of retained earnings for recovery and reconstruction as opposed to R&D and expansion. One failing of the standard model, however, is that improvements in safety are not necessarily viewed as a productivity improvement or even a gain in economic well-being. However, this is a problem associated with all modeling approaches and can be rectified by taking a broader view of societal objectives.

Finally, CGE models are superior to some other alternatives in modeling the role of lifelines and infrastructure, which include societal necessities such as electricity, water delivery, transportation, and communication. Many of these services are especially vulnerable because they are provided in linear links that disrupt all downstream extensions.[2] A CGE model can place a valuation on these services, even for public sector outputs that are typically unpriced. This is more than an academic exercise, since "shadow values" might serve as temporary prices to ration these services through the market rather than by administrative decree.

In sum, the standard I-O model is only able to analyze demand-side multipliers (*i.e.*, indirect and induced impacts on suppliers). A CGE model is able to analyze not only these, but also supply-side multipliers (impacts on customers), price effects, and limits on resources, and all within a non-linear framework. This does not mean that general equilibrium effects are larger than multiplier effects, in fact, quite the contrary since many of the considerations beyond demand-side multipliers may move in the opposite direction and hence offset the simple I-O impacts.

Basic CGE models have some limitations that would ideally be overcome in an application to natural hazards. First, CGE models require that the economy be in equilibrium. This poses less of a problem as the time horizon is extended, but is likely to be a serious shortcoming in the very short-run. Disequilibrium itself poses costs, *e.g.*, slow adjustment for cases where production is halted because of lack of critical inputs, or where input combinations used are not least cost. A constrained version of a CGE model might be best for short-run impact analyses.

CGE modeling lags behind more conventional macroeconometric models in handling financial variables. This is more of a problem, however, with national models, which need to include money supply variables and interest rates. Regional models do not require the former and can take the latter as given or as a local adjustment of the national rate. Also, it is possible to include an insurance industry, though it is crucial to specify the industry's asset base and location.

The final well-known shortcoming of CGE models is their lack of forecasting ability. This is important in assessing long-term losses, because it is often useful to compare the status of the affected economy several years hence from a projection of what the economy would have been like in the absence of the natural hazard event. This is surmountable with the help of exogenous forecasts and a recalibration of the CGE model for the projected future year.

Otherwise, most considerations associated with natural hazards can be modeled with some ingenuity relating to adjustment in input and import substitution elasticities, capital stock adjustments to reflect damage and reconstruction, and explicit constraints on resource availabilities.

[2] There are redundancies in many lifeline systems because they more resemble a web than a single thread, but this feature typically becomes less prevalent at the large (often wholesale, or transmission) scale level as opposed to small (retail, or local distribution) scale level. For example, contrast large transmission towers/lines for electricity with the network of neighborhood power poles/lines.

7.4 Electricity Disruption Response Stages

In the aftermath of an earthquake, a number of responses take place to minimize the ensuing economic losses. They range from advanced planning to ingenuity to reconfigurations of production processes. The responses are functions of time and cost, which are themselves interrelated. The longer the response period, the more options to choose from. Also, the longer the time lapse, the lower the cost of adjustment. The timeframe for each of the stages varies on a case-by-case basis, but some generalizations can be made.

Below, we distinguish several stages of earthquake responses with the intent of also identifying the appropriate economic modeling approach. The major characteristics of each stage are summarized in table 7.2. Note that the stages pertain to hazard recovery in general; recovery of lifelines, however, is not likely to last past the short-run stage. "Applicability" refers to the type of entity affected, "Best Modeling Approach" refers to our estimate of the most appropriate model to use at each stage, "Adjustment Mechanism" refers to the individual firm and economy-wide behavioral response, and "Adjustment Costs" refers to a combination of production losses and increased costs of doing business.

A. Emergency Response

This period refers to the immediate response for critical facilities (*e.g.*, hospitals, some utilities, public safety agencies) and for production processes that incur high costs of even a temporary outage (*e.g.*, fabricated metals, food production). These operations are the most likely to have low cost and dependable back-up measures in place, such as emergency generators, battery-powered equipment, etc. They are also likely to have an adequate supply of inventories of critical materials. Effectively, for short periods, there may be little or no disruption, and the costs are already incurred in the capital and carrying charges of preparedness activities. We have listed a time span for this stage as less than or equal to 7 days. In actuality, emergency equipment could be utilized for much longer, but these facilities are often the first to be repaired or to have power restored to them. It is especially important to evaluate these considerations on an individual sector basis. Hence, partial equilibrium analysis is needed, and, since business disruption is minimal, general equilibrium analysis is superfluous.

B. Immediate Response

This refers to more ordinary, immediate adjustments and includes some advanced planning, ad hoc responses, and even the short-term shutdown of business operations. It covers ordinary businesses and also some instances where emergency measures failed (*e.g.*, where backup generators were damaged by the earthquake as well). Adjustment mechanisms are limited and include rental of battery-operated systems, use of inventories, etc. The lack of flexibility in response means that an input-output model would be sufficient, though there are several subtleties associated with its appropriate use (see Rose *et al.*, 1997).

B'. Off-Hours

Many earthquakes take place in the middle of the night or on weekends (*e.g.*, the Northridge Earthquake took place at 4 a.m. on a national holiday). Since the vast majority of electricity disruptions last fewer than several hours, this may often be the pertinent stage. In this case, businesses are closed anyway and there is no cost or adjustment for the most part. Hence, no formal modeling is needed.

C. Very Short-Run

During this period, projected to last less than 7 days, businesses depend heavily on their ingenuity, inventories, and imports (including outside aid) of needed materials. They are often willing to pay higher prices and to utilize nonspecialized inputs where they are needed, thus they are likely to incur high costs of adjustment. A CGE model is appropriate here but with input and import elasticities set very low (probably less than 0.1).

D. Martial Law

In a case of especially devastating earthquakes and concern for public safety, governments may intercede in a manner that ranges from complete sociopolitical control or just economic control, such as the rationing of goods and services by central decree. This will definitely pertain to critical facilities and major businesses and, depending on the effectiveness of government, may extend to the remainder of the economy. Unlike the other cases discussed here, in which the market mechanism is essentially intact, and hence the behavior of individual producers and consumers still needs to be modeled, this case is better addressed by a linear programming model incorporating a region-wide objective function. We estimate the adjustment costs as moderate, because the deliberate reallocation of resources is likely to be more effective than grossly disrupted markets in minimizing region-wide losses.

E. Short-Run

This is a period where a greater range of resiliency measures can take place, including input substitution and new trading patterns, as well as making up lost production where possible (*i.e.*, manufacturing as opposed to services). A computable general equilibrium model would again be appropriate to this case with low input elasticities (generally less than 1.0), but very high import elasticities where transportation networks are not extensively damaged.

F. Medium-Run

We envision this stage as being less than or equal to a year and allowing normal input substitution while utilizing external resources. A computable general equilibrium model, with normal elasticities and mobile capital (reflecting the possibility of moving equipment across sectors), is appropriate here as well. Special attention, however, should be paid to inventory replenishment. Overall adjustment costs are estimated to be low except for this.

G. Long-Run

This is a case when reconstruction begins and can be adequately modeled by computable general equilibrium analysis with the capital stock restored. Adjustment costs to the disruption would be low except for rebuilding costs. Qualitative capital stock changes, especially those associated with hazard mitigation measures should, however, be incorporated. Ironically, many of these raise investment costs and would translate into capital productivity decreases, but only because the increase in safety is a conventionally nonmeasured output.

What are the implications of these various response stages for modeling economic losses? First, they emphasize that economic losses from natural hazards are doubly a "joint interaction" phenomenon. As on the stock side, the build-up of the human environment interacts with the physical ground shaking to determine the level of property damage, on the flow side of business interruption, human ingenuity makes the best use of the capital stock available. The longer the time span from the initial ground shaking, the greater is the ingenuity in terms of learning, adjustment options, making up lost production, and repair and reconstruction.

Second, this means one model does not fit all. Models, and analyses in general, need to be differentiated according to major characteristics, which are functions of time in a more sophisticated manner than the short-run/long-run bifurcation noted in table 7.1 above. Of course, table 7.2 is only a first approximation in two ways: a) time periods of applicability require confirmation, and b) adjustments do not always change in discreet jumps but often do so continuously. This is ominous for the modeler, as no one would want to respecify a model on a continuous basis for changes in elasticity values, for example. Some balance needs to be struck between accuracy and modeling cost. Those placing a greater emphasis on accuracy would want to make finer demarcations than are presented in table 7.2, or at least might interpolate elasticity values. In the analyses below, we simply utilize the various stages as midpoints of a continuous cost function. Unless this response function is very highly nonlinear, loss estimation accuracy is not sacrificed significantly.

Table 7.2 Post Earthquake Disruption Response Stages

Stage	Applicability	Best Modeling Approach	Adjustment Mechanisms	Adjustment Costs
A. Emergency ($\leq$ 7 days)	critical facilities	partial equilibrium	backup-generators, inventories	low in terms of previously incurred purchase costs
B. Immediate Response ($\leq$ 48 hours)	ordinary business	input-output	very limited	very high in terms of lost production
B'. Off-Hours ($\leq$ 16 hours or weekend)	non-critical business closed anyway	none needed	n.a.	none
C. Very Short-Run ($\leq$ 7 days)	all facilities	computable general equilib. (very low elasticities)	inventories, imports, outside aid	high in terms of cost increases & lost production
D. Martial Law ($\leq$ 30 days)	critical facilities & major businesses	linear programming (w/ some coefficient adj.)	centralized reallocation	moderate in terms of lost production
E. Short-Run ($\leq$ 6 months)	all facilities	computable general equilib. (low elasticities)	input & import substitution, new trading patterns, outside aid & insurance, make up lost production	moderate (except repair) in terms of cost increases & lost production
F. Medium-Run ($\leq$ 1 year)	all facilities except critical	computable general equilib. (normal elasticities, capital mobile)	input substitution, outside aid & insurance, inventory replenishment	low (except inventory replenishment) in terms of lost production primarily
G. Long-Run ($\geq$ 1 year)	all facilities	computable general equilib. (capital stock restored)	reconstruction, new equilibrium	none except rebuilding cost (including mitigation)

7.5 The Shelby County CGE Model

A prototype CGE model has been constructed for Shelby County, Tennessee. It is structured so as to be comparable to input-output and linear programming models previously used to estimate the indirect impacts of a New Madrid earthquake (see Rose *et al.*, 1997).[3] The CGE model consists of 22 production sectors, with an emphasis on those that are major users of electricity lifelines and those that are most crucial to the functioning of the regional economy (see table 7.A.1 in Appendix). It includes two primary factors of production (capital and labor). Production technology is specified as a 3-level constant elasticity of substitution function, with input flexibility across the aggregates of capital, labor, energy, and materials, as well as within the energy aggregate (*i.e.*, interfuel substitution), though the components of the materials aggregate (other intermediate goods) are modeled with fixed coefficients in the basic version. The prices of the aggregates are weighted averages of their individual components in the base year. Important parameter values of the model are presented in Appendix, table 7.A.2.

The demand for domestic goods relative to imports is also modeled by a CES function, thereby signifying that imports are imperfect substitutes for regionally produced goods. Overall production is modeled as a constant elasticity of transformation function of production for regional supply and for export. The region is treated as a small open economy, so that prices in the rest of the U.S. and rest of the world are purely exogenous.

Household behavior is modeled according to a Cobb-Douglas utility function for three household income bracket groupings. Factor incomes are distributed to households, government, and the rest of the world. Consumer expenditures are then divided among purchases of goods and services, payment of taxes, and savings. Governments earn additional income from taxes and their expenditures are split between purchases of goods and services and transfers to household and businesses. Investment expenditures are based on a regionally scaled national capital composition matrix.

The standard labor market accounting balance condition for CGE models is the Neoclassical closure rule, which consists of a fixed labor supply and a variable wage rate, and always results in full employment. This represents an unrealistic level of flexibility, and hence we drop it in favor of a Keynesian closure rule, which fixes the nominal wage rate and allows labor supply to vary in a way that represents unemployment and migration. In standard CGE modeling, in the short run, it is assumed that capital is fixed in each sector and that labor is mobile across sectors. In the long run, it is assumed that capital stock levels can change and capital is mobile across sectors. We modify these standard assumptions as required by the characteristics of the post-earthquake response stages depicted in table 7.2. Idealized versions of these modifications are presented in Appendix, table 7.A.3; the actual

[3] CGE should be viewed as a natural extension of I-O analysis rather than as an entirely different approach on both conceptual and empirical levels (Rose, 1995). For example, the input-output table of annual interindustry transactions serves as the major database for the CGE model.

parameter values used (as noted below) are somewhat different because of computational limitations of our software.[4]

7.6 Simulating Lifeline Disruption Impacts

7.6.1 Basic Considerations

We confine our simulations to the general equilibrium impacts of an electricity lifeline disruption in Shelby County, Tennessee, that would stem from a magnitude 7.5 Richter Scale earthquake. The analysis builds on previous efforts of researchers associated with the Multidisciplinary Center for Earthquake Engineering Research in modeling the fragility of the Memphis Light, Gas and Water (MLGW) system and translating substation outages in 36 electric power service areas (EPSAs) into reduced electricity availability on a sector by sector basis (see Shinozuka *et al.*, 1998; Rose *et al.*, 1997).

The CGE model simulates the partial and general equilibrium effects of the electricity curtailment. The former refers to adjustments such as the substitution of other primary fuels for electricity, the substitution of other inputs for energy in general, and energy conservation. These adjustments then cause changes in supply availabilities, product prices, and factor prices that set off general equilibrium impacts stemming from further input substitutions, including purchasing more goods from outside the region. In the more comprehensive version of the model, further losses would stem from production cutbacks associated with damage to capital stock outside the Electricity sector or unavailability of workers (through injuries or transportation problems), leading to lack of other inputs produced within the region. Production losses may also arise from shutdowns or slowdowns due to lack of customers (*e.g.*, transportation problems for consumers or the cancellation of orders by other businesses). These various effects typically manifest themselves in one of three forms of changes: parameters (*e.g.*, input elasticities), constraints (*e.g.*, labor mobility), or variables (*e.g.*, capital stock, or capacity, reductions).[5]

Note that we begin our simulations with the case of business disruption associated with a maximum predicted electricity outage of 44.8%. Restoration of service takes place in a progressive manner in the aftermath of the ground shaking (see Shinozuka *et al.*, 1998), and losses decline accordingly. Of course, the major point of simulating the various post-earthquake stages is to analyze how these losses are further decreased by resiliency measures, by implicit workings of the market, and by explicit policies. The original estimate of downtime for at least some aspects of MGLW for the simulated earthquake was up to 15 weeks (see Shinozuka *et al.*, 1997). The important thing to keep in mind is that, while ground shaking itself may last only a few seconds, the

---

[4] We are not able to obtain equilibrium solutions for our modified CGE model for very low substitution elasticities (<.01). Note also that the idealized elasticity values, which are assumed to be below 0.1 for the very short-run, are based on informed judgment. However, they would better be based on empirical estimation.

[5] The co-existence of damage among several lifelines or among lifelines and other structures is less likely to lead to the dangers of double-counting (see Chang *et al.*, 1996) in CGE models than in other approaches because of explicit sectoral capital stock variables and the disaggregation of input requirements.

economic disruption lasts over the entire period during which productive capacity is damaged, the regional economy is otherwise hindered in its operation, or until it achieves a new equilibrium (including changes in business location and trading patterns). Thus, the following stages are simulated below for our CGE model: Base Case (for purpose of comparison with the I-O results in Rose *et al.*, 1997), Very Short Run and Short Run (for purpose of comparison with the LP results of Rose *et al.*, 1997).

Note also an important distinction in CGE modeling of business interruption from lifeline damage vs. ordinary building damage. The latter can be readily modeled through sectoral capital stock reductions (with consideration of excess capacity). However, modeling the former depends on the type of damage. If large transmission lines coming into the region are damaged, or if a central power station serving all entities in the region is damaged, the same type of capital stock reduction (with an adjustment for load factor in the case of the latter) is appropriate. If damage is to substations or local distribution lines, electricity availability will differ by sub-region and hence by sector (no major region has uniformly distributed business activity). This necessitates explicit, differential constraints on electricity availability across all sectors. While this feature can be readily incorporated into I-O and LP models (see Rose *et al.*, 1997), it poses a significant problem in a CGE context. These models find it much more difficult to arrive at equilibrium solutions in the presence of intermediate input constraints. We have had difficulty obtaining solutions of this type using the General Algebraic Modeling System (GAMS) software, so we confine our presentation below to cases of uniform region-wide electricity disruptions.[6]

7.6.2 Basic Results

In the prototype Shelby County CGE model, we alter the input and import substitution elasticities to model the economy at different stages of recovery following an earthquake. The benchmark Shelby regional economy is replicated from the Social Accounting Matrix (SAM) by calibrating the model with a set of substitution elasticities for factor inputs, imports (CES), and exports (CET) shown in Appendix, table 7.A.2. This process initializes the endogenous parameters of the model, and allows us to exogenously introduce a modified set of elasticities to evaluate the production possibilities of the regional economy in different scenarios.

The reaction of the regional economy has been classified into eight stages, as depicted in table 7.2, with the assumption that resiliency is built into the recovery process. Of these, three stages have been simulated below (Base, Very Short Run, Short Run), each beginning with an equal reduction in the capital stock of the electricity sector but different modifications of the substitution elasticities. We run each case at the maximum outage level for purposes of comparison of impacts under

---

[6] In a related study for Los Angeles, we have been able to model sectorally differentiated electricity availability by switching to the GAMS/MPSGE system. Applied GE models may be generalized as square system of nonlinear equations, which are ordinarily solved as constrained NLPs, with rationality bounds on variables. Alternatively, the CGE model may be solved as a mathematical complimentarity problem (which uses a combinatorial logic to select a solution set that satisfies the constraints). This is done using the MPSGE formulation, which is solved by the GAMS/PATH solver. In this chapter, we have used the GAMS/MINOS NLP solver, which is likely to be of more interest to the majority of CGE modelers who use GAMS/MINOS.

different conditions of the flexibility of the response. In actuality, the time path of outages would decrease as the power system is restored so that economic impacts would decrease as well over time. Thus, the results are presented in percentage terms for the maximum disruption period and not as dollar values.[7]

Base Case

The capital stock of the electricity sector is reduced by a proportion corresponding to the peak outage to induce supply reduction. The substitution elasticities are kept the same as the benchmark, since this serves as a control case.

For the Base Case, which reflects the average parameter values in the basic CGE model unmodified for earthquake analysis,[8] overall Gross Output (production or sales) in Shelby County declines by only 1.4% (see table 7.3). The 44.8% disruption translates into a 44.8% reduction in electricity sector capital stock but only a 13% reduction in electricity output because labor and other inputs are substituted for capital. (Note that the Gross Output reduction using an I-O model would be at least 44.8%, see Rose *et al.*, 1997.) The price of electricity is increased by 39.5%, but electricity use is reduced by only 9.6% for intermediate goods production and only 23.7% by final demand (primarily households, or residential customers), both reflecting the price in elasticity of demand for this necessity. These results indicate there is too much flexibility in an ordinary CGE model for direct application to indirect loss estimation. First, in the real world, electric power plants cannot so readily substitute other inputs for capital, especially in the short timeframe of an earthquake-induced disruption. Second, the relatively largest decreases in electricity use are absorbed by residential customers, who are in fact likely to receive priority treatment. Third, it is unlikely that the price of electricity would be allowed to rise by 39% (even if this rate represents its true value in an emergency situation). We make adjustments for the first of these considerations in our very short-run and short-run simulations, and adjustments for the latter two in separate sets of simulations in progress.

Very Short-Run (VSR) (< 7 days)

Capital stock of the Electricity Sector is reduced as before to reflect supply loss. Reducing the substitution elasticities close to 0 for the Electricity Sector would reflect the true reaction of the system. However, solver limitations only permit a more modest reduction to 0.26 in the second tier of the production function, and to 0.13 in all other tiers. Import elasticities are held at 0.1 for the Electricity Sector across all cases.

For all sectors other than Electricity, the interfuel (4th and subsequent tier) elasticities, which range from 0.6 through 0.8 in the base case, are reduced by 80% in the VSR case. Import elasticities range from 0.75 to 2.0 for trade with the rest of the world, and from 0.9 to 3.4 for trade with the rest of the US.

---

[7] We admit to an inconsistency in this approach in that the Very Short-Run and Short-Run stages might more properly be considered sequentially with a lower level of outage during the latter. However, this would make comparisons, like we emphasize in this chapter, more difficult.

[8] The Base Case most closely resembles the Long-Run Case (G) in table 7.2.

Table 7.3 Economic Impact of Electricity Lifeline Disruption in Shelby County: Base Case (Keynesian) Region-Wide Electricity Availability Constraints (percentage change from the benchmark economy)

Sector	Output	Capital	Labor	Output Price	Capital Price	Electricity Use	Exports	Imports	Household Demand
Agriculture	-1.4	0.0	-1.3	0.0	0.0	-16.8	-1.4	-1.0	-1.2
Mining	-1.6	0.0	-1.6	0.0	0.0	-4.2	-1.5	-2.9	-2.3
Construction	-5.0	0.0	-5.8	-1.0	-9.5	-3.8	-3.9	-6.2	-100.0
Food Processing	-0.8	0.0	-0.9	0.0	0.0	-7.6	-0.8	-0.9	-1.2
Manufacturing	-1.5	0.0	-1.5	0.0	0.0	-7.3	-1.3	-2.3	-1.2
Petroleum Products	-2.4	0.0	-2.6	0.0	-6.7	-2.5	-1.9	-3.2	-1.1
Transportation	-0.6	0.0	-0.7	0.0	0.0	1.8	-0.3	-2.4	-0.8
Communication	-0.6	0.0	-0.9	0.0	0.0	-11.9	0.3	-1.7	-0.9
Electricity Services	-14.7	-44.8	4.8	39.6	216.7	-100.0	-40.0	-11.8	-23.6
Gas Distribution	-2.5	0.0	-3.0	-1.0	0.0	-33.3	0.0	-4.8	-1.1
Water & Other Utilities	-0.5	0.0	-0.8	0.0	0.0	0.0	-0.1	-100.0	-0.7
Wholesale Trade	-2.1	0.0	-2.4	0.0	-6.3	-10.1	-1.0	-4.6	-0.3
Retail Trade	-1.5	0.0	-1.5	0.0	0.0	-12.9	-1.3	-1.6	-1.2
Fin., Ins., & Real Estate	-0.3	0.0	-0.5	-1.0	0.0	-11.6	0.4	-1.8	-0.9
Personal Services	-1.5	0.0	-1.4	0.0	0.0	-13.0	-1.7	-1.2	-1.4
Bus. & Prof. Services	-0.9	0.0	-1.1	0.0	0.0	-5.4	-0.2	-1.7	-0.9
Entertainment	-1.2	0.0	-1.1	0.0	0.0	-11.8	-1.2	-1.2	-1.2
Health	-0.9	0.0	-1.1	0.0	-7.1	-10.9	-0.6	-1.5	-1.0
Education	-0.4	0.0	-0.4	0.0	-6.7	-10.1	-0.4	-0.5	-1.2
Government	-0.4	0.0	-0.2	0.0	0.0	-16.9	-0.7	0.0	-1.6
Total	-1.4	-0.7	-1.4	0.1		-9.6	-0.8	-2.1	-1.4
Total Except Electricity	-1.3	0.0	-1.4	-0.2		-9.6	-0.8	-1.9	-1.0

Table 7.4 Economic Impact of Electricity Lifeline Disruption in Shelby County: Very Short Run (Keynesian) Region-Wide Electricity Availability Constraints (percentage change from the benchmark economy)

Sector	Output	Capital	Labor	Output Price	Capital Price	Electricity Use	Exports	Imports	Household Demand
Agriculture	-3.9	0.0	-3.7	0.0	0.0	-41.3	-4.0	-3.0	-3.3
Mining	-5.8	0.0	-6.2	-1.0	-6.7	-12.5	-4.9	-15.8	-4.5
Construction	-14.1	0.0	-16.1	-2.9	-23.8	-19.0	-11.2	-17.1	-100.0
Food Processing	-2.7	0.0	-2.6	0.0	0.0	-26.2	-2.7	-2.4	-3.3
Manufacturing	-4.6	0.0	-4.7	0.0	-6.7	-20.9	-4.4	-6.6	-3.2
Petroleum Products	-7.4	0.0	-8.0	-1.0	-13.3	-9.1	-6.2	-9.5	-2.9
Transportation	-1.9	0.0	-2.2	0.0	0.0	-5.1	-1.1	-7.0	-2.2
Communication	-1.6	0.0	-2.5	-1.0	0.0	-31.1	0.9	-4.6	-2.5
Electricity Services	-34.3	-44.8	-15.4	136.5	658.3	-100.0	-72.0	-28.3	-51.5
Gas Distribution	-9.0	0.0	-10.6	-3.0	-14.3	-33.3	-12.5	-16.4	-2.9
Water & Other Utilities	-2.6	0.0	-3.7	-2.0	-7.1	0.0	-0.8	-100.0	-1.1
Wholesale Trade	-5.9	0.0	-6.8	-1.0	-12.5	-26.4	-3.2	-12.5	-0.8
Retail Trade	-4.1	0.0	-4.2	0.0	-7.1	-32.6	-4.0	-4.3	-3.3
Fin., Ins., & Real Estate	-0.9	0.0	-1.5	-1.0	-6.7	-31.2	1.2	-4.9	-2.3
Personal Services	-4.3	0.0	-4.1	1.0	-7.1	-31.5	-5.0	-3.2	-3.8
Bus. & Prof. Services	-2.6	0.0	-3.2	-1.0	-6.7	-20.5	-0.7	-4.9	-2.5
Entertainment	-3.3	0.0	-2.9	0.0	0.0	-32.4	-3.4	-3.2	-3.4
Health	-2.6	0.0	-2.9	0.0	-7.1	-28.7	-1.8	-4.0	-2.8
Education	-1.2	0.0	-1.2	0.0	-6.7	-24.9	-1.1	-1.3	-3.3
Government	-1.3	0.0	-0.7	1.0	0.0	-41.1	-2.1	-0.1	-4.2
Total	-4.1	-0.7	-4.1	0.5		-26.4	-2.5	-6.1	-3.5
Total Except Electricity	-3.8	0.0	-4.1	-0.5		-26.4	-2.5	-5.8	-2.8

Table 7.5 Economic Impact of Electricity Lifeline Disruption in Shelby County: Short Run (Keynesian) Region-Wide Electricity Availability Constraints (percentage change from the benchmark economy)

Sector	Output	Capital	Labor	Output Price	Capital Price	Electricity Use	Exports	Imports	Household Demand
Agriculture	-2.2	0.0	-2.1	0.0	0.0	-26.5	-2.2	-1.6	-1.9
Mining	-2.8	0.0	-3.0	0.0	-6.7	-6.2	-2.4	-6.9	-2.3
Construction	-7.9	0.0	-9.2	-1.9	-14.3	-10.4	-6.2	-9.7	-100.0
Food Processing	-1.4	0.0	-1.4	0.0	0.0	-13.6	-1.4	-1.3	-1.9
Manufacturing	-2.4	0.0	-2.4	0.0	0.0	-11.8	-2.2	-3.6	-1.8
Petroleum Products	-4.0	0.0	-4.3	0.0	-6.7	-4.1	-3.2	-5.2	-1.6
Transportation	-0.9	0.0	-1.1	0.0	0.0	-0.2	-0.5	-3.8	-1.3
Communication	-0.9	0.0	-1.4	-1.0	0.0	-19.2	0.5	-2.6	-1.4
Electricity Services	-21.7	-44.8	-0.7	66.7	350.0	-100.0	-56.0	-17.6	-34.1
Gas Distribution	-4.5	0.0	-5.4	-2.0	-7.1	-33.3	-12.5	-8.5	-1.7
Water & Other Utilities	-1.1	0.0	-1.7	-1.0	0.0	0.0	-0.3	-100.0	-0.8
Wholesale Trade	-3.3	0.0	-3.8	-1.0	-6.3	-15.0	-1.7	-7.2	-0.5
Retail Trade	-2.3	0.0	-2.3	0.0	0.0	-19.7	-2.1	-2.5	-1.8
Fin., Ins., & Real Estate	-0.5	0.0	-0.8	-1.0	-6.7	-19.0	0.7	-2.8	-1.3
Personal Services	-2.3	0.0	-2.2	0.0	0.0	-18.7	-2.7	-1.8	-2.1
Bus. & Prof. Services	-1.4	0.0	-1.7	-1.0	0.0	-11.1	-0.3	-2.7	-1.4
Entertainment	-1.8	0.0	-1.6	0.0	0.0	-19.3	-1.8	-1.8	-1.9
Health	-1.5	0.0	-1.6	0.0	-7.1	-17.1	-1.0	-2.3	-1.6
Education	-0.7	0.0	-0.7	0.0	-6.7	-14.8	-0.6	-0.7	-1.8
Government	-0.7	0.0	-0.4	1.0	0.0	-25.6	-1.2	0.0	-2.4
Total	-2.3	-0.7	-2.2	0.1		-15.3	-1.3	-3.3	-2.0
Total Except Electricity	-2.0	0.0	-2.2	-0.4		-15.3	-1.3	-3.2	-1.6

For our Very Short-Run Case, the economic impact on Regional Gross Output almost triples to 4.1% (see table 7.4). Here, with less flexibility in input substitution in the electricity sector, the 44.8% reduction in capital turns into a 34% decrease in electricity output. This reduction is also stimulated by the price increase of 136.5%. Intermediate sector energy use decreases by 26.4%, but again household demand absorbs more of the impact, in this case 51.5%.

Short-Run (SR) (< 6 months)

At this stage, greater factor substitution possibilities are envisaged. For the Electricity Sector, we increase substitution elasticities to the average of the base and VSR cases. For all other sectors, we double the elasticities of interfuel substitution from the VSR case.

The results for the Short-Run Case are intermediate to those of the two previous cases, though closer to the Base Case (see table 7.5). This is consistent with intuition and helps validate the model. Still, the results for all three cases indicate rather low levels of economic losses when compared to the I-O results of 44.8% or more, or to the LP results of nearly half that much (see Rose *et al.* 1997). Simulations of additional restrictions are likely to lead to higher indirect losses, and will therefore likely be more accurate.

7.7 Policy Implications

A major advantage of CGE modeling over more standard approaches to economic loss estimation for natural hazards is its emphasis on prices and markets. The function of the market institution is to effectively allocate scarce resources, and there is no more dire challenge than in the aftermath of a disaster. The question is whether the market can be counted on to function properly in emergency situations?

First, we should step back and note the important role of resource allocation as a recovery measure. Economic losses are often couched in terms of property damage, and recovery and reconstruction couched in terms of emergency service and construction efforts. However, if we focus on "production flows" that are stifled by the property damage, we can gain a valuable perspective—the extent of losses is a variable depending on the speed and pattern of recovery. The same damage level to an electricity lifeline can yield vastly different outcomes depending on service restoration. Obviously, output and employment losses can be affected by shifting lifeline services between customers (with due regard to high priorities for providers of health care and other basic necessities). This reallocation essentially represents a *costless* post-earthquake loss mitigation measure.[9]

[9] For the sake of brevity, we overlook two important issues. First is the political sensitivity of the reallocation decision and the associated political cost. Second, the potential for reallocation of lifeline services is limited by engineering constraints and is more appropriate in situations where brownouts are the problem or when distribution/transmission lines are down, than for the case of a central power station, which affects all customers equally and allows for reallocation only when certain sectors can be connected to non-damaged substations.

This measure can be implemented by central decree, such as a public official shutting down non-essential businesses in the face of a supply shortage. On the other hand, it can be accomplished through various price mechanisms, such as interruptible service contracts or price rationing. Elsewhere, Rose and Benavides (1999) have simulated these two possibilities with use of a linear programming model and have noted one of the shortcomings of interruptible service contracts—the fact that in choosing whether to pay the premium, an individual firm will only consider its own potential damage and neglect the damage to its suppliers or customers if its operations are curtailed. The simulation here presents a first, though not comprehensive assessment of the situation, since we have not explicitly modeled such contracts before the fact, but instead modeled Electricity sector price changes after the fact.

Essentially, the price changes of our simulations reflect the increased scarcity value of electricity (between 36% and 145%), and customers respond accordingly. The assumption within a CGE model is that decision-makers make optimal responses to price changes, including substituting away from electricity and using less energy overall. Moreover, the decision is made in a general equilibrium context so that implicitly the value of electricity to indirect customers and suppliers is taken into account by the workings of the market, and is reflected in price changes. Hence, our CGE analysis may be superior to alternative ways of establishing non-interruptible service premiums. Note these would be the probability of occurrence multiplied by the percentage price increases. Note also that the socially optimal premium is the same constant value for all customers regardless of individual direct business losses.

To summarize, our model is able to simulate a major lifeline recovery measure associated with market pricing. Moreover, we were able to quantify the emergency value of electricity by estimating the change in its price under alternative assumptions regarding the flexibility of the economy. This contrasts with the conventional situation where prices are fixed by contract or fixed by government decree in the aftermath of an earthquake. Although some of these options and proportional cutbacks in electricity service are viewed as equitable, since they are applied equally across the board, they may not be as effective in reducing output and employment losses as the measure proposed here.[10] Our results indicate that the market has the potential to reduce hazard losses through reallocation. In reality, crisis conditions strain the market's abilities (*e.g.*, they inhibit information flows and delay adjustments). Moreover, the present system of interruptible service contracts may lead the market to a private optimum that differs from the social one.

[10] There is no universal consensus on the best definition of equity, and even an across-the-board interpretation of no price changes is tenuous because it pertains only to the situation confronting electricity users and does not take into account the actual impact on their operations, *i.e.*, percentage differences in their potential business losses.

7.8 Conclusions

This chapter has presented a new approach to modeling business disruption losses and recovery efforts associated with lifeline utility damage caused by earthquakes. The approach is especially adept at analyzing the role of markets and prices in the extent of mitigation of economic losses. A progression of simulations was performed indicating that the model yields reasonable results given the input data and assumptions used. However, the main result of the simulations is that the typical CGE model, even based on short-run (vs. long-run) substitution elasticities, is far too flexible and is likely to greatly understate earthquake impacts (in our case by more than a factor of ten). Deliberate efforts must be made to incorporate real world rigidities as well as resiliencies.

Several other areas of further research immediately present themselves. In relation to engineering, the first is to integrate more engineering features of electricity lifelines into the model (*e.g.*, the ability to distinguish between substation, transmission tower, and distribution line damage). The second is to improve the spatial dimensions of the model so that it can better address engineering aspects of damage and of service restoration efforts. The third is to handle multiple lifeline failures without double-counting.

As to economic features, several extensions need to be incorporated to account for complications. For example, when recovery starts, there may be massive goods and services available in the damaged area through donation or some other means with reduced cost. These may distort the market mechanism, and thus the model may over- or under-estimate price changes. This may become more significant when the damages are extensive and devastating. Also, for the long-run impact, the insurance premium might increase; therefore, the production/living cost may increase to some extent.

Finally, we note the applicability of this modeling approach to other lifelines and other hazards. CGE modeling has recently been applied to a water service disruption from an earthquake, with only modest additional refinement (see Rose and Liao, 2002). Application to floods is also being explored by the senior author. Given the times, CGE might also be used to examine the regional economic impacts of terrorist attack and other manmade disasters.

References

Boisvert, R. 1992. "Indirect Losses from a Catastrophic Earthquake and the Local, Regional, and National Interest," in *Indirect Economic Consequences of a Catastrophic Earthquake*. Washington, DC; FEMA.

Brookshire, D. and M. McKee. 1992. "Other Indirect Costs and Losses from Earthquakes: Issues and Estimation," in *Indirect Economic Consequences of a Catastrophic Earthquake*, Washington, DC: FEMA.

Chang, S.E., H. Seligson, and R. Eguchi. 1996. *Estimation of the Economic Impact of Multiple Lifeline Disruption: Memphis Light, Gas, and Water Case Study*. Buffalo, NY; National Center for Earthquake Engineering Research.

Cochrane, H. *et al.* 1997. "Indirect Economic Losses," in *Development of Standardized Earthquake Loss Estimation Methodology Vol. II*, Menlo Park, CA; RMS, Inc.

Cole, S. 1995. "Lifelines and Livelihood: A Social Accounting Matrix Approach to Calamity Preparedness," *Journal of Contingencies and Crisis Management,* 3: 1-11.

Deardoff, A.V. and R.M. Stern. 1986. *Michigan World Trade Model.* Cambridge, MA; MIT Press.

Ellson, R., J. Milliman, and R. Roberts. 1984. "Measuring the Regional Economic Effects of Earthquakes and Earthquake Prediction," *Journal of Regional Science.* 24: 559-79.

Gordon, P., H. Richardson, and B. Davis. 1998. "Transport-related Impacts of the Northridge Earthquake," *Journal of Transportation Statistics,* 1: 22-36.

Guimaraes, P., F. Hefner, and D. Woodward. 1993. "Wealth and Income Effects of Natural Disasters," *Review of Regional Studies,* 23: 97-114.

Hanson, K. and A. Rose. 1997. "Factor Productivity and Income Inequality: A General Equilibrium Analysis," *Applied Economics,* 29: 1061-71.

Hashimoto, K. and J. Heath. 1995. "Estimating Elasticities of Substitution by the CDE Production Function: An Application to Japanese Manufacturing Industries," *Applied Economics,* 27: 167-72.

Li, P.C. 1994. *Environmental Policy, Coal, and the Macroeconomy: The Impact of Air Quality Regulations on the Pennsylvania Economy.* Ph.D. Dissertation, University Park, PA; The Pennsylvania State University.

McKibbin, W. and P. Wilcoxen. 1998. "The Theoretical and Emprical Structure of the G-Cubed Model," Brookings Institution, Washington, DC.

Minnesota IMPLAN Group. 1998. *Impact Analysis for Planning System (IMPLAN).* Stillwater, MN.

Prywes, M. 1986. "A Nested CES Approach to Capital-Energy Substitution," *Energy Economics,* 8: 22-28.

Rose, A. 1995. "Input-Output Economics and Computable General Equilibrium Models," *Structural Change and Economic Dynamics,* 6: 295-304.

Rose, A. and J. Benavides. 1999. "Optimal Allocation of Electricity After a Major Earthquake: Market Mechanisms vs. Rationing," in K. Lawrence *et al.* eds., *Advances in Mathematical Programming and Financial Planning.* Greenwich, CT: JAI Press.

Rose, A. and S. Liao. 2002. "Modeling Regional Economic Resiliency to Earthquakes: A Computable General Equilibrium Analysis of Water Service Disruptions," in Proceedings of the 7th National Conference on Earthquake Engineering, Oakland, CA; EERI.

Rose, A. and S.M. Lin. 1995. "Regrets or No Regrets—That is the Question: Is Conservation a Costless CO_2 Mitigation Strategy?" *Energy Journal,* 16: 67-87.

Rose, A., J. Benavides, S.E. Chang, P. Szczesniak, and D. Lim. 1997. "The Regional Economic Impact of an Earthquake: Direct and Indirect Effects of Electricity Lifeline Disruptions," *Journal of Regional Science,* 37: 437-58.

Rose, A., J. Carmichael, and G. Oladosu. 1999. "Modeling the Economics of Natural Hazard Impacts and Policy Responses Using Computable General Equilibrium Analysis," University Park, PA; The Pennsylvania State University.

Shinozuka, M., A. Rose, and R. Eguchi eds. 1998. *Engineering and Socioeconomic Analysis of a New Madrid Earthquake.* Buffalo, NY; Multidisciplinary Center for Earthquake Engineering Research.

Shoven, J. and J. Whalley. 1992. *Applying General Equilibrium.* New York, NY; Cambridge University Press.

Tierney, K. 1997. "Impacts of Recent Disasters on Businesses: The 1993 Midwest Floods and the 1994 Northridge Earthquake," in B. Jones ed. *Economic Consequences of Earthquakes: Preparing for the Unexpected.* Buffalo, NY; National Center for Earthquake Engineering Research.

Appendix

Table 7.A.1 Structure of Shelby County Economy

Sector	Output ($ millions)	Output (%)	Employment (persons)	Employment (%)	Capital ($ millions)	Capital (%)
Agriculture	128	0.25	4,934	0.83	228	0.36
Mining	20	0.04	92	0.02	35	0.05
Construction	2,782	5.53	32,274	5.42	2,245	3.50
Food Processing	1,999	3.98	5,652	0.95	1,552	2.42
Manufacturing	8,151	16.22	43,617	7.32	8,234	12.83
Petroleum Products	511	1.02	469	0.08	327	0.51
Transportation	6,235	12.40	56,151	9.42	4,990	7.77
Communication	654	1.30	2,953	0.50	1,489	2.32
Electricity Services	530	1.05	2,133	0.36	989	1.54
Gas Distribution	21	0.04	35	0.01	21	0.03
Water & Sanitation	57	0.11	442	0.07	71	0.11
Wholesale Trade	4,581	9.11	41,293	6.93	4,438	6.91
Retail Trade	4,148	8.25	100,489	16.86	4,612	7.19
Finance, Insurance & Real Estate	6,931	13.79	40,719	6.83	20,095	31.31
Personal Services	1,546	3.08	37,933	6.36	1,520	2.37
Business & Professional Services	3,917	7.79	85,814	14.40	5,445	8.48
Entertainment	293	0.58	6,395	1.07	369	0.57
Health	3,143	6.25	41,765	7.01	3,164	4.93
Education	1,324	2.63	37,083	6.22	52	0.08
Government	3,291	6.55	55,741	9.35	4,313	6.72
Total	50,263	100.00	595,986	100.00	64,186	100.00

Table 7.A.2 Elasticities Used in the Shelby County CGE Model

Sector	Elasticities of Transformation: Exports vs. Regional		Elasticities of Substitution: Imports vs. Regional		Elasticities of Input Substitution			
	ROW	RUS	ROW	RUS	KEL - M	KE - L	K - E	Interfuel
Agriculture	0.90	1.50	1.90	3.40	0.70	0.85	1.50	0.60
Mining	1.10	1.50	1.10	1.20	0.65	0.65	0.65	0.65
Construction	0.75	0.90	0.75	0.90	0.65	0.65	0.65	0.65
Food Processing	0.75	2.00	1.15	2.10	0.65	0.65	0.80	0.80
Manufacturing	0.75	2.00	1.30	2.50	0.65	0.65	0.70	0.80
Petroleum Products	0.75	2.00	2.00	2.00	0.65	0.65	0.65	0.80
Transportation	0.50	2.00	0.75	2.00	0.65	0.65	0.65	0.65
Communication	1.10	2.00	0.90	2.00	0.65	0.65	0.65	0.65
Electricity Services	0.75	1.10	0.10	0.10	0.65	0.65	0.65	0.65
Gas Distribution	0.75	1.10	0.75	3.00	0.65	0.65	0.65	0.65
Water & Sanitation	0.75	1.10	0.75	3.00	0.65	0.65	0.65	0.65
Wholesale Trade	2.00	2.00	1.10	2.00	0.65	0.65	0.65	0.65
Retail Trade	2.00	2.00	1.10	2.00	0.65	0.65	0.65	0.65
Finance, Insurance & Real Estate	2.00	2.00	0.85	2.00	0.65	0.65	0.65	0.65
Personal Services	2.00	2.00	0.75	1.10	0.65	0.65	0.65	0.65
Business & Professional Services	2.00	2.00	0.75	1.10	0.65	0.65	0.65	0.65
Entertainment	2.00	2.00	1.10	2.00	0.65	1.15	0.65	0.65
Health	0.90	2.00	0.90	2.00	0.65	0.80	0.65	0.65
Education	0.90	2.00	0.90	2.00	0.65	0.80	0.65	0.65
Government	0.75	0.90	0.75	0.90	0.85	0.65	1.50	0.65

ROW = Rest of the World; RUS = Rest of the United States

Sources: Elasticities are based on the following studies: Li (1994), Mckibbin and Wilcoxen (1998), Hashimoto and Heath (1995), Deardoff and Stern (1986), and Prywes (1986), and well as modifications on expert judgement from Li (1994), which was based on Robinson *et al.* (1992).

Table 7.A.3 CGE Model Adjustments for Electricity Lifeline Disruptions

Stage	Electricity Capacity	Electricity Elasticities		All Other Sectors		Substitution & Distribution	Power Plant & Transmission	Capital Mobility	Closure Rule
		Input	Import	Input	Import				
O. Base Case	maximum outage	base	base	base	base	sectoral availability constraint	K stock reduction	fixed	Keynesian & Neoclassical
C. Very Short-Run ($\leq$ 7 days)	maximum outage or some recovery	$\leq$ 0.1	10X of base	$\leq$ 0.1	10X of base, elec imp $\leq$ 0.1	sectoral availability constraint	K stock reduction	fixed	Keynesian
E. Short-Run ($\leq$ 6 months)	various stages of recovery	0.5 of base	5X of base	0.5 of base	5X of base, elec imp=base	sectoral availability constraint	K stock reduction	fixed	Keynesian
F. Medium-Run ($\leq$ 1 year)	full recovery	base	base	base	base	n.a.	n.a.	sectoral	Keynesian
G. Long-Run ($\geq$ 1 year)	new equilibrium	base adjust for mitig	base or new	base adjust for resil	base or new	n.a.	K stock augmented	economy-wide	Keynesian & Neoclassical

8 The Fall of the Iron Curtain and the Evolution of German Regional Labour Markets: A Self-Organized Criticality Perspective*

Aura Reggiani[1] and Peter Nijkamp[2]

[1] Department of Economics, Faculty of Statistics, University of Bologna, Italy
e-mail: reggiani@economia.unibo.it
[2] Department of Economics, Faculty of Economics, Free University, The Netherlands
e-mail: pnijkamp@feweb.vu.nl

8.1 Systems in Transition

Modelling dynamic systems has in the past decades become an integral part of the analysis of complex behaviour of evolutionary phenomena. After a period of linear dynamics, we have in the 1990s observed an avalanche of interest in non-linear dynamic systems, as witnessed inter alia in catastrophe theory, chaos theory, synergetics and learning theory. These advances have demonstrated that modelling such systems calls for a variety of approaches (see *e.g.* Nijkamp and Reggiani, 1998).

An important question turns out to be whether a system that is brought out of balance after being exposed to an external force or an endogenous shock will return to its original equilibrium, and if not, whether it will move towards a new equilibrium or whether it will continue as an unstable system. The literature on these issues is abundant, not only in the natural sciences, but increasingly also in the social sciences. Given the complicated and uncontrolled nature of human-made systems, the application of non-linear dynamic models is still fraught with many hurdles.

In recent years we have seen a great popularity of contingency studies, which aims to study the conditions under which a dynamic system might return to an equilibrium after an exogenous shock. Such shocks may comprise wars, revolutions, earthquakes, volcanic eruptions, floods or fast political transformations. As an illustration studied in the present chapter, we will focus on the labour market shifts in West Germany after the fall of the iron curtain. For this purpose we have deployed here a recent technique for studying shifts in dynamic systems, viz. self-organised criticality.

The present chapter aims to review insights and reflections originating from the novel concept of Self-Organised Criticality (SOC), which is based upon the idea that complex behaviour can develop spontaneously in various many-body systems whose dynamics may vary abruptly and unexpectedly. This concept is an outgrowth of recent advances in dynamic systems analysis in which attempts are made to investigate the

* The authors wish to thank Simonetta Longhi (Tinbergen Institute, Amsterdam) and Giovanni Russo (University of Utrecht) for their computer assistance, as well as Rainer Andergassen (University of Bologna and Siena) for his useful suggestions. The authors also wish to acknowledge the statistical support by the Institute for Employment Research of the Federal Employment Services (IAB) in Nürnberg (Germany), in particular Uwe Blien and Erich Maierhofer, for providing the data set used in the empirical analysis.

transition trajectories of evolutionary systems. The dynamics of such systems is not only determined by external shocks, but also by internal response mechanisms, at both micro and macro levels. SOC is a recently developed scientific approach that aims to examine which critical conditions emerging out of the internal dynamics of the system itself lead to significant and structural transformations of the system's behaviour.

The present chapter is organised as follows. We will offer first some observations on the study of evolutionary dynamics in the past decade, by focussing on the methodological steps which have led to the formulation of the concept of emergence (section 8.2). Then, in section 8.3 the concept of SOC will be dealt with, by highlighting in particular various relevant definitions and characteristics, while section 8.4 will be devoted to the analysis and modelling of SOC in spatial-economic systems. Finally, in section 8.5 we will offer an illustration of SOC in the labour market sector, while we will conclude, in section 6, with some retrospective and prospective remarks.

8.2 Complex Dynamics: Sources and Processes

8.2.1 Preface

The interaction and balance between order and disorder, or equilibrium and disequilibrium, has always created a great deal of interest among scientists. The 'adventure' of nonlinear dynamics, and particularly of chaos and complexity theory, has offered – in the past decades – a powerful and promising perspective on evolutionary spatial systems and processes (see Nijkamp and Reggiani, 1998; Reggiani, 2000). It is noteworthy that the so-called complex dynamics, where chaos and complexity appear to be strictly interlinked, is still catching a lot of scientific attention and debate. On the one hand, chaos shows fascinating properties and dynamic patterns, as is witnessed by the following quotation: "*simple rules might produce arbitrarily complex patterns which appear quite random*" (Bossomaier and Green, 2000, p. 2). Thus, chaos is one of the causes leading to complexity, although this does not hold vice-versa. Complexity, on the other hand, shows numerous definitions on record: "*As yet, there is no agreed theory of complexity. Much of the mathematics is intractable and computer simulation plays a major part*" (see, again, Bossomaier and Green, 2000, p. 2). The new 'systems complexity approach' is still prompting much scientific debate, mainly in terms of its definition, potential, modelling and empirical significance, not only in bio-ecological systems, but also in spatial-economic systems. Such intriguing problems are also witnessed in the following statement: "*... complex systems have the inconvenient property that they can sometimes generate elaboration upon elaboration in patterns in state space, literally forever. This makes measurement a problem*" (Barrett *et al.*, 1998, p. 73).

Table 8.1 Fundamental Elements Identifying the Complexity Approach

SOURCES	PHASES	COMPLEX OUTCOME	PROCESSES
	Simple building blocks		Self-organisation
	Interaction		Emergence
	Iteration		Criticality

However, as argued by Bossomaier and Green (2000, p. 5), even tough there is still no clear universal theory, there are certainly common paradigms and methods. They try to identify and describe in a meaningful way the fundamental feature of complexity, viz., the outcome should not be obvious from the simple building blocks. Consequently, we may define a sort of methodological structure, in order to identify the fundamental elements in the complexity approach, as depicted in table 8.1.

In analysing the sources of complexity, which can be found together in any complex system, we have to start from the details of the constituents, which then can interact and iterate, so as to produce complex entities. We can distinguish the following components or factors:

- Simple building blocks: fundamental components of the system;
- Interaction: interdependence of these basic elements;
- Iteration: repetition of simple operations, *i.e.*, generating rules which produce complex phenomena (*e.g.*, chaos, fractals).

All together they lead to a new paradigm shift in the development of scientific thinking, which can be characterised as from 'top down' to 'bottom up'. This new paradigm, the so-called emergence phenomenon, and its connection with the self-organisation principle, will be outlined in the next subsection.

8.2.2 Emergence and Self-Organisation

The emergence phenomenon is strictly connected to the fact that the interacting simple 'entities' (the simple building blocks) give rise to interesting global properties, which are more than the 'sum of the parts'. An illustrative statement of this new thinking can be found in Aleksić (2000, p. 99): "*The main characteristic of an emergent phenomenon is that it cannot exist and thus cannot be interacted with or observed at lower generating levels (Baas, 1994). It exists as the result of the aggregate behaviour of more elementary parts. In some cases the emergent phenomena can be deduced or computed knowing the generating level, but the most interesting emergence is that one which is computationally intractable or even undecidable. The connection of interesting emergent phenomena at different organisational levels is not a trivial one.*"

In this context, the connections of emergence with self-organisation are certainly crucial. If we consider the following definition of self-organisation: "*...the ability by certain equilibrium systems to develop structures and patterns in the absence of control or manipulation by an external agent (Nicolis, 1989)*" (quoted in Jensen 1998, p. 3), we can identify a first difference among the two approaches. Particularly, we may argue that self-organisation is produced in the absence of external shocks, while emergence can also encapsulate the results from external shocks.

On top of it, the emergence principle outlines the 'old' economic issue of the relationships among micro- and macro-structures, since, as Seeley (2000, p. 86) states: "*...society does not antedate the roles of individuals, but rather emerges from the rich interaction between them. Hence, complex systems can be seen to be upholding a strong role for the individual while examining the emergence of high-level behaviours globally for the ensemble of individuals. It is taking both a global and local perspective to whole systems.*"

Consequently, both the macro- and micro-level interactions as well as their continuous interplay are fundamental for the maintenance of the emergent, self-organised structure (see figure 8.1). This has also been emphasised by Walleczek (2000, p. 2) as follows: "*Emergent or self-organizing properties can be defined as properties that are possessed by a dynamical system as a whole but not by its constituents parts. In this sense, the whole is more than the sum of its parts. Put in different terms, emergent phenomena are phenomena that are expressed at higher levels of organization of the systems but not at the lower levels.*" This local-global aspect is in fact the 'key point' concerning emergence, as was also stressed by Zimmermann (2002, p. 2): "*The point is here that we visualize processes as acting on two levels relative to each other, one the microlevel on which a large number of relevant processes take place being performed by individual agents associated with specific actions, the other the macrolevel on which a collective phenomenon can be observed which can be understood as a kind of superposition of all the individual actions being performed on the microlevel.*"

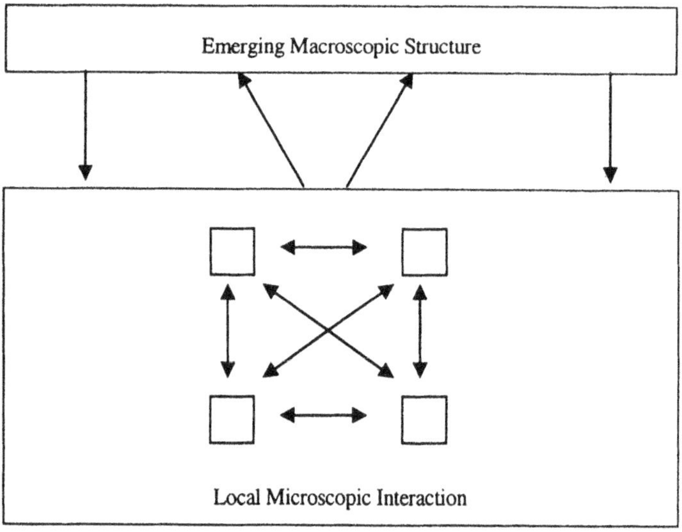

Figure 8.1 Emerging Macroscopic Structure as a Result of the Dynamic Interdependence between Macro and Micro-Levels of the System

By synthesising, according again to Walleczek (2000), the reductionist approach is giving space to the concept of self-organisation, that is based on two fundamental features:

(a) living systems as open systems, *i.e.*, in a permanent state of exchanging energy;
(b) complex organisation, *i.e.*, vast networks of interactions involving a high degree of nonlinearity.

Thus, the combination of these two features can allow complex systems to exhibit the emergent or self-organising properties defined above.

It is noteworthy that again a growing interest is coming to the fore in the theory of self-organising systems, after the worldwide recognition, in the 1970s, of the importance of the studies carried out by Nobel laureate Ilya Prigogine (see Nicolis and Prigogine, 1977). Particularly, Prigogine paid profound attention to the fact that systems far from thermodynamic equilibrium tend to self-organise by exchanging entropy and form. He focussed on 'dissipative structures'[1], by showing that the entropy (in the form of disorder) can decrease[2]. It should also be noted that, on the basis of Nicolis and Prigogine's work, the concept of evolution implying fundamental changes in form, character or behavioural strategy has emerged (see, *e.g.*, Allen, 1988, and, for a review on related applications in economic and social sciences, Nijkamp and Reggiani, 1998).

In this context, it is interesting to observe that synergy and synergetics[3] play a crucial role for the study and identification of the emergent phenomena. It should also be noted here that self-organisation, connected nowadays with the concept of emergence and complexity (as previously outlined), is not only related to the dissipative structures studied by Prigogine, but also to conservative structures, like, for example, in the concept of self-organised criticality (see next section).

The 'emergence' perspective also highlights a new principle in the study of complexity, the so-called 'organised complexity', in contrast to 'disorganised complexity', addressed by many scientists in the past century. This principle has clearly been pointed out by Stuart A. Kauffman in his notable book 'The Origins of Order', where he argues (1993, p. 173): "*First of all, contrary to our deepest intuition, massively disordered systems can spontaneously "crystallize" a very high degree of order. Much of the order we see in organisms may be the direct result not of natural selection but of the natural order selection was privileged to act on. Second, selection achieves complex systems capable of adaptation. Moreover, I shall suggest that there are general principles characterizing complex systems able to adapt: They achieve a "poised" state near the boundary between order and chaos, a state which optimizes the complexity of tasks the system can perform and simultaneously optimizes evolvability.*"

In this context, Kauffman defines a complex regime as the transition region, on the edge between order and chaos: "*In the ordered regime, many elements in the system freeze in fixed states of activity. These frozen elements form a large connected*

[1] Dissipative systems are those for which volume elements in phase space shrink with increasing time. This property leads to attracting domains in phase space, such as fixed points, limit cycles and strange attractors (see, among others, Nijkamp and Reggiani, 1992).

[2] This forms an apparent contradiction to the second law of the thermodynamics, stating that entropy always increases within closed, *i.e.*, conservative, systems. However, living systems are open systems, since everything interacts with its surroundings, and hence the law does not apply locally (see Green, 2000).

[3] We should recall here the fundamental work by the 'father' of synergetics (*i.e.*, the science of co-operation), Herman Haken (1973 and 1983).

cluster, or frozen components, which spans, or percolates, across the system and leaves behind isolated islands of unfrozen elements whose activities fluctuate in complex ways. In the chaotic regime, there is no frozen component." (1993, p. 174)[4].

The ordered regime arises, because a large fraction of the elements falls to fixed (active or inactive) states. Thus, the complex regime corresponds to a phase where the frozen islands begin to melt and percolate, while the unfrozen component is just ceasing to percolate. In Kauffman's view this is the most complex but controllable behaviour.

In parallel, Kauffman focuses his attention on co-evolving systems[5], by arguing that complex co-evolving systems tend towards a state in which each system internally is poised at the edge of chaos. Particularly, all such systems may co-evolve to the edge of chaos (the so-called self-organised criticality) as will be discussed in the next section.

8.3 Self-Organised Criticality: A Summary

8.3.1 The SOC Concept

Bak and his co-authors (1987) used the example of the sand pile in order to illustrate the empirical evidence of SOC. They noted that, as soon as a drop of sand is added to the pile of sand at its steady state[6], a chain of many tiny and few large avalanches occurs. They coined this evolutionary state as critical[7]. Moreover, Bak and Chen (1991, p. 26) defined the critical state as follows: "*Many composite systems naturally evolve to a critical state in which a minor event starts a chain reaction that can affect any number of elements in the system. Although composite systems produce more minor events than catastrophes, chain reactions of all sizes are an integral part of the dynamics. According to the theory, the mechanism that leads to minor events is the same one that leads to major events. Furthermore, composite systems never reach equilibrium but instead evolve from one meta-stable to the next. Self-organized criticality is a holistic theory: the global features, such as the relative number of large and small events, do not depend on the microscopic mechanisms.*" Surprisingly, these authors found a characteristic power-law distribution relating the frequencies and the size of avalanches.

The criticality concept has subsequently been described by Jensen as follows: "*The system becomes critical in the sense that all members of the system influence each other*" (1998, p. 3). Thus the author emphasises the crucial role of the interaction

[4] In particular, in the chaotic regime unfrozen elements percolate across the system, leaving behind isolated frozen islands.

[5] Co-evolving systems reflect the property that entities do not merely evolve, but co-evolve with other entities and with a changing environment. Explorations of co-evolutionary behaviour in spatial systems have been carried out by Batten (1998), in the context of traffic networks.

[6] In this steady state the mean rate of dropping sand onto the pile equals the mean rate at which sand falls over the edges. This is a conservative law. SOC is also related to conservative systems (see footnotes 1 and 2).

[7] "*The word criticality has a precise meaning in equlibrium thermodynamics (Binney et al., 1992)...at the transition temperature the local distortion will propagate throughout the entire system...*" (Jensen, 1998, p. 3).

among the micro-elements. However, the same author raises some doubts on the precise definition, as well as on the formalism concerning SOC: "*Connected with this lack of definition of SOC is a lack of mathematical formalism. Despite an enormous interest from people working in many different fields, SOC still lacks a well-established mathematical framework*" (Jensen, 1998, p. 2).

It is now worthwhile to describe the main characteristics inherent in SOC systems. These are depicted in table 8.A.1, in Appendix. From table 8.A.1 we can deduce the difficulties in proving analytically the SOC existence. Even the power law distribution – which seems the clearest indicator – is likely dependent on the data clustering and aggregation methods (see, *e.g.*, Gabrielov *et al.*, 1999; Reggiani *et al.*, 2002a, 2002b). On top of it, the most fundamental methodological question, *i.e.*, whether SOC is a unifying theme or just a special case of more general turbulence phenomena, is still under debate and emphasises evoking the necessity of more thorough studies and empirical applications.

However, despite these difficulties in formally identifying the analytical properties as well as the empirical evidence of SOC, the idea that SOC appears to be the 'culmination' of adaptive evolutionary systems certainly opens new perspectives in the analysis of evolutionary complexity. In this light, it is interesting to focus the attention on the similarities and differences between SOC and chaos. A first attempt will be offered in the next subsection.

8.3.2 SOC and Chaos

In section 8.2 we pointed out the relevance of the 'edge of chaos', mainly outlined by Kauffman (1993). This author hypothesises that systems between order and chaos are the natural culmination of selective evolution. This fascinating issue is based on the possibility that "*Adaptive evolution achieves the kind of complex systems which are able to adapt*" (Kauffman, 1993, p. 235). In other words, a selection causes the systems able to adapt, while the complex region may be the attractor of that selective dynamics.

This fascinating principle of 'adaptation to the edge of chaos' directs the attention towards the state – between order and chaos – of self-organised criticality (SOC), to which co-evolving systems seem to move. According to Kauffman, in this domain some entities are frozen for long periods, such that elements in the frozen component do not change, while other species continue to evolve. In other words, in this region some, but not all, partners cease changing and form a percolating frozen component, leaving isolated islands of partners which continue to co-evolve and change. The co-evolving system reaches then a structure where all partners attain the highest sustained fitness. Here then, avalanches of change at local points propagate by causing speciation and extinction events. Consequently, several authors identify criticality also with the 'edge of chaos', *i.e.*, the transition from ordered to chaotic behaviour (Kauffmann 1993; Langton, 1990).

The main question is then whether or not SOC is characterised by unpredictability. On the one hand, according to Aleksić (2000, p. 116): "*The fact that the system operates at or near the critical state means that the smallest possible perturbations can cause drastic changes over large areas of the system. Small perturbations can cause changes very far within the system's space and time, practically infinitely far.*" And consequently, the author states: "*Treating large systems in the critical state means that computational intractability is likely to occur.*"

On the other hand, given the fixed laws of the power law distribution of the avalanches described by Bak and his co-authors, SOC can be controllable. In this context, Seeley (2000) outlines the relevance of connectivity – instead of the capacity of the system – by arguing that in the experiment of the sand pile adding a grain of sand is similar to adding a new connection. He then summaries: "*The phase change which occurs at the connectivity avalanche is from a system sparsely connected to a system which is richly connected*" (p. 65).

We should also add here another fundamental difference, namely the following. While chaos is deterministic and sensitive to very small changes in initial conditions, SOC does not necessarily show these characteristics. On top of it, the test on a power law distribution makes a SOC system more transparent than a chaotic system.

It is therefore worthwhile to analyse in more detail the SOC concept, particularly with a view to meaningful interpretations and applications of those phenomena, which may imply waves diffusion and evolution, in a continuous interplay between micro- and macro-structures with substitution and complementarity effects, like the so-called disaster cases.

A first attempt to identify SOC from a methodological viewpoint, particularly in relation to the 'universal' logistic law that is able to map out cycles and chaos patterns, will be presented in the next section. We will then try to show the connections among cycles, chaos and criticality, by using the formal representation of a dynamic logistic function.

8.4 Identification of Self-Organised Criticality in Spatial Economic Systems

8.4.1 Introduction

It can be argued from the previous section, that SOC is sometimes seen as the typical behaviour of interacting many-body systems, where the external forces are not shocks. We may refer here to the following statement by Jensen (1998, p. 126): "*We will expect SOC behaviour in slowly driven, interaction-dominated threshold systems.*" Jensen suggests then to use the abbreviation: SDIDT, in order to emphasise its constructive definition, since in his opinion SOC is a phenomenological definition.

According to the SDIDT definition, we have then the following properties:

- many degrees of freedom are interacting;
- "*the dynamics of the system must be dominated by the mutual interaction between these degrees of freedom, rather than by the intrinsic dynamics of the individual degrees of freedom*" (Jensen, 1998, p. 126).

In summary, SOC highlights the tendency of self-organisation towards a critical state. These phenomena emerge in both natural and man-made systems. It is then interesting to evaluate the possibility of SOC events in conventional dynamic models deployed in spatial economics, such as the logistic equation expressing the self-organisation dynamics of several spatial phenomena, such as innovation diffusion, transport competition, species substitution, network organisation, etc. (see for a study on the dynamics of spatial logistics in various forms of complexity, Nijkamp and Reggiani, 1998). There is a wealth of mathematics on this issue.

We will consider here as an illustration the well-known (Verhulst) logistic equation in discrete terms:

$$y(t+1) = ry(t)(1-y(t)) \tag{8.1}$$

Its properties can be analysed by performing the related phase-space analysis (y vs. r). In equation (8.1), y is the dynamic variable (*e.g.*, the percentage of persons adopting a certain innovation) and r is the growth-rate parameter. It is well-known that for the range of values $3 < r < 3.824...$ the logistic equation gives rise to cycles, before starting the chaotic period for $r = 3.824$. This logistic behaviour may be considered as the envelope curve of micro-logistic interacting behaviours.

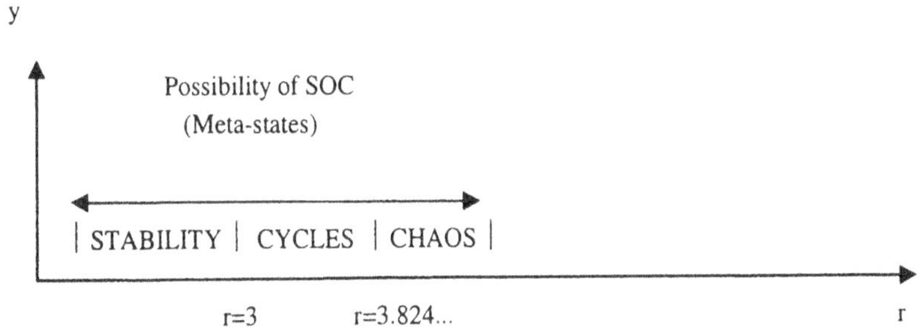

Figure 8.2 Possible Configurations of SOC in the Logistic Growth Equation in Discrete Terms

In this context, it seems reasonable to identify the possibility of SOC events in the range of the parameters ranging from stability to cycles (see figure 8.2). This is a domain at the border of chaos, although robust. In addition, it can easily display meta-states (local attractors where the system tends to and self-organises). The external driving forces can then lead the system (in a short or long time period) to meta-states, and hence to the state 'edge of chaos'. This phenomenon can be represented in a more general equation as follows:

$$y(t+1) = ry(t)(k-y(t)) \tag{8.2}$$

If we assume different capacities/thresholds k and k_1, both the slope of the curves and the threshold values appear to be crucial for the criticality configuration (see figure 8.3). In figure 8.3 various conditions of SOC, by varying the threshold values and the curve slopes in the logistic growth in discrete terms, are mapped out.

In particular, the slope (growth parameter) is relevant for reaching in a rapid way the criticality, if the strength of the interaction among the micro-components – leading to the same threshold – is the same (see the b, d curves for the threshold k, as well as the a, c curves for the threshold k_1). If not, it is certain that the strength of interaction can determine the threshold's critical state independently or in conjunction with the slope (a, b curves at time $t-1$; c curve at time t; d curve at time $t+1$).

In conclusion, in order to identify SOC in a spatial-economic network, we have to consider the following features:

- the spatio-temporal behaviour of the systems;
- the highly interconnected and interacting webs of agents/entities;
- the threshold dynamics; particularly, the local thresholds, which allow for a large number of static metastable configurations;
- intermittency (unstable-stable phases);
- chain reactions;
- robustness of the critical state; in a way, the critical state is a sort of 'attractor state';
- tuning parameters; particularly, there is still an open debate concerning the relevance of fine tuning for reaching critical regimes (see also next subsection).

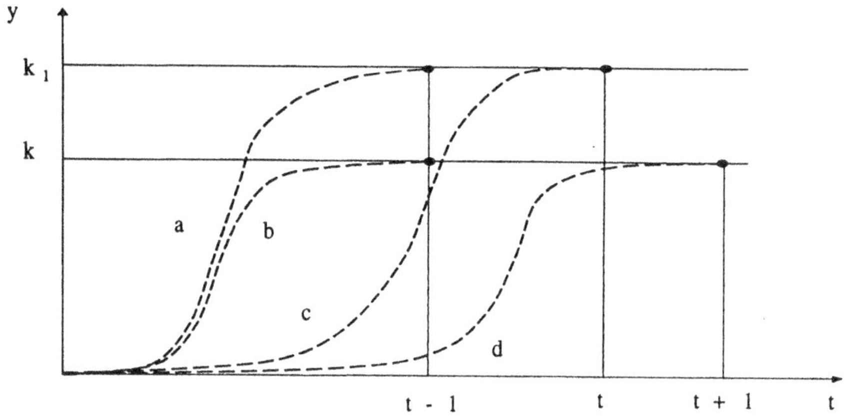

Figure 8.3 Conditions of SOC by Varying the Threshold Values and the Logistic Growth Curve Slopes in Discrete Terms

Clearly, the main question is how to measure SOC. In this context, various modelling efforts have been carried out in recent years, mainly in the context of cellular automata approaches. In the next subsection we will offer a brief overview on SOC simulations, with particular attention to the prototypes models in physical systems (sand pile, earthquake, forest fire). Since these models clearly present characteristics of negative avalanches, negative chain reactions and collapses, we can also coin them, in general, as disaster cases. This will be the focus of the next sections, although we have to point out that SOC can also model positive avalanches and chain reactions (see Reggiani *et al.*, 2002a).

8.4.2 Modelling Disaster Cases by means of SOC: A Brief Overview

The SOC principle has inspired a great number of empirical studies, with particular reference to physical systems. An informal survey of applications, where simple adjustment rules have been identified to generate SOC, can be found in Bak (1996). In parallel, recent applications to regional and economic science have also come to the fore (see for a brief review, Reggiani *et al.*, 2002a).

In this section we will pay attention to SOC models, which refer to disaster cases. In this context, the book by Jensen (1998) shows a series of prototype models, like the forest fire and earthquake models[8], mainly based on the sand pile models. The models are based on a dynamic variable, *e.g.*, the local slope of the sand heap or the stress in the earthquake fault. In this context, several of these analytical experiments shared with SOC models the slow external driving and the avalanche response. Particularly, in the sand pile model by Bak *et al.* (1987) a simple cellular automaton was introduced in order to illustrate numerically the SOC principle. The first necessary differentiation between sand pile and earthquake models is that the former can model conservative systems (the number of grains are conserved as they flow down), while the latter refer to nonconservative systems. The dynamic variable is a quantity here, viz. the local force on the tectonic sheets; thus conservation may be broken (see, *e.g.*, the prototype model proposed by Olami *et al.*, 1992). Surprisingly, also these models with no apparent conservation law display a power law distribution of avalanche sizes. However, self-organisation here is due to synchronisation, a mechanism very different from that in the conserved models (Middleton and Tang, 1992).

It is noteworthy that much debate arose in identifying the role of self-organisation in SOC models (see Portugali, 2000; for a review, Vespignani and Zapperi, 1998), as well as the role of tuning parameters. In the latter context Vespignani *et al.* (2000) pointed out that the SOC state can be ascribed to the presence of two infinitely separated time scales, which correspond to the above mentioned external energy input and the microscopic evolution (avalanches). These authors showed the relevance of fine tuning parameters in contrast to Bak's idea; this is certainly an ongoing debate which deserves more scientific research (see, *e.g.*, Nagel and Paczuski, 1995; Nagel and Rasmussen, 1994). It is interesting to note here that Vespignani and Zapperi (1998) also emphasised that the language of cellular automata (CA) with absorbing states can be employed for describing SOC models.

Another group of disaster case models is the so-called 'critical forest fire model.' The forest fire models are a sort of diffusion models, used for modelling the spreading of measles in the population of the Faroe and Bornhlom islands (Jensen, 1998; Rhodes and Anderson, 1996).

This kind of model is indeed controlled by a critical point, however the critical point is repulsive, *i.e.*, the model is not self-organised critically, but it can be tuned to criticality (see again Jensen, 1998). Drossel and Schwabl (1992, 1993) developed this forest fire model, by defining a d-dimensional cubic lattice, on the basis of the model by Bak *et al.* (1990). The lattice is updated according to four simple rule algorithms. It is interesting to know that if the lattice is updated only to the first three rules conceived by the authors, the evolution becomes periodic rather than critical. This finding supports our methodological representation of SOC (see figure 8.2), as well as the previous arguments on the relationships between SOC and chaos (see section 8.3.2),

[8] See also Chapter 5 in Bak (1996).

by showing how the periodic cycles and criticality are closely related phenomena, since a rule algorithm is able to shift the system from one domain to the other.

In the case of the forest fire model, Drossel and Schwabl (1992) find that the critical state corresponds to the one where the number of growing trees and of burned trees is at maximum. In addition, they find a separation of the time scale, analogously to those emerging in the sand pile and earthquake models; particularly, they find that the time in which a forest cluster burns is much shorter than the time in which a tree grows.

Concerning the application of these models to the epidemic diffusion of measles cases, Rhodes and Anderson (1996) found a similarity between the temporal variation in the number of fires and the temporal records of measles in the islands under analysis.

Given this brief overview, it is clear that much research can be imagined[9], especially with reference to the study of the conditions needed for self-organisation towards criticality. In the context of disaster case SOC models, the threshold dynamics seems to play a crucial role. In systems where the dynamic evolution is the result of a fight against various types of barriers, the action will occur when the difference between the total load applied locally and the local barriers is at a minimum. Thus, in the earthquake model: "*the site with the largest force (and therefore the smallest distance to the threshold for slipping) is identified. The global load on the systems increased precisely enough to make this site critical, and then the system is relaxed*" (Jensen, 1998, p. 69).

A final point of interest is the ability of SOC models in predicting evolutionary systems. In this context, the debate is still going on, since several scientists argue that such models are formulated on heuristic grounds, which makes difficult to assess the relevance of predictions extracted from them (see again, Jensen, 1998). Sensitivity analyses and studies in this respect are then certainly necessary. However, it should be noted here that the empirical tests concerning the presence – in disaster cases and in general in complex dynamics – of a power law distribution are already a good instrument for detecting a SOC state, and hence to infer behavioural rules, in contrast to linear and conventional modelling, where the complex and dynamic behaviour cannot be modelled and diagnosed with certainty.

In conclusion, the first simulations and applications clearly show the great potential of SOC as a new approach that is able to identify the dynamics of large interacting systems which are governed by threshold dynamics. In this context, the study of self-organisation and criticality is certainly crucial for disaster case models, in order to contribute to a better understanding and forecasting of these disaster phenomena. On top of it, by deploying the SOC terminology, we are able to define – in broad terms – as disaster cases a SOC state presenting two infinitely separated time scales, which correspond to the above mentioned external (slow) energy input and the microscopic evolution leading to avalanches (collapse). Also social-economic complex events leading to disaster cases or collapses (*e.g.*, war, negative economic indicators, etc.) may then be analysed by means of SOC. An example in this respect will be illustrated in the next section, where we will show the dynamic pattern of negative fluctuations in German labour market, mainly as a result of the fall of the iron curtain in the 90s. This is an interesting case of (negative) micro-economic evolution, which can be analysed by using the SOC principle.

---

[9] According to Bak (1996, p. 93): "*By 1995 there were more than 100 articles in the literature supporting the view of earthquakes as an SOC phenomenon.*"

8.5 Does the German Regional Labour Market Exhibit SOC?

The use of SOC analysis in social science systems is still in its infancy. In this sector we will investigate the dynamic adjustment behaviour of German labour markets after the fall of the iron curtain and the subsequent political and economic transformations in this country. Given a large data base on the employment evolution of regional labour markets in West-Germany in the 1990s, we will be able to trace and investigate the complex dynamics involved. In order to illustrate a case of 'possible' SOCs in an empirical setting, we will apply here the SOC concept to a statistical analysis of emerging shifts in West-German regional labour market data (for the related analysis see Longhi *et al.*, 2002). The available data consist of the absolute volume of employment in the 327 districts (the so-called Kreisen) in the former West Germany for ten years (1987-1997). These districts can also be grouped in 9 regions/clusters according to economic criteria (see table 8.2, and, for details, Blien and Tassinopoulos, 1999), as well as in 10 regions (administrative subdivisions).

Table 8.2 Economic Regions

Type of district (BfLR/BBR-typology):	Type of region
A. Regions with urban agglomeration	1. Central cities
	2. Highly urbanised districts
	3. Urbanised districts
	4. Rural districts
B. Regions with tendencies towards agglomeration	5. Central cities
	6. Highly urbanised districts
	7. Rural districts
C. Regions with rural features	8. Urbanised districts
	9. Rural districts

Source: IAB (Institute for Employment Research), Nürnberg, Germany, 2000
Note: BBR is the Bundesanstalt für Bauwesen und Raumordnung, Bonn, whose former name was Bundesforschungsanstalt für Raumordnung und Landeskunde (BfLR).

Figures 8.4 and 8.5 show the evolution of the employment growth rates in these economic clusters and administrative regions of West Germany. From figures 8.4 and 8.5 it can be clearly seen that avalanches in the labour market emerged after the 1990s and propagated among all the districts until the year 1997, as is witnessed by the negative growth rates in both types of regional clustering. The conclusion is thus clear: the fall of the iron curtain has been followed by drastic regional labour shifts in West Germany, which lasted at least until 1997 before a new – less favourable – stable situation was reached. By considering a random sample of districts (see figure 8.6), we can again observe the same negative growth rate pattern.

This behaviour might in principle be analysed by means of the SOC concept. In this framework we can guess that the German unification in 1990, as well as the wide-scale introduction of new technologies in the 90s, has led to small idiosyncratic shocks at the micro-level (districts), by producing a propagation of negative fluctuations in the growth rates, as depicted in the above figures. This may mean that the labour market in the 90s was likely at its criticality state.

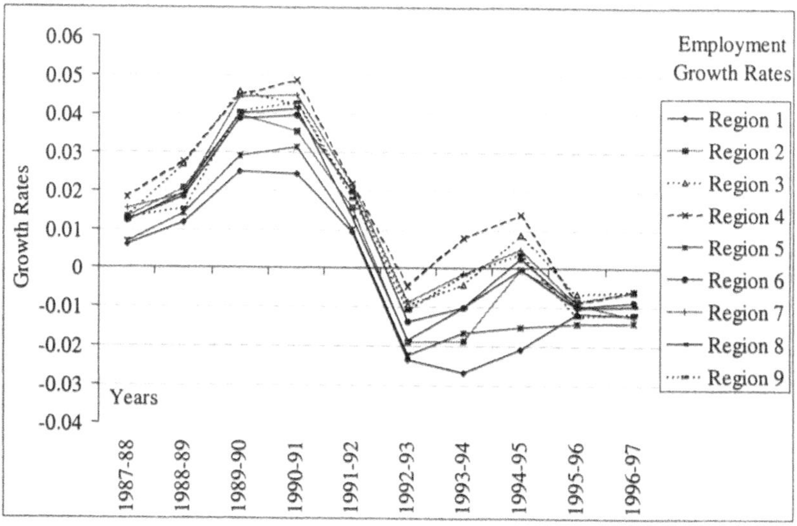

Source: Andergassen *et al.*, 2003

Figure 8.4 Evolution of Employment Growth Rates in the 9 Economic Regions of West Germany (see table 8.2 for the type of economic clustering) (1987-1997)

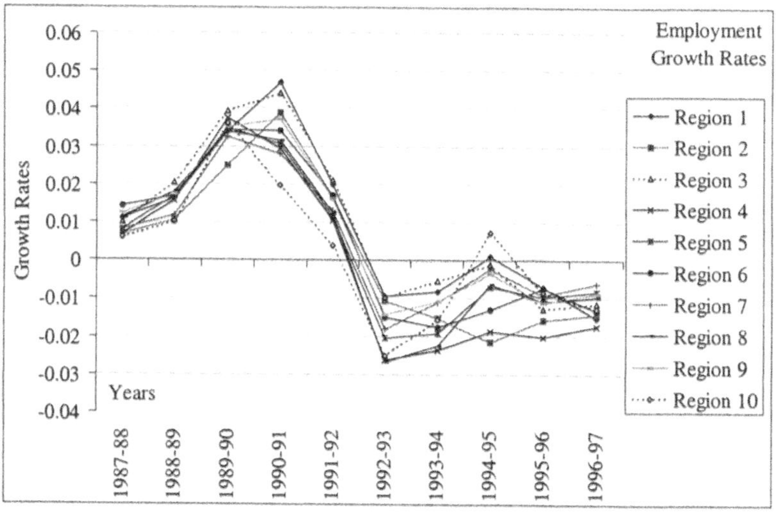

Figure 8.5 Evolution of Employment Growth Rates in the 10 Administrative Regions in West Germany (1987-1997)

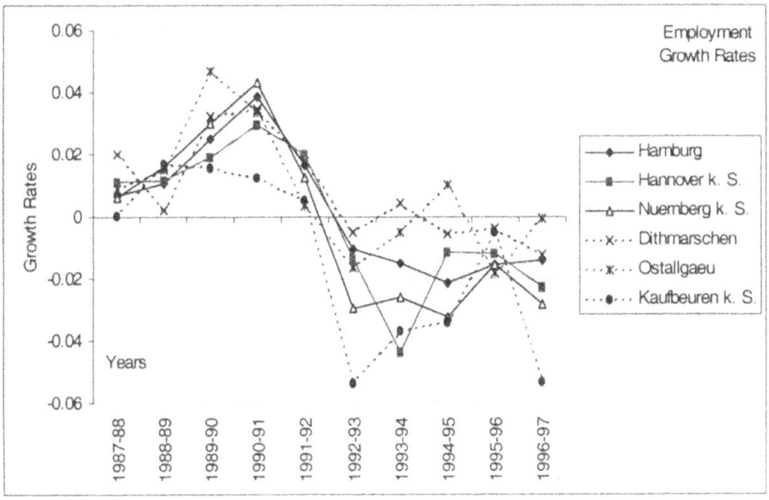

Note: The first 3 districts of Hamburg, Hannover and Nuemberg belong to the region of type 1 in table 8.2 (*i.e.*, central cities); the last 3 districts of Dithmarschen, Ostellgaeu, Kaufbeuren belong to the region of type 9 in table 8.2 (*i.e.*, rural districts).

Figure 8.6 Evolution of Employment Growth Rates in 6 Districts (1987-1997)

Clearly, such considerations might be demonstrated by testing the power law distribution of the avalanches. This will be our next research endeavour in the empirical analysis of SOC (see also Andergassen *et al.*, 2003; Reggiani *et al.*, 2002b). In addition, it is interesting to note that SOC in this context might explain economic fluctuations (Scheinkman and Woodford, 1994), in particular the development of business cycles (see also a study by Andergassen, 2001, where it is argued that economic evolution, growth and business cycles occur along the critical state).

8.6 Concluding Remarks

The aim of this study was the analysis of the potential of a recent methodological tool concerning the dynamics of systems, viz., the self-organised criticality (SOC) approach. Particularly the possibility of modelling disaster cases by means of SOC has been outlined. In this context we were able to define – in broad terms – as disaster cases a SOC state, essentially constituted of two infinitely separated time scales (viz. the external (slow) energy driving and the microscopic evolution (avalanches)).

It is interesting to note that, in parallel to the sand pile model, which represents the prototype SOC models in conservative systems, we can model and simulate – by means of SOC – non-conservative systems, that nevertheless present the time-scale separation, like the earthquake model or forest fire model.

The possibility for these systems to reach the critical state is then tested by the presence of a power law distribution. This is undoubtely a significant characteristic, which can offer additional information that is useful for the prediction and forecast of these disaster phenomena.

Also social complex events leading to disaster cases or collapses (*e.g.*, wars, devastating economic shifts, etc.) may be analysed by means of SOC, by emphasising the novel capability of the SOC principle.

A first empirical attempt to investigate the SOC state was carried out with reference to the dynamics of West German labour markets. The 'post iron-curtain' shifts appear to be significantly present here. In this context, particular attention would have to be devoted, in the next phase of our research, to the elaboration of statistical tests in order to demonstrate – also from an analytical viewpoint – the existence of a power law distribution in these systems. A major effort would be to separate out the effects of different drivers (*e.g.*, innovation breakthroughs, international economic downswings), in order to identify the single impact of the fall of the iron curtain.

As already pointed out, SOC may explain the dynamics of sudden jumps like earthquakes, up- and downswings in economic markets and fast changes in ecosystems, by showing that large interactive systems naturally evolve toward a critical state in which small shocks can induce the system to start a chain reaction which can lead to a catastrophe. We may argue that SOC seems to be the 'culmination' of the adaptive evolutionary systems, *i.e.*, an 'attractor' state – positioned at the edge of chaos – where the system, by means of adaptation and selection mechanisms, reaches its highest fitness. An essential element is the strong spatio-temporal interaction among the constituent parts of the system, in a continuous interplay with the dynamic emergent structure.

SOC appears to be an interesting appealing tool of analysis for economic and spatial sciences, since it can also explain catastrophes and avalanches usually considered as the result of aggregate shocks. In this context, future research may address the following themes: the interrelation between micro and macro states of a dynamic system (including specification problems), the transition points from stationary developments to sudden changes (including their endogenous explanation), and the impact of different time and spatial scales on the stability conditions of multi-layer evolutionary systems.

References

Aleksić, Z. 2000. "Artificial Life: Growing Complex Systems," in T.R.J. Bossomaier and D.G. Green eds. *Complex Systems*, Cambridge, UK; Cambridge University Press: 91-126.

Allen, P.M. 1988. "Evolution: Why the Whole is Greater than the Sum of the Parts," in W. Wolff, C.J. Soeder, and F.R. Drepper eds. *Ecodynamics*. Berlin, Germany; Springer-Verlag: 3-30.

Andergassen, R. 2001. "Economic Growth and Aggregate Fluctuations," Quaderni del Dipartimento di Siena, N. 318, University of Siena (Italy).

Andergassen, R., P. Nijkamp, and A. Reggiani. 2003. "Analysis of Regional Labour Market Dynamics: In Search of Indications for Self-Organised Criticality," *Journal of Geographical Systems*, 5 (3): 275-290.

Baas, N. 1994. "Emergence, Hierarchies, and Hyperstructure," in C.G. Langton ed. *Artificial Life III*. Reading, MA; Addison Wesley: 515.

Bak, P. 1996. *How Nature Works: The Science of Self-Organised Criticality*. New York, NY; Copernicus/Springer.

Bak, P. and K. Chen. 1991. "Self-Organised Criticality," *Scientific American*. January: 26-33.

Bak, P., C. Tang, and K. Wiesenfeld. 1987. "Self-Organised Criticality: An Explanation of 1/f Noise," *Physical Review Letters*, 59 (4): 381-384.

Bak, P., K. Chen, and C. Tang. 1990. "A Forest-Fire Model and some Thoughts on Turbulence," *Physics Letters A*, 147: 297-300.

Barrett, C., R. Thord, and C. Reidys. 1998. "Simulations in Decision Making for Socio-Technical Systems," in M.J. Beckmann, B. Johansson, F. Snickars, and R. Thord eds. *Knowledge and Networks in a Dynamic Economy*. Berlin Germany; Springer-Verlag: 59-82.

Batten, D. "Co-Evolutionary Learning on Networks," in M.J. Beckmann, B. Johansson, F. Snickars, and R. Thord eds. *Knowledge and Networks in a Dynamic Economy*. Berlin, Germany; Springer-Verlag: 311-332.

Binney, J.J., N.J. Dowrick, A.J. Fisher, and M.E.J. Newman. 1992. *The Theory of Critical Phenomena*. Oxford, UK; Oxford University Press.

Blien, U. and A. Tassinopoulos. 1999. "Forecasting Regional Employment with the Entropy Method," Paper Presented at the European Regional Science Association Conference, Dublin, Ireland.

Bossomaier, T.R.J. and D.G. Green. 2000. "Introduction," in T.R.J. Bossomaier and D.G. Green eds. *Complex System.*, Cambridge, UK; Cambridge University Press: 2-9.

Drossel, B. and F. Schawbl. 1992. "Self-Organised Critical Forest-Fire Model," *Physical Review Letters*, 69 (11): 1629-1632.

Drossel, B. and F. Schawbl. 1993. "Forest-Fire Model with Immune Trees," *Physica A*, 199: 183-197.

Gabrielov, A., W.J. Newman, and D.L. Turcotte. 1999. "Exactly Soluble Hierarchical Clustering Model: Inverse Cascade, Self-Similarity, and Scaling," *Physical Review E*, 60 (5): 5293-5300.

Green, D. 2000. "Self-organisation in Complex Systems," in T.R.J. Bossomaier and D.G. Green eds. *Complex Systems*, Cambridge, UK; Cambridge University Press: 11-50.

Haken, H. 1973. *Synergetics*. Berlin, Germany; Springer-Verlag.

Haken, H. 1983. *Advanced Synergetics*. Berlin, Germany; Springer-Verlag.

Jensen, H.J. 1998. *Self-Organised Criticality*. Cambridge, UK; Cambridge University Press.

Kauffman, S.A. 1993. *The Origins of Order*. New York, NY; Oxford University Press.

Langton, C.G. 1990. "Computation at the Edge of Chaos," *Physica D*, 42 (1-3): 12-37.

Longhi S., P. Nijkamp, A. Reggiani, and U. Blien. 2002. "Forecasting Regional Labour Markets in Germany: An Evaluation of the Performance of Neural Network Analysis," *CD Proceedings of the European Regional Science Association Conference*, Dortmund.

Middleton, A. and C. Tang. 1992. "Self-Organised Criticality in Nonconserved Systems," *Physical Review Letters*, 74 (5): 742-745.

Nagel, K. and M. Paczuski. 1995. "Emergent Traffic Jams," *Physical Review E*, 51: 2909-2918.

Nagel, K. and S. Rasmussen. 1994. "Traffic at the Edge of Chaos," in R.A. Brooks and P. Maes eds. *Artificial Life IV*. Cambridge, MA; MIT Press: 222-235.

Nicolis, G. 1989. in P. Davids ed. *The New Physics*. Cambridge, UK; Cambridge University Press.

Nicolis, G. and I. Prigogine. 1977. *Self-organisation in Nonequilibrium Systems*. New York, NY; J. Wiley.

Nijkamp, P. and A. Reggiani. 1992. *Interaction, Evolution and Chaos in Space*. Berlin, Germany; Springer-Verlag.

Nijkamp, P. and A. Reggiani. 1998. *The Economics of Complex Spatial Systems*. Amsterdam, The Netherlands; Elsevier.

Olami, Z., H.J.S. Feder, and K. Christensen. 1992. "Self-Organised Criticality in Continuous, Nonconservative Cellular Automaton Modeling Earthquakes," *Physical Review Letters*, 68 (8): 1244-1247.

Portugali, J. 2000. *Self-Organization and the City*. Berlin, Germany: Springer-Verlag.

Reggiani, A. ed. 2000. *Spatial Economic Science: New Frontiers in Theory and Methodology*. Berlin, Germany; Springer-Verlag.

Reggiani, A., P. Nijkamp, and R. Andergassen. 2002a. "Firm Dynamics and Self-Organised Criticality," Paper Presented at the Workshop "Entrepreneurship in the Modern Space-Economy", Tinbergen Institute, June 10-11, Amsterdam.

Reggiani, A., P. Nijkamp, and R. Andergassen. 2002b. "Self-Organised Criticality: An Operational Framework for Spatial Economic Dynamics," Paper Presented at the European Regional Science Association Conference, Dortmund.

Rhodes, C.J. and R.M. Anderson. 1996. "Power Law Governing Epidemics in Isolated Populations," *Nature*, 381 (June): 600-602.

Scheinkman, J.A. and M. Woodford. 1994. "Self-Organised Criticality and Economic Fluctuations," *American Economic Review*, 84: 417-421.

Seeley, D.A. 2000. "Network Evolution and the Emergence of Structure," in T.R.J. Bossomaier and D.G. Green eds. *Complex Systems*. Cambridge, UK; Cambridge University Press: 51-89.

Vespignani, A., R. Dickman, M.A. Muñoz, and S. Zapperi. 2000. "Absorbing-state Phase Transititions in Fixed-Energy Sandpiles," *Physical Review E*, 62 (4): 4564-4582.

Vespignani, A. and S. Zapperi. 1998. "How Self-Organized Criticality Works: A Unified Mean-Field Picture," *Physical Review E*, 57 (6): 6345-6362.

Walleczek, J. 2000. "The Frontiers and Challenges of Biodynamics Research," in J. Walleczek ed. *Self-Organised Biological Dynamics & Nonlinear Control*. Cambridge, UK; Cambridge University Press: 1-11.

Zimmermann, R.E. 2002. "Decentralization as Organizing Principles of Emergent Urban Structures," Paper Presented at the AKSOE Meeting, Section 14, Spring Conference of DPG, Regensburg.

Appendix

Features of SOC

In this Appendix, we will summarise the characteristics of SOC (see table 8.A.1), also in order to better identify its possible applications in spatial-economic systems. In doing so, we will mainly refer to Jensen (1998, p. 3-5).

Table 8.A.1 Characteristics of the Self-Organising Criticality Principle

CHARACTERISTICS OF SOC
The process connected with the external driving of the system needs to be much slower than the internal relaxation process; see, *e.g.*, the earthquake phenomenon, where the driving force may take years, but the relaxation only a few seconds/minutes.
There is evolution with meta-stable states and threshold dynamics.
There is presence of the marginally stable states. They occur when a slight increase in the applied force can lead to almost any response: lack of any typical timescale or length scale. These are a sort of 'intermediate states' towards SOC.
SOC is related to both conservative systems (*e.g.*, sandpile) and nonconservative systems (*e.g.*, earthquakes).
SOC is characterised by the response of a system to external perturbation: the same perturbation applied at different positions or at the same position at different times can lead to a response of any size.
A small memory effect exists – in SOC systems at a micro-level – depending on the strength of the exogenous driving force (the larger is this one, the lower will be the related memory effect).
SOC can possibly be an explanation of 1/f (flicker) noise and fractals.
The distribution functions describing the frequency with which various events occur in the SOC state exhibit power laws.
SOC refers also to systems displaying weak chaos/border of chaos; here no uncertainty grows more slowly than it does for chaotic systems (in particular, by means of a power law rather than an exponential law; see the previous point). Weakly chaotic systems lack a time scale beyond which it is impossible to make predictions (in contrast to chaotic systems). They allow for long-term predictions.

9 Risk Perception, Location Choice and Land-use Patterns under Disaster Risk: Long-term Consequences of Information Provision in a Spatial Economy

Hirokazu Tatano,[1] Kentaro Yamaguchi,[2] and Norio Okada[1]

[1] Disaster Prevention Research Institute, Kyoto University, Japan
e-mail: tatano@imdr.dpri.kyoto-u.ac.jp and okada@imdr.dpri.kyoto-u.ac.jp
[2] Mitsubishi Research Institute Co., Japan
e-mail: yamaken@mri.co.jp

9.1 Introduction

It appears that the number of death and magnitude of damage due to natural disaster is increasing in the last decades. Worldwide economic losses due to natural disasters are increased exponentially: the measured worldwide economic loss in 1990s is US$608.5 billion and is nine times larger than in 1960s according to Munich Re (2003). Although the number of the catastrophic events in the 1990s is three times as large as in 1960s, the economic loss per event has increased in the three times faster than the growth of the number. There is no doubt that accumulation of population and assets into the disaster prone area is one of the main factors which cause such a large increase in economic losses due to natural disasters.

In order to reduce the damages caused by natural disasters, it is necessary to control accumulation of population and assets in the disaster prone area as well as to increase the capacity of social infrastructure against natural disasters by developing "disaster prevention and mitigation countermeasures." It has been pointed out that the investments on disaster prevention facilities cause accumulation of population and assets in vulnerable areas against a huge catastrophic disaster. Therefore, countermeasures to decrease damages give incentives for individuals to choose safer locations and appropriate mitigation countermeasures by themselves.

A location pattern of population and assets in a region is determined by the location choice behavior of individuals as equilibrium. Individuals collect information of lands in the considered region. Based on the information they obtained, they decide a specific location and size of land to locate, considering its quality and its rental price under the budget constraint. Vulnerability of the area is one of the major factors of quality of the land. Therefore, public information on vulnerability of the land against natural disaster has potential to achieve the appropriate land use patterns and to contribute redaction of the damage potential of the region.

The argument is based on the assumption which households decide rationally their location by obtaining information of lands. In the former literature (for example, Palm, 1981), a little supportive empirical result was found on the assumption, but recent ones (Berknoph *et al.*, 1997, and Nakagawa *et al.*, 2002) seem to obtain more supportive results. It can be explained by the reason why the reliability of the information is getting more reliable because of extension of scientific knowledge upon natural disaster processes.

However, a large-scale natural disaster is an extreme event. It is difficult to assume that individuals have achieved an appropriate risk perception in which perceived risk level is identical with the objective risk level. We call the difference in perceived risk and objective one as "perception bias" upon risk. Since individual actions are chosen based on their own belief, individual investment for anti-disaster mitigation cannot be efficient because of this perception bias.

Viscussi (1990) formulated a risk perception process of each individual for events with low frequencies as a rational learning process using the Bayesian learning process. Based on his framework, we will describe household's belief formation process upon vulnerability in a linear mono-centric city.

Some theoretical studies have been done to investigate location choice behavior of households and equilibrium land use patterns under uncertainty (*e.g.*, DeSolvo *et al.*, 1982, Turnbull, 1991, and Papageorgiou, 1991). However, these studies did not take information provision into account. In the past study, the authors (Tatano *et al.*, 1998) investigated the conditions for public information about the disaster risk to contribute the reduction of the damage potential of the region.

Given the learning model, the chapter analyzes effects of information provision in terms of vulnerability against disaster. Equilibrium land use patterns are derived by a residential choice model with different perception upon vulnerability against disasters. Optimal land use model is also formulated. By comparing equilibrium land use conditions with optimal land use conditions, the chapter examines the influence of perception biases in terms of vulnerability against disaster.

Welfare analysis is also conducted as to derive effective policy implications. In conclusion, we discuss the obtained two main results: 1) the optimal land use pattern is not attained by the market mechanism even if the government introduces economic incentive systems, e.g., taxes and/or subsidies, 2) an appropriate land-use planning is necessary as to realize the optimal land-use pattern.

9.2 The Model

9.2.1 Assumptions

A linear mono centric city is concerned to analyze the effect of providing vulnerability information against disaster risks on its land use patterns and its damage potential against natural disasters. Figure 9.1 demonstrates abstract conditions of the land in the city. The city has a central business district (CBD) in the center of the city.

All the residents have the same preference and commute to the CBD. The city is assumed large enough to neglect immigration from other cities in our model, *i.e.*, the closed-city assumption is adopted. The total population of the city is fixed to N. All the land in the city is owned by absent landowners who live in the outside of the city (absent landownership is assumed).

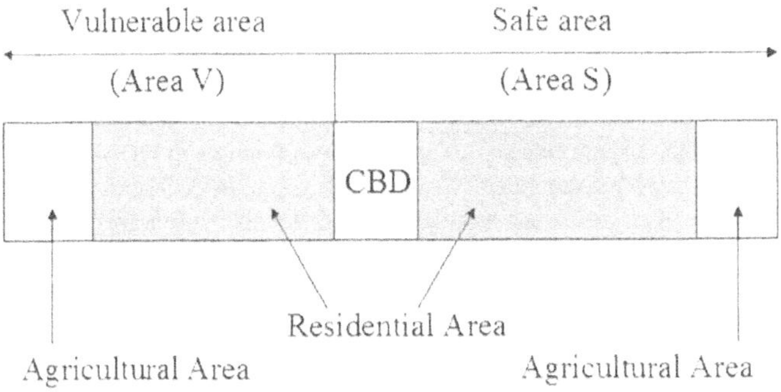

Figure 9.1 Physical Conditions of the City Concerned

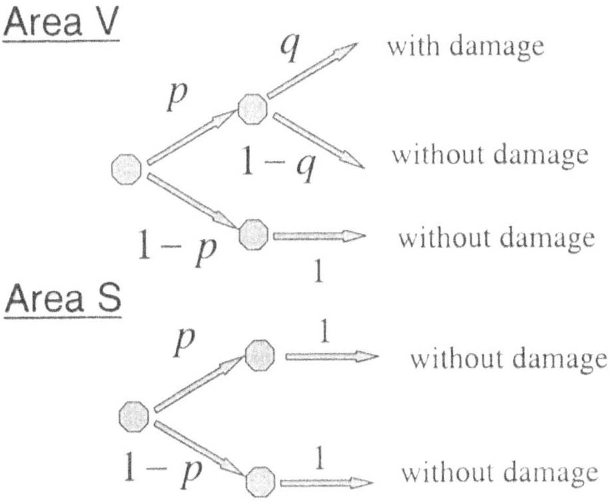

Figure 9.2 Actual Vulnerability in the City

Conditions of land in the city are assumed to be homogeneous except for the condition of "vulnerability against natural disaster." The left-hand side of the CBD (area V) is vulnerable area against natural disasters. The damage caused by the disaster is assumed to be very severe in the area V. When a disaster attacked the city, a resident may suffer severe damage with probability q and affected victims cannot survive. On the other hand, the right side of the CBD (area S) is a safe area against the disaster. Even after any disaster attack the city, residents in the area can enjoy the same amenity as before. This assumption is summarized in the figure 9.2.

The probability of the occurrence of a disaster is assumed to be same in all areas in the city. Let p denote the (objective) probability. It is assumed that all the residents and landowners know the exact value of p. Namely, p is *the common knowledge* among all the households in the city. But, it is also assumed that the households do not know how vulnerable their own land in the city is without disaster risk information on the vulnerability of land, *i.e.*, all the households have no information upon vulnerability of each piece of land in the city. Although they have no information upon vulnerability, they can develop belief in terms of the vulnerability of each piece of land but the belief should be the same in each piece of the land. This is because there is no information which can differ the vulnerability of a certain area from the other area in the city. These assumptions in terms of availability of information are made to reduce complexity. But, it is general enough to treat essential problems derived from perception biases of disaster risk. In the context of some disasters, such as earthquakes and floods, this assumption means that their return period are known but exact vulnerability of buildings and land against them are often not known by owners of the property because it is necessary for them to ask investigation of safety of their building to professionals and the owners are not necessary to believe the result or to understand its meaning.

Residents in the city maximize their utility defined over market goods and amenities subject to a budget constraint. Market goods are composed of the city's own traded (composite) good z produced in the CBD and housing service s which represents the size of the residential lot size. Landowners who live in the outside of the city own the houses. Residents pay rents for the houses to the absent landowners. The composite good is purchased from the same market in the CBD. The rental prices of housing vary spatially associated with the level of amenities at the location of the housing.

Utility function of a representative resident is defined by the function, $u(s,z,e)$. The amenity level of the land, e, assumed to take values of e_0 or e_1. The amenity level e_0 represents an amenity level in normal periods. Residents both in area S and area V enjoy their life at the amenity level e_0 before the disaster occurs. The amenity level e_1 represents an amenity level in disaster periods. When the disaster occurs and the residents may suffer damage depending on their locations in the city, the residents in area V face with the amenity level e_1 but the residents in the area S enjoy the same level of amenity e_0 as before. As we mentioned before, if a household suffers damage, he/she can not survive after the disaster. Hence, we set $u(s,z,e_1)=0$ and $u(s,z,e_0)=u(s,z)>0$ as to reflect the assumptions on the amenity levels in areas V and S, respectively, on our model.

9.2.2 Learning Vulnerability from Low Frequent Events

Risk Learning Processes and Perceived Risk

As we have assumed, in this chapter, despite of that the occurrence of a certain scale of natural disaster, p, is the common knowledge, all the households in the city have no information about vulnerability of each piece of land in the city although they can form belief in terms of the vulnerability. In this section, belief-forming processes in terms of the vulnerability are investigated.

Viscussi (1990) has described a model which could describe learning processes on risk. Based on the Bayesian learning theory, he obtained the following formula which represents the perceived risk level, $\tilde{q}$, given an objective risk information, q:

$$\tilde{q} = \frac{a + \tau q}{1 + \tau} \tag{9.1}$$

where a is individual's initial perception of the risk, and τ is a parameter which represents relative reliability of the information over initial perception of the risk a. $q - \tilde{q}$, the difference between the objective risk level, q, and perceived risk level, $\tilde{q}$, represents perception bias caused by the ambiguity of the information. From equation (9.2) the bias is formulated as:

$$q - \tilde{q} = \frac{q - a}{1 + \tau} \tag{9.2}$$

If the individuals very frequently experience the event on which the risk is defined, τ takes large value and the bias, $q - \tilde{q}$, is small. If it puts in another way, such perception bias is inevitable for the risk that has low frequency and of which scientific estimation is believed to have some noise. Since a large-scale natural disaster has these features, it is not realistic to assume that all individuals have a non-biased perception about disaster risk.

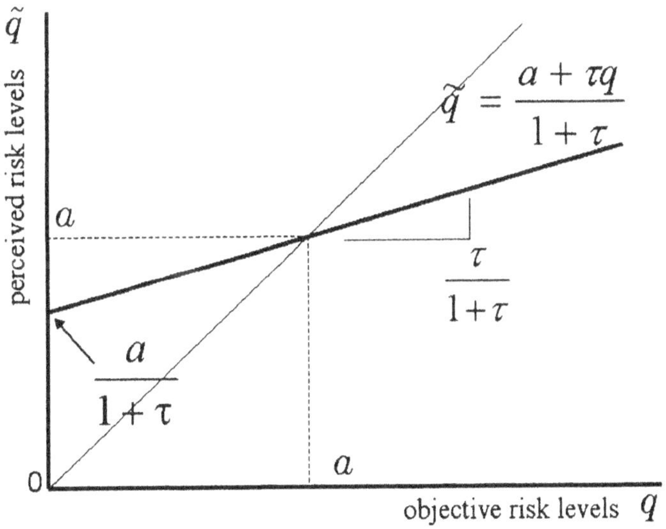

Figure 9.3 Relationship between Objective Risk Levels and Perceived Risk Levels

Figure 9.3 illustrates the relationships between the objective risk level, q, and perceived risk level, $\tilde{q}$. From this figure, it is found that, for smaller objective risk, objective risk is smaller than perceived risk and marginal perceived risk with respect to objective risk, $\frac{\partial \bar{q}}{\partial q} = \frac{\tau}{1+\tau}$, is smaller than one. The latter feature of the perceived risk means that individuals judge marginal changes of the risk smaller than actual. This feature of perceived risk causes non-sensitive reactions upon new information.

Representation of Perceived Vulnerability

Based on the Viscussi's work, let us formulate the belief-forming process as it relates to vulnerability. Vulnerability is defined as conditional probability of occurrence of a certain level of damage due to the occurrence of a certain scale of disaster. In our context, since a single type of disaster is assumed to possibly occur in the city and the only possible outcomes are state "without damage", e_0, and state "with damage", e_1, vulnerability for each individual is specified as the probability of occurrence of the state with disaster e_1 conditional on the occurrence of the disaster. Let a denote the household's initial belief regarding his/her vulnerability in the city and $Q_\delta(\tau,a,q)$ be the perceived vulnerability of the area, δ $(\delta = S,V)$, after the following information is provided: the vulnerability in area S is zero and the one in area V is q, where the information is assumed to be identical with the objective estimate of vulnerability of each area. $Q_\delta(\tau,a,q)$, where δ $(\delta = S,V)$ and the perceived vulnerability after obtaining the information, is formulated by:

$$Q_S(\tau,a,q) = \frac{a}{1+\tau}, \text{ and } Q_V(\tau,a,q) = \frac{a+\tau q}{1+\tau} \tag{9.3}$$

where τ refers again to the subjective reliability measure of the information. If the household believe the information perfectly, $\tau \to \infty$, $Q_S(\tau,a,q) = 0$, and $Q_V(\tau,a,q) = q$. If no information is available or if households do not believe the information at all, $\tau \to 0$, $Q_S(\tau,a,q) = a$, and $Q_V(\tau,a,q) = a$. Biases of the perceived vulnerabilities are also defined as the difference between the objective vulnerabilities and the perceived vulnerabilities. Therefore, the biases decrease as the reliability of information τ increases.

For the finite number of τ, a location change from area V to area S is believed to contribute to the reduction of vulnerability in $\frac{\tau q}{1+\tau}$, although the actual contribution is q. This means that the existence of a perception bias causes a less sensitive response in location choice. This idea will be developed in the following section.

9.2.3 Location Choice with Perceived Vulnerability

Let y, t, and $R_\delta^\tau(r)$ denote income, travel cost per a unit distance and land rent at the place where r is distant from CBD in area δ(=S,F), respectively. Suppose composite good z to be a *numeraire*. Demand for the housing lot and the composite good of the household who lives at the place (δ, r) is determined by the following expected utility maximization problem subject to a budget constraint:

$$V_\delta^\tau(R_\delta^\tau(r), y-tr) = \{\max_{s,z}\ (1-pQ_\delta(\tau,a,q))u(s,z) \mid R_\delta^\tau(r)s+z+tr = y\} \tag{9.4}$$

where $V_\delta^\tau(R_\delta^\tau(r), y-tr)$ is the indirect expected utility function. The first order conditions are:

$$\frac{\partial u(s,z)}{\partial s} / \frac{\partial u(s,z)}{\partial z} = R_\delta^\tau(r)\text{, and } R_\delta^\tau(r)s+z+tr = y \tag{9.5}$$

From these conditions, it is easy to verify that the demand functions of housing lot and composite good are the functions of land rent and income after spending of transportation cost:

$$s_\delta^\tau = \hat{s}(R_\delta^\tau(r), y-tr)\text{, and } z_\delta^\tau = \hat{z}(R_\delta^\tau(r), y-tr) \tag{9.6}$$

The location choice behavior is formulated as $\max_{\delta,r} V_\delta^\tau(R_\delta^\tau(r), y-tr)$.

9.2.4 Land Use Equilibrium

In order to derive land-use equiliburia in the city, let us define the bid rent function as:

$$\Psi_\delta^\tau(r,u;p,q,a) = \max_s \left\{ \frac{y-tr-z}{s} \mid (1-pQ_\delta(\tau,a,q))u(s,z) = u \right\} \tag{9.7}$$

The bid rent function is the function of the representative household's willingness-to-pay for a unit of land at distance r from the CBD in order to maintain his/her utility level equal to or above the reserved utility level, u.

Since we assume the homogeneity of preference of households, bid rent at each point of the city equals to market land rent at equilibrium:

$$R_\delta^\tau(r) = \Psi_\delta^\tau(r,u^\tau;p,q,a) = R_A \quad (r \le \bar{r}_\delta^\tau, \delta = S,V) \tag{9.8}$$

Let u^τ and $\bar{r}_\delta^\tau$ denote the equilibium utility level and distance from CBD to the city border in area δ. By their definitions, they should satisfy the following conditions at equilibrium:

$$R_\delta^\tau(\bar{r}_\delta^\tau) = \Psi_\delta^\tau(\bar{r}_\delta^\tau, u^\tau; p,q,a) = R_A \quad (\delta = S,V) \tag{9.9}$$

$$N = \sum_{\delta=S,V} \int_0^{\bar{r}_\delta^\tau} h / \hat{s}(\Psi_\delta^\tau(\bar{r}_\delta^\tau, u^\tau; p,q,a), y-tr)dr \tag{9.10}$$

where R_A is the land rent of farm land and N is the total number of population. By solving equations (9.9) and (9.10), the values of and u^τ are $\bar{r}_\delta^\tau$ determined.

9.2.5 Comparative Static Analysis

In this subsection, features of the model are studied through use of comparative static analysis. The main focus of the analysis closely investigates the information provision effect on equilibrium land-use.

For purpose of this analysis, the indirect utility function and the Solow type bid rent function have been introduced. Theses functions are defined by:

$$v(R_S^\tau(r), y-tr) \equiv \{\max_{s,z} u(s,u) \mid R_S^\tau(\overline{r}_S^\tau)s + z = y - tr\} \tag{9.11}$$

$$\psi(u^\tau, y-tr) \equiv \left\{ \max_{s,z} \frac{y-tr-z}{s} \mid u(s,z) = u \right\} \tag{9.12}$$

By comparing equations (9.4) and (9.7), and (9.11) and (9.12), the following relationships among the utilities and bid rents are found to hold:

$$v(R_S^\tau(r), y-tr) \equiv V_S^\tau(R_S^\tau(r), y-tr)/(1-pQ_S^\tau(\tau,a,q)) \tag{9.13}$$

$$\psi(\frac{u^\tau}{1-pQ_S^\tau(\tau,a,q)}, y-tr) = \Psi_S^\tau(r,u^\tau;p,q,a) \tag{9.14}$$

Proposition 1. For any $r,u^\tau,p,q,a,\tau \geq 0$, $\Psi_S^\tau(r,u^\tau;p,q,a) \geq \Psi_V^\tau(r,u^\tau;p,q,a)$. The equality holds in the condition if and only if $\tau q = 0$.

Proof. From the assumption, $1-pQ_S^\tau(\tau,a,q) \leq 1-pQ_V^\tau(\tau,a,q)$ always holds. By definition of u^τ and equation (9.13), the following condition is derived:

$$u^\tau = (1-pQ_S^\tau(\tau,a,q))v(\Psi_S^\tau(r,u^\tau;p,q,a), y-tr) \tag{9.15}$$

From the assumption of utility function $u(s,z)$, $\partial v(R(r), y-tr)/\partial R(r) < 0$. Hence, $\Psi_S^\tau(r,u^\tau;p,q,a) \geq \Psi_V^\tau(r,u^\tau;p,q,a)$. The equality holds if and only if $1-pQ_S^\tau(\tau,a,q) = 1-pQ_V^\tau(\tau,a,q)$, *i.e.*, $\tau q = 0$. (Q.E.D)

Proposition 1 means that when information regarding vulnerability against the disaster is provided, the land rent of a piece of land in area S is higher than in area V, if both pieces of land are located the same distance from CBD. It also means that land rent curves in each area are exactly the same when $\tau q = 0$. The condition holds in the following situations: (1) information regarding vulnerability against the disaster is not available; (2) all households do not believe the available information at all; or (3) vulnerability in area V is also zero, *i.e.*, no damage will be caused in the city by the disaster.

Proposition 2. For any $q > 0, \tau \geq 0$, $\frac{\partial \Psi_S^\tau(r,u^\tau;p,q,a)}{\partial \tau} \geq 0$ and $\frac{\partial \Psi_V^\tau(r,u^\tau;p,q,a)}{\partial \tau} \leq 0$.

Proof. See Appendix

Proposition 2 means that as the relative reliability of the information increases, land rent in the area S increases but land rent in the area V decreases. This result implies that more reliable information expands the difference between land rents in area S and area V.

Theorem 1. If land in the city is owned by absent land-owners and the city is closed, providing information regarding vulnerabilities against disaster for each piece of land in the city increases land rents in the safer area and decreases them in the more vulnerable area. More reliable information increases land rents in the safer area and decreases them in the more vulnerable area. Hence, the gap in land rents between two pieces of land in the same size and same distance from CBD increases the greater the reliability of information.

9.3 Welfare Analysis

In this section, a welfare analysis is conducted in order to analyze the effect that providing vulnerability information will have on the social welfare in the city.

9.3.1 Welfare Measure When a Perception Bias Exists

In this chapter, all the households are assumed to have no information regarding the vulnerability of the land before information on vulnerability is provided by the government. Therefore, their perception of vulnerability may originally be biased. Even after information is provided, such bias remains except in the case where they have perfect (perceived) information. Except in this situation, the equilibrium utility levels are not identical with their actual welfare levels. The equilibrium utility is the same among all households in the city. If there is some level of perception bias, households who have lower perceived vulnerability than actually exists locate in a more vulnerable place than they thought. Therefore, their welfare levels are smaller than the equilibrium utility level. On the other hand, households who are living safer area enjoy actually higher welfare levels than the equilibrium utility level. Hence, it is important not to use the equilibrium utility as the measure of welfare for the households, but instead to use another appropriate measure which can exactly reflect the actual vulnerability of households.

This chapter proposes an "objective welfare level" as the measure of the actual welfare of households. The objective welfare level is the actual expected utility which is enjoyed by households. In our model, a household in the area S can enjoy utility $v(\Psi_S^\tau(r,u^\tau;p,q,a),y-tr)$ with probability one. Therefore, their actual expected utility is $v(\Psi_S^\tau(r,u^\tau;p,q,a),y-tr)$. On the other hand, households in the area V also enjoy utility $v(\Psi_V^\tau(r,u^\tau;p,q,a),y-tr)$ while they survive but they can survive with probability of $1-q$ after occurrence of the disaster. The probability of occurrence of the disaster is p. Therefore, the actual expected utility they can enjoy is $(1-pq)v(\Psi_V^\tau(r,u^\tau;p,q,a),y-tr)$. We define the objective welfare function as:

$$w_\delta^\tau(\Psi_\delta^\tau(r,u^\tau;p,q,a),y-tr)=\begin{cases} v(\Psi_S^\tau(r,u^\tau;p,q,a),y-tr) & (\delta=S) \\ (1-pq)v(\Psi_V^\tau(r,u^\tau;p,q,a),y-tr) & (\delta=V)\end{cases} \tag{9.16}$$

The objective welfare function has the following features:

Proposition 3. For any $q>0$ and $\tau\geq 0$;

$$\frac{\partial w_S^\tau(\Psi_S^\tau(r,u^\tau;p,q,a),y-tr)}{\partial\tau}<0\text{, and }\frac{\partial w_V^\tau(\Psi_V^\tau(r,u^\tau;p,q,a),y-tr)}{\partial\tau}>0 \tag{9.17}$$

and

$$\lim_{\tau\to\infty} w_\delta^\tau(\Psi_\delta^\tau(r,u^\tau;p,q,a),y-tr)=u^*\text{, where } u^*=\lim_{\tau\to\infty}u^\tau \tag{9.18}$$

Proof. From proposition 2,

$$\frac{\partial w_S^\tau(\Psi_S^\tau(r,u^\tau;p,q,a),y-tr)}{\partial\tau}=\frac{\partial v(\Psi_S^\tau(r,u^\tau;p,q,a),y-tr)}{\partial R_S^\tau(r)}\frac{\partial\Psi_S^\tau(r,u^\tau;p,q,a)}{\partial\tau}<0,$$

$$\frac{\partial w_V^\tau(\Psi_V^\tau(r,u^\tau;p,q,a),y-tr)}{\partial\tau}=$$

$$=(1-pq)\frac{\partial v(\Psi_V^\tau(r,u^\tau;p,q,a),y-tr)}{\partial R_V^\tau(r)}\frac{\partial\Psi_V^\tau(r,u^\tau;p,q,a)}{\partial\tau}<0.$$

Since $\lim_{\tau\to\infty}Q_S(\tau,a,q)=0$ and $\lim_{\tau\to\infty}Q_F(\tau,a,q)=q$,

$\lim_{\tau\to\infty}u^\tau=\lim_{\tau\to\infty}w_\delta^\tau(\Psi_\delta^\tau(r,u^\tau;p,q,a),y-tr)$ (Q.E.D).

The first part of Proposition 3 means that the objective welfare of a household located in area S (area V) is an increasing (decreasing) function of the subjective reliability of information. The second part of the proposition means that welfare levels of households in both areas S and V take the same value if the information is perfectly believed. Namely, if no perception bias exists, a representative household has indifferent preference of location choice between area S and area V.

From this proposition, it is easy to verify the following inequality:

$$w_V^\tau(\Psi_V^\tau(r,u^\tau;p,q,a),y-tr)<w_S^\tau(\Psi_S^\tau(r,u^\tau;p,q,a),y-tr) \tag{9.19}$$

Therefore, the objective welfare level in area S is always higher than the one in the area V. From the Proposition 3, it is clear that the gap in objective welfare levels between the area S and area V is the largest at null information situation and decreases as the reliability of information increases.

Theorem 2. If a perception bias regarding vulnerability exists, the objective welfare level is always larger than the one in the vulnerable area. The difference between the objective welfare levels decreases with the increase of the reliability of the information. If perception bias does not exist, the objective welfare level is identical with the equilibrium expected utility level.

9.3.2 Optimal Land-Use When a Perception Bias Exists

We also examine the concept of optimal land-use when a perception bias exists. The Herbert-Stevens model (HS model) is the standard model for analyzing optimal land-use patterns. The model derives optimal land use patterns by maximizing total surplus subject to reserved utility conditions. If the reserved utility level is set at the equilibrium utility level, optimal land-use is *identical* with the equilibrium land-use pattern.

The reason why the model has such reserved utility constraints is to treat all equals equal. Due to the non-convexity introduced by space the spatial economy, Mirrlees' inequality or unequal treatment of equals easily occurs (Fujita, 1989). As we discussed in the last section, welfare level of each household should be evaluated by its objective value but not its subjective utility if there exist perception biases.

Natural extension of the HS model is to use the objective welfare function instead of the subjective utility:

$$\max_{\bar{r}_\delta, n_\delta, s_\delta(r)} S_S + (1-pq)S_V \tag{9.20}$$

$$\text{s.t.} \qquad \sum_{\delta=S,V} \int_0^{\bar{r}_\delta} h/s_\delta(r)dr = N \tag{9.21}$$

where S_δ, total surplus in the area δ is $S_\delta = \int_0^{\bar{r}_\delta} \{\frac{y-tr-Z_\delta(s_\delta(r),w)}{s_\delta(r)} - R_A\}h\,dr$; and $Z_\delta(s_\delta(r),w)$, the compensated demand for the composite good at area δ, is $Z_S(s_S(r),w) = \arg\{z \mid u(s_S(r),z) = w\}$ and $Z_V(s_V(r),w) = \arg\{z \mid (1-pq)u(s_V(r),z) = w\}$, respectively.

According to Fujita (1989), the solution of the extended HS model is described as:

$$S_\delta^*(r) = \begin{cases} s(w, y-tr-G^*) & (\delta = S) \\ s(\frac{w}{1-pq}, y-tr-G^*) & (\delta = V) \end{cases} \tag{9.22}$$

$$z_\delta^*(r) = \begin{cases} z(w, y-tr-G^*) & (\delta = S) \\ z(\frac{w}{1-pq}, y-tr-G^*) & (\delta = V) \end{cases} \tag{9.23}$$

$$R_\delta^*(r) = \begin{cases} \max\{\Psi(w, y-tr-G^*), \mathrm{R_A}\} & (\delta = S) \\ \max\{\Psi(\frac{w}{1-pq}, y-tr-G^*), R_A\} & (\delta = V) \end{cases} \tag{9.24}$$

$$\bar{r}_\delta^*(r) = \begin{cases} \arg\{r \mid \Psi(w, y-tr-G^*), \mathrm{R_A}\} & (\delta = S) \\ \arg\{r \mid \Psi(\frac{w}{1-pq}, y-tr-G^*), R_A\} & (\delta = V) \end{cases} \tag{9.25}$$

$$G^* = \arg\{G \mid \int_0^{\bar{r}_S} h/s(w, y-tr-G)dr + \int_0^{\bar{r}_V} h/s(\frac{w}{1-pq}, y-tr-G)dr = N\} \tag{9.26}$$

where $\Psi(w, I) = \arg\{\max_{s,z} \frac{I - z}{s} \mid u(s, z) = w \}$.

Theorem 3. If households have perfect information and beliefs consistent with actual vulnerability, the optimal land-use pattern is brought about through a market mechanism. If a perception bias exists, the optimal land use pattern cannot be attained through a market mechanism.

Proof. If households have perfect information, the objective welfare level is identical with the equilibrium utility level from Proposition 3. From the definition of equilibrium utility, it is clear that the lump-sum tax G^* should be zero. In this case, the extended HS model is equivalent to the ordinary HS model. Therefore, the optimal land-use pattern is attained through a market mechanism.

If a perception bias exists, the expected utility level which corresponds to the target objective welfare level will be never identical with the equilibrium utility level. To see this, let $\tilde{u}^\tau$ denote a equilibrium expected utility level. The objective welfare levels at each area are:

$$w_S^\tau = v(\Psi(\frac{u^\tau}{1 - pQ_S(\tau, a, q)}, y - tr), y - tr) = \frac{u^\tau}{1 - pQ_S(\tau, a, q)},$$

$$w_F^\tau = (1 - pq)v(\Psi(\frac{u^\tau}{1 - pQ_V(\tau, a, q)}, y - tr), y - tr) = \frac{(1 - pq)u^\tau}{1 - pQ_V(\tau, a, q)}.$$

If the market mechanism can attain the optimal allocation, there is an allocation which corresponds to the equilibrium expected utility $\tilde{u}^\tau$ and the market mechanism attains the optimal allocation. If the allocation is the optimal one, the objective welfare levels should be the same value, *i.e.*, the target objective welfare level, because of the definition of the extended HS model, *i.e.*, $w_S^\tau = w_F^\tau = w$. However, $w_S^\tau - w_F^\tau = \frac{pq(1 + pq)u^\tau / (1 + \tau)}{(1 - pa/(1 + \tau))(1 - p(a + \tau q))} < 0$. This leads to a contradiction. Hence, any allocation through a market mechanism is not optimal. (Q.E.D)

This result also means that optimal land-use will not be stable unless there is some intervention by the government. In order to attain the optimal land-use, the government should control land-use in a direct manner. Therefore, the existence of a perception bias in vulnerability against natural disaster supports the need for land-use regulation, *e.g.*, land-use planning and management.

9.4 Discussions

This chapter focuses on the perception bias which is brought about by the nature of natural disasters, *i.e.*, their low-frequency. Theoretical analysis has been conducted to analyze its effect of this perception bias on equilibrium land-use patterns and the welfare levels of households. The major results obtained are summarized as follows:

(1) If land in the city owned by absent land-owner and the city is closed, providing information regarding vulnerability against disaster of each piece of land in the city increases land rents in the safer area and decreases them in the more vulnerable area. More reliable information increases land rents in the safer area and decreases them in the more vulnerable area. Hence, the land rents gap between two pieces of land which have the same attributes except for the vulnerability against natural disaster become greater with increase in the reliability of the information provided.
(2) If a perception bias of vulnerability exists, the objective welfare level in the safer area is always greater than the objective welfare level in the vulnerable area. This difference in the objective welfare levels decreases with an increase in the reliability of the information. If a perception bias does not exist, the objective welfare level is identical with the equilibrium utility level.
(3) If all households have perfect information and beliefs consistent with actual vulnerability, the optimal land use pattern can be attained through a market mechanism. If a perception bias exists, the optimal land-use pattern cannot be attained through a market mechanism.

The first and second results reflects the importance of risk communication because it is not necessary to discuss problems brought about by a perception bias if we can find an appropriate way for households to form an consistent belief regarding the actual vulnerability.

The final result of our study suggests that the existence of a perception bias in vulnerability against natural disaster may support land-use regulation such as land-use planning. It is necessary to closely investigate the role of the land-use regulation in order to find the functional method of land-use regulation to implement when a perception bias exists. The result also can be interpreted to illustrate the importance of risk communication since risk communication is key activity needed in order to decrease perception biases.

In this chapter, *ex post* transfer of losses is assumed not possible. In reality, it is possible to transfer losses by insurance and/or other risk transferring measures. It is also possible for households to implement some mitigation measures. Integration of such risk management countermeasures and location choice behavior are two possible future directions in which to extend this research. In order to investigate it, the model should incorporate the dynamic behavior of households regarding their choice of risk management countermeasures (Yokomatsu and Kobayashi, 2001).

References

Berknoph, R.L., S. Brookshire, M. McKeeand, and D.L. Soller. 1997. "Estimating the Social Value of Geologic Map Information: A Regulatory Application," *Journal of Environmental Economics and Management*, 32: 204-218.

DeSolvo, J.S. and L.R. Eechhoudt.1982. "Household Behavior under Income Uncertainty in a Monocentric Urban Area," *Journal of Urban Economics*, 11: 98-111.

Frame, D.E. 1998. "Housing, Natural Hazards and Insurance," *Journal of Urban Economics*, 44: 93-109.

Fujita, M. 1989. *Urban Economic Theory*. Cambridge, UK; Cambridge University Press.

Herbert, J.D. and B.H. Stevens. 1960. "A Model of the Distribution of Residential Activity in Urban Areas," *Journal of Regional Science*, 2: 21-36.

Mirrlees, J.A. 1972. "The Optimum Town," *Swedish Journal of Economics*, 74: 114-135.

Munich Re. 2003. *Annual Review: Natural Catastrophes* 2002, available at http://www.munichre.com/pdf/topics_2002_e.pdf.

Nakagawa, M., M. Saito, and H. Yamaga. 2002. "On Risk Aversion toward Earthquake Risks, Part I: Land Pricing in Tokyo Metropolis," Proceedings of the Second Annual IIASA-DPRI Meeting on *INTEGRATED DISASTER RISK MANAGEMENT: Megacity Vulnerability and Resilience*, available at http://www.iiasa.ac.at/Research/RMS/dpri2002/proceedings.html.

Palm, I.R. 1981. "Public Response to earthquake Hazard Information," *Annals of the Association of American geographers*, 71 (3): 389-399.

Papageorgiou, Y.Y. 1991. "Residential Choice When Place Utility Is Not Precisely Known in Advance," working paper, McMaster University, Canada.

Tatano, H., N. Okada, N. Tanaka, and K. Yamaguchi. 1998. "Public Information Provision on Disaster Risk and Equilibrium Land Use in a Mono-Centric City," in *Risk Research and Management in Asian Perspective*, Proceedings of the First China-Japan Conference on Risk Assessment and Management, International Academic Publishers: 470-477.

Turnbull, G.K. 1991. "The Spatial Demand for Housing with Uncertain Quality," *Journal of Real Estate Finance and Economics*, 4: 5-17.

Viscussi, K.W. 1990. "Sources of Inconsistency in Social Responses to Health Risks," *American Economic Review*, 80 (2): 527-554.

Viscussi, K.W. 1998. *Rational Risk Policy*. Oxford University Press.

Yokomatsu, M. and K. Kobayashi. 2001. "Physical Asset Loss and Economic Benefits of Disaster Risk Mitigation," the First Annual IIASA-DPRI Meeting on Integrated Disaster Risk Management: Reducing Socio-Economic Vulnerability, IIASA, Laxenburg Austria, available at http://www.iiasa.ac.at/Research/RMS/dpri2001/papers.html.

Zenou,Y. and L. Eeckhoudt. 1997. "Bid Rent under Unemployment Risk: Delayed versus Timeless Uncertainty," *Journal of Urban Economics*, 42: 42-63.

Appendix

Proof of Proposition 2

From equation (9.14),

$$\frac{\partial \Psi_{\delta}^{\tau}(r,u^{\tau};p,q,a)}{\partial \tau}=\frac{\partial \psi(U_{\delta},y-tr)}{\partial U_{\delta}}\frac{\partial U_{\delta}}{\partial \tau} \tag{9.A.1}$$

where $U_{\delta}=\dfrac{u^{\tau}}{1-pQ_{\delta}^{\tau}(\tau,a,q)}$, $(\delta=S,V)$.

According to Fujita (1989), $\dfrac{\partial \psi(U_{\delta},y-tr)}{\partial U_{\delta}}<0$ always holds. Therefore, it is sufficient to show $\dfrac{\partial U_{S}}{\partial \tau}\leq 0$ and $\dfrac{\partial U_{V}}{\partial \tau}\geq 0$ as to prove Proposition 2.

At the equilibrium, the equilibrium utility u^{τ} should take same value for the representative household in the different location, *i.e.*:

$$u^{\tau}=V_{\delta}^{\tau}(\Psi_{\delta}^{\tau}(r,u^{\tau}),y-tr) \tag{9.A.2}$$

$$\partial u^{\tau}/\partial r=0 \tag{9.A.3}$$

Hence, differentiate equation (9.A.2) with r, and the following equation is derived:

$$\frac{\partial V_\delta^\tau(\Psi_\delta^\tau(r,u^\tau), y-tr)}{\partial R_\delta^\tau}\frac{\partial \Psi_\delta^\tau(r,u^\tau)}{\partial r} - t\frac{\partial V_\delta^\tau(\Psi_\delta^\tau(r,u^\tau), y-tr)}{\partial (y-tr)} = 0 \tag{9.A.3}$$

The following Roy's identity is also holds:

$$-\frac{\partial V_\delta^\tau(\Psi_\delta^\tau(r,u^\tau), y-tr)}{\partial R_\delta^\tau} \Bigg/ \frac{\partial V_\delta^\tau(\Psi_\delta^\tau(r,u^\tau), y-tr)}{\partial (y-tr)} = \hat{s}(\Psi_\delta^\tau(r,u^\tau), y-tr) \tag{9.A.4}$$

From the equations (9.A.3) and (9.A.4), we obtain:

$$\frac{1}{\hat{s}(\Psi_\delta^\tau(r,u^\tau), y-tr)} = -\frac{\partial \Psi_\delta^\tau(r,u^\tau)}{\partial r}\frac{1}{t} \tag{9.A.5}$$

Substitute the $\hat{s}(\Psi_\delta^\tau(r,u^\tau), y-tr)$ of the equation (9.A.5) for it in equation (9.10), the equation (9.10) is rewritten as:

$$N = (\Psi_S^\tau(0,u^\tau) + \Psi_V^\tau(0,u^\tau) - 2R_A)h/t \tag{9.A.6}$$

From assumption, since total number of household is fixed, differentiating equation (9.A.6) by τ, the following condition is obtained:

$$\frac{\partial \Psi_S^\tau(0,u^\tau)}{\partial \tau} + \frac{\partial \Psi_V^\tau(0,u^\tau)}{\partial \tau} = 0 \tag{9.A.7}$$

By use of equation (9.A.1), the equation (9.A.7) can be reduced to:

$$\frac{\partial U_S}{\partial \tau} + \frac{\partial U_V}{\partial \tau} = 0 \tag{9.A.8}$$

If equation (9.A.8) holds, then:

$$\frac{\partial U_S}{\partial \tau}\frac{\partial U_V}{\partial \tau} \le 0 \tag{9.A.9}$$

Suppose $\frac{\partial U_S}{\partial \tau} > 0$, then $\frac{\partial U_V}{\partial \tau} < 0$. If:

$$\frac{\partial U_S}{\partial \tau} = \frac{\frac{\partial u^\tau}{\partial \tau}(1-pQ_S^\tau) + u^\tau p\frac{a}{1+\tau}}{(1-pQ_S^\tau)^2} > 0 \quad \text{and} \quad \frac{\partial U_V}{\partial \tau} = \frac{\frac{\partial u^\tau}{\partial \tau}(1-pQ_V^\tau) + u^\tau p\frac{a-q}{1+\tau}}{(1-pQ_V^\tau)^2} < 0 \ ,$$

then, $\frac{pa}{(1+\tau)(1-pQ_S^\tau)} < \frac{\partial u^\tau}{\partial \tau}\frac{1+\tau}{u^\tau} < \frac{p(q-a)}{(1+\tau)(1-pQ_V^\tau)}$. However;

$$\frac{pa}{(1+\tau)(1-pQ_S^\tau)} = \frac{p(q-a)}{(1+\tau)(1-pQ_V^\tau)} + \frac{q(1-pa)}{(1+\tau)(1-pQ_S^\tau)(1-pQ_V^\tau)} \ge \frac{p(q-a)}{(1+\tau)(1-pQ_V^\tau)}.$$

This leads a contradiction. Hence, $\frac{\partial U_S}{\partial \tau} \le 0$ and $\frac{\partial U_V}{\partial \tau} \ge 0$. (Q.E.D)

Part III: Integrative Models

10 The Dynamics of Recovery: A Framework*

Stephanie E. Chang[1] and Scott B. Miles[2]

[1] School of Community and Regional Planning, and Institute for Resources, Environment, and Sustainability, University of British Columbia, Canada
e-mail: stephanie.chang@ubc.ca
[2] Department of Geography, University of Washington, USA
e-mail: smyles@u.washington.edu

10.1 Introduction

Each disaster reminds us that, from an economic standpoint, losses do not occur instantaneously, but are accumulated over the course of a sometimes long and complex recovery process. Moreover, disasters are spatial events that impact some places and some groups within those places more heavily than others. These effects can be observed in both natural disasters and human-induced events. Thus in the September 11th tragedy of 2001, the loss of the World Trade Center towers and the thousands of human lives do not in themselves constitute the economic impact of the disaster. Neither does the loss of gross regional product (GRP) on the 11th itself. Rather, the economic impact of the disaster is strongly influenced by the multitude of decisions made in the days and months following – decisions regarding whether to relocate an office to New Jersey and for how long, whether to lay off workers and how many, and whether to inject stimulus spending into the New York City economy and how much. Similar post-event decisions strongly influence recovery in natural disasters such as floods or earthquakes. In other words, the process of disaster recovery is critical to understanding the spatial economic impacts of disasters, yet the recovery process itself is extremely complex and uncertain.

Perhaps as a result, no comprehensive framework or model of disaster recovery currently exists in the literature. Many studies touch upon facets of recovery, but none take it as their focus. A comprehensive recovery model is needed in order to evaluate the potential consequences of decisions that affect disaster losses and recovery timepaths. This chapter provides an initial attempt at such a model (for a more detailed treatment, see Miles and Chang (2003)). Because disaster recovery is a very complex process, model development should be guided in part by empirical observations of actual disasters. Section 2 reviews the modeling literature on disaster economic losses, as well as the empirical literature on disaster recovery. Section 3 develops a conceptual model for community disaster recovery. The simulation is implemented in Section 4 for a simple, hypothetical city struck by an earthquake. Section 5 concludes with a discussion of further research needs.

* This work was supported by the Earthquake Engineering Research Centers Program of the National Science Foundation under Award Number ECC-9701471 to the Multidisciplinary Center for Earthquake Engineering Research.

10.2 Literature Review

A substantial literature has been emerging in recent years on modeling the economic impacts of natural disasters. Early studies proposed relatively simple applications of Input-Output methodologies (Cochrane, 1974; Kawashima and Kanoh, 1990), while recent research has developed increasingly sophisticated approaches including econometric models (West and Lenze, 1994), economic rebalancing models (Brookshire *et al.*, 1997), sequential Input-Output models (Okuyama *et al.*, 2000), computable general equilibrium models (Rose and Guha, 1999), and integrated infrastructure-economy models (Rose *et al.*, 1997; Gordon *et al.*, 1998; Cho *et al.*, 2001; Chang *et al.*, 2002). In all of these studies, impacts are primarily driven by damage to various economic sectors and inter-industry linkages. All of these studies focus on the urban or regional scale.

While the literature on loss modeling has been growing rapidly, modeling of recovery processes has been largely neglected. The significance of this distinction can be illustrated by the schematic diagram in figure 10.1. Loss models generally focus on the initial loss caused by a disaster, treating the recovery timepath in a summary fashion; in the extreme case, losses are assumed to be incurred over one year, after which the economy returns to normal. Yet the recovery timepath itself clearly makes a great difference in determining loss, as illustrated by comparing the different disaster cases in the figure. Losses (from damage) and gains (from reconstruction stimulus) are measured as the difference between the disaster case and the without-disaster baseline. In Case A, the economy suffers a substantial initial loss followed by a small gain, before returning to the baseline trend. By contrast, the economy in Case B eventually reaches a new equilibrium that is lower than the without-disaster baseline, indicating a much larger net loss. In disaster Case C, the reconstruction stimulus pushes the economy to a new trend that is above the baseline, so that in the long run, gains may completely offset losses.

The extent to which the recovery timepath can be influenced by decision variables will be of great interest to policy-makers. To date, models incorporating temporal processes have focused on the temporal distribution of reconstruction spending, production chronology factors (Okuyama *et al.*, 2000), reconstruction borrowing and debt repayments over time (Brookshire *et al.*, 1997), and prioritizing lifeline reconstruction to minimize economic disruption (Rose *et al.*, 1997; Cho *et al.*, 2001).

In contrast, the empirical literature on disaster recovery as a process – while lacking any comprehensive models – provides many useful insights. The classical work by Haas, Kates and Bowden (1977) provides a generalized framework for disaster recovery in which a community undergoes four post-disaster stages in regular, predictable sequence. Subsequent case studies have cast doubt on this idea of an orderly, inevitable progression of recovery stages (Hogg, 1980; Rubin and Popkin, 1990; Rubin, 1991; Berke *et al.*, 1993; Bolin, 1993). Instead, this more recent literature has been concerned with disparities and inequalities in recovery, and with conceptualizing disaster recovery as a social process involving decision-making, institutional capacity, and conflicts between interest groups. These themes resonate with the development of social vulnerability theory in disaster studies, which suggest that marginal groups may not only be especially vulnerable to suffering losses, but they are likely to have more difficulty in recovering (Hewitt, 1997; Blaikie *et al.*, 1994). They may, for example, have lesser access to insurance, loans, relief aid, or government bureaucracies and decision-making, or face shortages in low-income housing (*e.g.*, Bolin and Bolton, 1986; Bolin and Stanford, 1991; Hirayama, 2000).

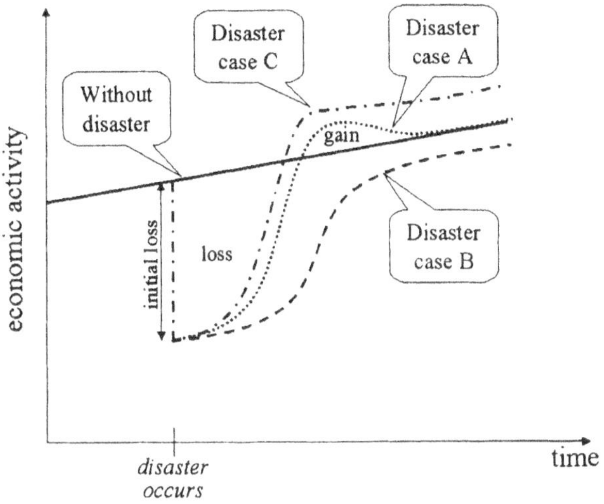

Figure 10.1 Schematic of Disaster Recovery

The importance of disparities has also been borne out by empirical studies of businesses in disasters. In various California earthquakes, researchers have found that small businesses and those that were generally marginal even before the disaster had the most difficulty in recovering (Durkin, 1984; Kroll *et al.*, 1991; Tierney and Dahlhamer, 1998; Alesch and Holly, 1998). One study of the 1994 Northridge earthquake identified four main factors that significantly influenced the survival and recovery of small businesses: entrepreneurial skill of the business owner, post-event demand for the business' products, pre-event business characteristics such as financial condition, and availability of resources for recovery (Alesch and Holly, 1998).

Further, spatial effects have been found to be important in disaster recovery. Decentralization of population and economic activity may be accelerated (Chang, 2001), business losses are correlated with disaster severity in the neighborhood (Tierney and Dahlhamer, 1998), and retail and other locally-oriented businesses generally lag in recovery (Alesch and Holly, 1998; Kroll *et al.*, 1991).

Developing a conceptual framework and model of disaster recovery should build on the major insights from this empirical literature. This means challenging some of the conventions of regional economic modeling: emphasizing dynamic processes over static equilibrium models, recognizing differences between individual agents within a sector, addressing spatial feedback effects, and acknowledging model limitations by allowing for randomness and uncertainty.

10.3 Conceptual Model

In order to translate findings from the empirical literature into a simulation of the urban disaster recovery process, a conceptual model is needed. The conceptual model of disaster recovery is derived here more by induction than by deduction – it is, in other words, guided by empirical observations rather than distilled from rigorous theory. It is especially interested in acknowledging how differences between business types and between household types help to determine recovery prospects. Thus, while the recovery simulation is intended to *measure* recovery at the community and neighborhood levels, it *models* recovery at the scale of the individual business and household. Moreover, the simulation models the influence of agents' environments on their recovery processes.

Figure 10.2 provides an overview of the conceptual model of earthquake disaster recovery. A more detailed description of the conceptual model and its development is presented in the following section. A key feature of the conceptual model is the emphasis on relationships between various units of analysis, as highlighted in the figure. The basic units of analysis are individual households and businesses ("agents"), lifelines such as the water delivery system, neighborhoods that contain agents and lifelines, and the community as a whole, which is comprised of individual neighborhoods. Recovery is assessed at the neighborhood level in terms of household and business recovery in that neighborhood. The changes that take place at each level are influenced to some extent by characteristics and changes in other units of analysis. For example, the housing damage suffered by an individual household is influenced by whether or not the community has adopted seismic provisions in its building code. An individual retail business's recovery is influenced by neighborhood-level household recovery because households purchase and consume the business's products.

The conceptual model reflects numerous simplifications, which may be addressed at later stages of development. For instance, it allows agents to exit the region – businesses can fail and households can leave – but it does not capture how they might relocate within the region following the disaster. Also, while it distinguishes between locally-oriented and export-oriented production sectors, a distinction shown to be important in the empirical literature, it does not disaggregate these into specific industries. In terms of community-level and policy variables, the framework emphasizes the role of lifeline infrastructure restoration.

10.3.1 Design Methodology

The methodology adopted for developing the conceptual model is based on the Object Modeling Technique (OMT) introduced in Rumbaugh *et al.* (1991). OMT is a methodology for conceptual modeling, originally used in software engineering. With OMT, the conceptual model is comprised of (1) the object (or static) model, (2) the dynamic model, and (3) the functional model, which together describe the real world system, and comprise an implementation-independent design.

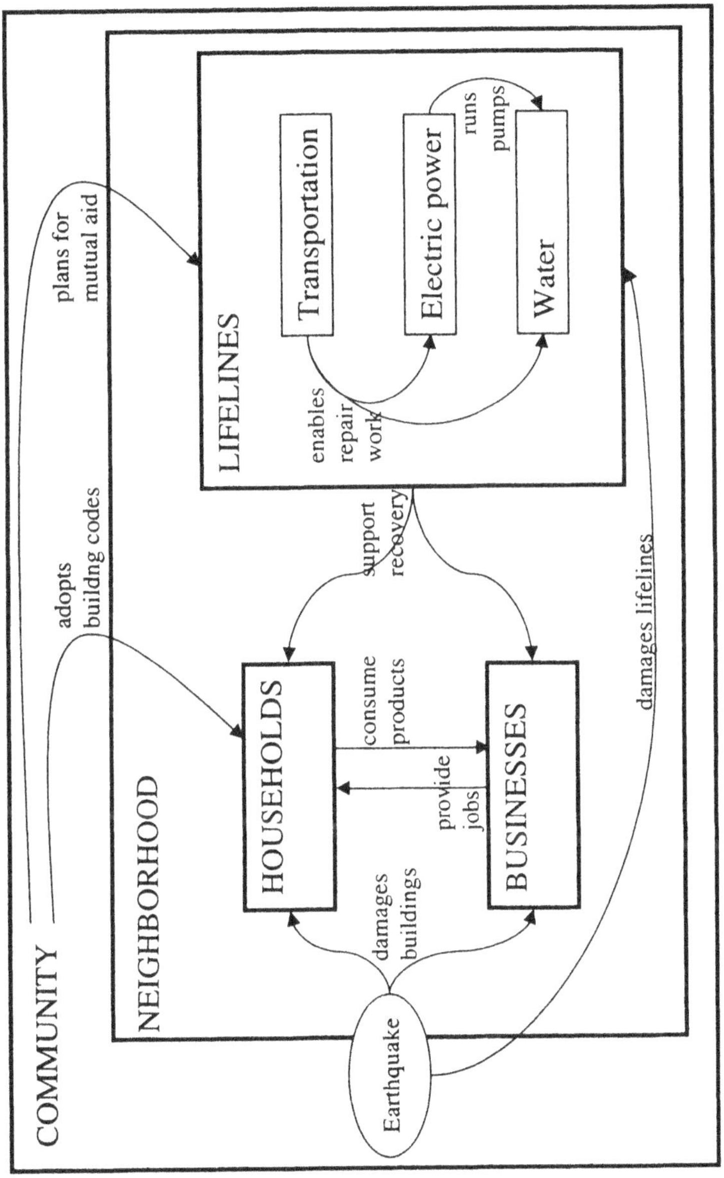

Figure 10.2 Conceptual Model Overview

There are several reasons why OMT is appealing as a means of designing the disaster recovery simulation model. Perhaps most obvious is the paucity of numerical data that can be used in developing a simulation of such high detail and broad scope. Similarly, as described above, there is a rich body of knowledge (including various theories and conceptual models, qualitative data, and quantitative data) on which to base a model. OMT provides an effective way of incorporating this available knowledge and is not hampered by the lack of data (or, more specifically, data describing some relationships and not others). Another significant reason for using OMT is the desire for an implementation-independent design. That is, it is important to have a sound conceptual framework that is founded in the disaster recovery literature that can serve as a guide for multiple approaches to computer simulation. A strong conceptual model will also serve other likely objectives, such as constructing a related database system or providing educational aids for students or decision makers. An object-oriented design may also facilitate an object-oriented implementation, which has benefits of modularity and encapsulation. With modularity, once the simulation objects, such as communities or neighborhoods, are designed and implemented, it is simple to create multiple objects and model the interactions (*e.g.*, multiple interacting communities). With encapsulation, the function of an object can be modified, say as new data or equations become available, with little or no revision to the overall simulation.

The steps of OMT are (1) problem definition, (2) object (static) modeling, (3) dynamic modeling, and (4) functional modeling. The goal of defining the problem is to identify all of the objects and relationships that exist within the system, which can be abstracted during subsequent stages of analysis. An object model captures the static structure of a system by showing the objects in the system, relationships between objects, and the attributes and operations that characterize each class of objects. Dynamic modeling is not needed for purely static systems (*i.e.*, a database) or computational systems, but rather for interactive software systems. (The dynamic model is not presented or discussed here because it is intended to assist later design of a spatial decision support system and has no effect on the simulation functionality.) The functional model describes the computations within a system in a general way.

The two most effective approaches to developing a detailed, unambiguous problem statement is to either write a requirements document of the model or compose a narrative of the real world system being modeled. The former approach is primarily suited to software development. The latter approach was adopted here based on the literature review, which was distilled into a short narrative describing the events and interactions during an earthquake disaster.

10.3.2 Static Model

An object model captures the static structure of a system by showing the objects in the system, relationships between objects, and the attributes and behaviors that characterize each class of objects. An object can be anything that makes sense to the particular application: typically a concept, abstraction, or physical thing with well-defined boundaries. Objects should be chosen to promote understanding of the modeled system (*i.e.*, disaster recovery) and provide a sound basis for computer implementation.

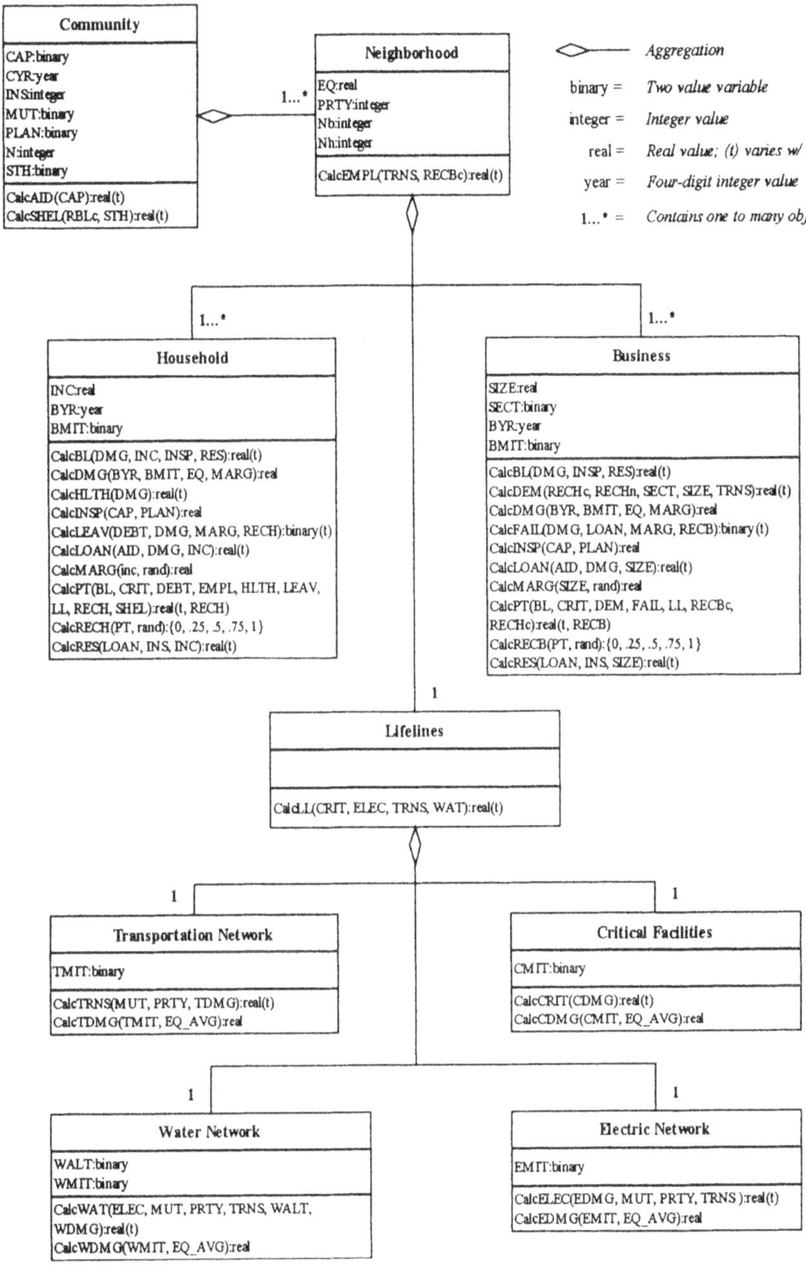

Figure 10.3 Main Objects in Conceptual Model (The three parts of each box respectively indicate the object's name, attributes, and behaviors or functions)

The initial step in creating the object model was identifying important objects from the problem narrative. Once potential objects were listed, some were discarded if they were beyond the current scope or better represented as attributes of or relationships between other objects. The potential objects were analyzed to determine what, if any, associations exist among the objects (for example, a community contains one or more neighborhoods, which contain one or more households and businesses). With a short list of potential objects and their associations, the problem narrative was used to help determine important attributes of each object (*e.g.*, the size and market sector of a business). Additional attributes of objects were obtained by considering likely decision variables, which may not necessarily be associated with any particular object. One design choice involved representing associated physical and economic objects (*e.g.*, electric network and electric company, respectively) as a single economic object with attributes and functions that represent the important aspects of the associated physical object.

The static aspects of the conceptual model of disaster recovery are represented by the diagram in figure 10.3. The diagram describes the important object types of the conceptual model and lists the attributes and functions of each type of object. So, for example, an object of type "household" has attributes of income (*INC*), year building of residence was built (*BYR*), and whether any building mitigation has been done (*BMIT*). Within an implementation of the conceptual model there may be any number of households having the same data structure, but with different values for the respective attributes (and thus different output for the respective functions). The functional dependencies listed in figure 10.3 were determined in the functional modeling step described below. The diagram shown is a simplified version of the object model developed and eventually implemented. Some secondary objects and associations are not shown. For example, the diagram does not show the inheritance association between households and businesses, which are both economic agents and have many similar attributes and functions.

10.3.3 Functional Model

The functional model shows how output values are derived from input values, without regard to the order in which these values are computed. The functional model specifies the meaning of the functions in the static model. Where the static model is represented using an object diagram, a function model is typically represented using a data flow diagram. The data flows represent object attributes that are passed between the functions of the different objects.

To develop the functional model, the most important top-level inputs (*e.g.*, decision variables) and outputs (*e.g.*, recovery indicators) were identified. The overall inputs that were not already specified as attributes were assigned to an appropriate object in the object model. As in the static modeling step, an effort was made to reduce the number of top-level inputs and outputs. This was done primarily by eliminating similar variables or creating proxy variables in place of related variables. Intermediate outputs that are necessary to map the top-level inputs to the top-level outputs were then considered. This was greatly assisted by referring to the potential attribute list of each object and finding those that vary with time (*e.g.*, a household's health after a disaster). These intermediate outputs were then assigned as functions to the appropriate objects (*e.g.*, *CalcHLTH* to household objects). Functional dependencies (*i.e.*, what inputs are needed to derive a particular output) were arrived at with reference to the literature, the problem narrative, and common sense.

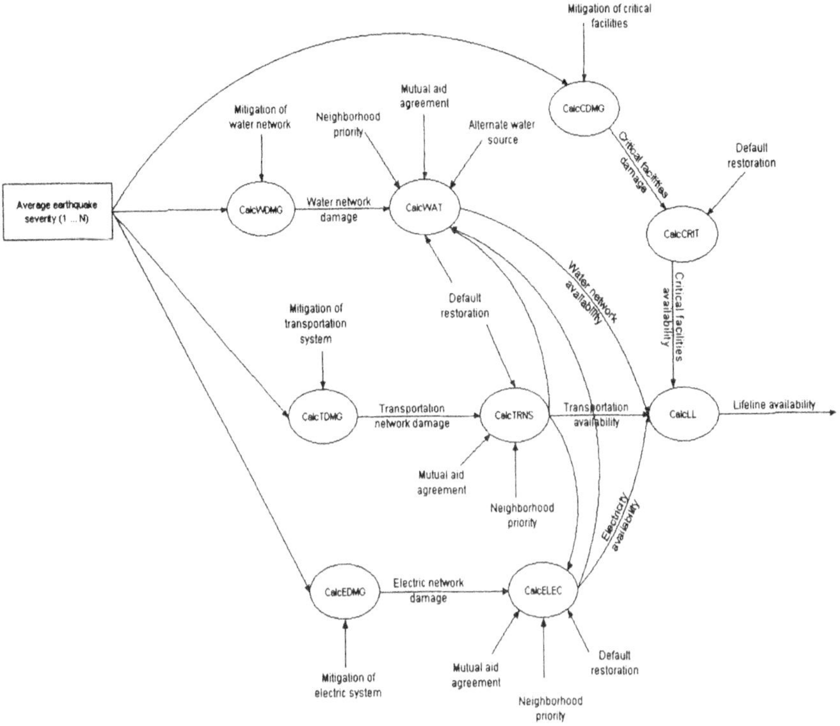

Figure 10.4 Flow Diagram for Lifeline Availability Showing Relationships of Function Inputs and Outputs.

The functional dependencies within the disaster recovery conceptual model are illustrated in the flow diagrams of figures 10.4 and 10.5. For demonstration purposes, these figures show only portions of the overall flow diagram. Figure 10.4 illustrates the data flow required in assessing lifeline availability within a given neighborhood. The flow diagram clearly conveys the interconnectedness of the components that make up a community's lifeline network. Figure 10.5 describes the data flow requirements for determining the recovery of an individual business within a given neighborhood. The blocks related to lifeline mitigation are in bold to help illustrate the potential effects of lifeline availability on business recovery. (This is referred to later.) The flow diagram for household recovery is similar to the one for business recovery, with respective differences in variables. Notice that the flow diagrams (and, thus, the conceptual model) do not specify the numerical equations for each function (ellipses). This modularity is important so that it can be carried over to the implementation of the simulation model. In this way, existing equations or algorithms can be used or experimented with without affecting the overall structure or function of the simulation.

The functional model describes five principal types of recovery influences and processes that are useful in organizing and explaining the relationships expressed by the numerical framework. The five types are: (1) dynamic processes; (2) agent-attribute influences; (3) interaction effects; (4) spatial feedbacks; and (5) policy effects. Dynamic processes refer here to changes over time. In true dynamic processes, a variable's current level depends upon its level in a previous period. What can be called pseudo-dynamic processes – changes over time that can proceed independently of variable levels in previous periods – also play an important role.

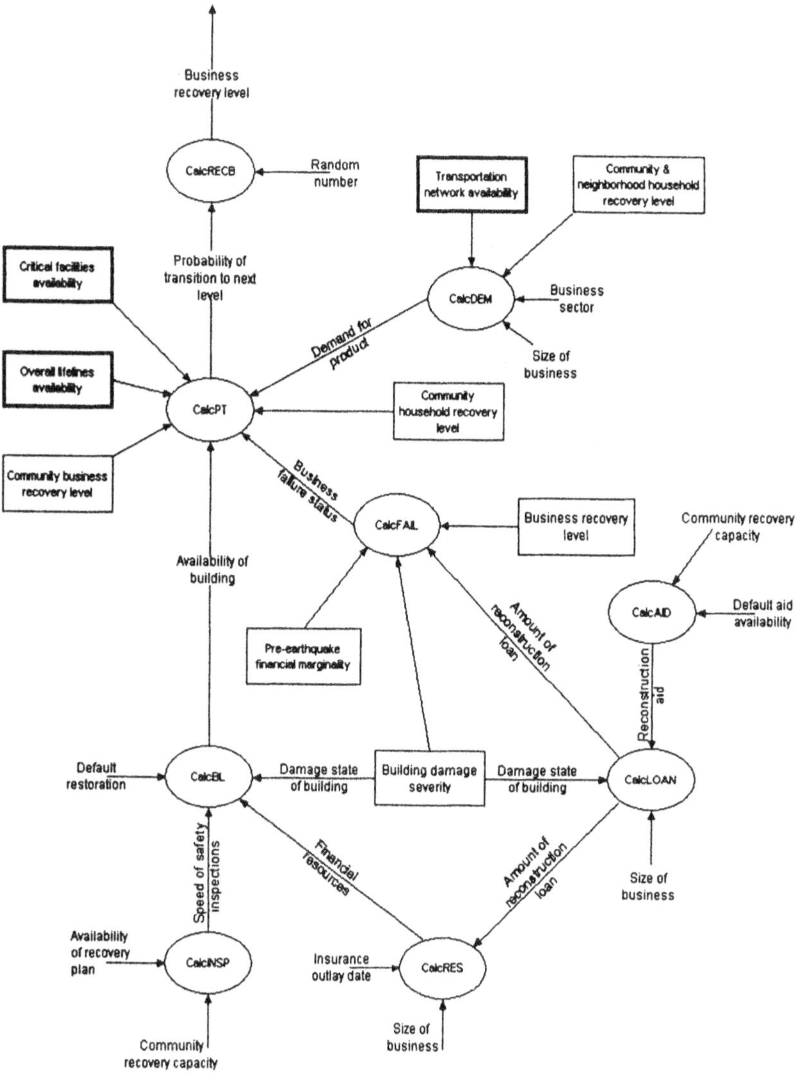

Figure 10.5 Flow Diagram of Lifeline Mitigation Influences on Business Recovery (Bold boxes indicate lifeline mitigations. Rectangular boxes indicate relationships not described within this diagram.)

In addition to dynamic processes, a second main type of recovery influence consists of agent-attribute effects. Attributes of a business or a household influence its respective recovery trajectory. A good example is the influence of a business's attributes on the post-earthquake demand for its products (figure 10.5). Product demand depends upon whether the particular business is in a locally-oriented or export-oriented sector and whether it is a large or small business. In particular, if locally-oriented, then the recovery of households in the neighborhood and community matters because these are its customers. Similarly, local transportation conditions will influence locally-oriented business's product demand. However, if a business is export-oriented, these local variables do not play a role and the demand for its product remains unchanged by the disaster.

A third type of recovery influence consists of interaction effects. For example, in figure 10.4, water availability is influenced by the survival of the electric power and transportation systems. Electric power may be needed to drive pumps that enable the water system to function; transportation disruption can impede the ability of the water utility to make repairs in a timely manner. Similarly, the relationships driving business product demand described in figure 10.5 demonstrate some of the ways in which households, businesses, neighborhoods, and the community as a whole interact. Households influence business recovery through consumption demand. The availability of lifelines and critical facilities influence business recovery, as does the overall recovery level of households and businesses in the economy.

The fourth type of influence, spatial feedbacks, can similarly be seen in the examples presented so far. Households and businesses do not exist aspatially, but are affected by conditions in their specific neighborhoods, whether in terms of water availability, transportation conditions, or local consumer demand. Thus, the same type of household or business may recover differently depending upon which neighborhood it is located in.

The final type of influence consists of policy or decision effects. These are community-level decisions made either before the event, such as emergency planning and mitigation measures, or afterwards, such as recovery policy decisions. Figure 10.4 illustrates the influence of decisions regarding mutual aid and neighborhood prioritization on lifeline availability. Others decisions that are modeled include the year that the community put into effect a seismic design code for its buildings (if it did); emergency planning for alternative water supplies such as water trucks; whether or not seismic mitigations had been conducted for lifeline systems; the availability of a restoration and recovery plan; reliance on short-term housing in this plan; and a broadly defined variable indicating the community's capacity for recovery (a proxy for the community's degree of integration and consensus). Modeling the influence of these decisions is critical in that implementation of the conceptual model will enable "what if" explorations of the recovery consequences for different policy decisions.

Further information on the model can be found in the appendices. Appendix 10.A.1 lists all the variables in the model. Appendix 10.A.2 outlines the model structure and functional relationships. In a series of generalized equations, it shows relationships between variables at various levels of analysis. It also identifies policy and other exogenous variables that can be considered as the "drivers" of the model. Details on the mathematical equations that make up the simulation can be found in Miles and Chang (2003).

10.4 Prototype Application

10.4.1 Quantifying the Conceptual Model

Prior to actual implementation, the dependencies specified in the functional modeling step were quantified. The need for quantification in the absence of data or (quantitative) theory from which to develop a mathematical relationship is often an insurmountable obstacle for complex systems with many objects and relationships. The objective of the work at this point is to demonstrate that such a model is both feasible and useful. For this reason, we employed a simple numerical scheme to facilitate implementing the many relationships of the conceptual model.

The functional model is quantified using a relative index framework in which every variable varies between 0 and 1, rather than real world metrics, such as dollars. In many cases, the variables do not have a common metric; for example, financial marginality and health. The approach taken is useful for integrating many metrics that would otherwise be difficult to mathematically combine. With each variable varying between 0 and 1, it was relatively simple to create basic first-order algebraic relationships based on the functional model. This numerical framework clearly simplifies the nature and relationships of disaster recovery; for example, in actual disaster situations, lifeline restoration curves typically follow a logistic-type rather than linear functional form. Note, however, that while certain equations are linear in form, the overall model is highly nonlinear in behavior. After this prototype stage, further refinements can be made to better address the complexity and non-linearity of disaster recovery.

The complete set of model equations are not listed here due to space constraints. The general information is already described in figures 10.3 through 10.5 and Appendices 10.A.1 and 10.A.2. However, the use of recovery levels in the conceptual model is not well represented in the associated diagrams. It is worthwhile, then, to list the equations describing the likelihood (*PT*), at each simulation time step, that individual businesses and household move on to each of the four levels of recovery. (These recovery stages or levels are represented within the 0~1 numerical scheme as 0, 0.25, 0.5, 0.75, and 1). The equations will also give some idea of how the conceptual relationships have been quantified, though the other equations are simpler (*i.e.*, linear and/or non-conditional) because they vary less with the state of the simulation.

The general structure of the functions for calculating *PT* are similar for both businesses and households, the difference being the particular variables in the function and what particular variable is assumed relevant for each recovery level. Overall, the expected recovery of any particular business is a function of building availability (*BL*), critical facilities availability (*CRIT*), overall lifeline availability (*LL*), demand for product (*DEM*), whether or not the business has failed (*FAIL*), community-wide business recovery (*RECBc*), and community-wide household recovery (*RECHc*) (see figure 10.3 "Business" and figure 10.5). Equation (10.1) describes in more detail the relationship of each input variable to the likelihood of reaching each recovery level.

$$PT(t) = \begin{cases} 0 & , \; if \quad FAIL = 1 \quad and \quad REC_b > 0 \\ 0.5 * (CRIT(t) + LL(t)) & , \; if \quad REC_b = 0 \\ 0.333 * \sqrt{DEM(t)}(\sqrt{BL(t)} + \sqrt{LL(t)} + \sqrt{CRIT(t)}) & , \; if \quad REC_b = 0.25 \\ 0.2 * \sqrt{DEM(t)}(BL(t) + LL(t) + CRIT(t) + & , \; if \quad REC_b = 0.5 \\ \quad + RECB_c(t) + RECH_c(t)) & \\ 0.25 * DEM(t)(0.45BL(t)^2 + LL(t)^2 + & , \; if \quad REC_b = 0.75 \\ \quad + RECB_c(t)^2 + RECH_c(t)^2) & \\ 1 & , \; if \quad REC_b = 1.0 \end{cases} \tag{10.1}$$

The business recovery level (REC_b) is computed by adding 0.25 each time that *PT* exceeds the actualized value of some uniformly distributed random number until the recovery level reaches a maximum value of one or the simulation times out. This is shown in the following equation:

$$REC_b(t) = \begin{cases} 0.25 * \left((PT(t) \geq rand) + REC_b(t-1)\right) & , \; if \quad REC_b < 1 \\ 1 & , \; otherwise \end{cases} \tag{10.2}$$

For a household, the overall recovery is a function of its health (*HLTH*), building availability (*BL*), availability of temporary shelter (*SHEL*), household indebtedness (*DEBT*), critical facilities availability, (*CRIT*), overall lifelines availability (*LL*), community employment (*EMPL*), and whether or not the household leaves the area (*LEAV*) (see figure 10.3 "Household"). The role of the input variables within each recovery level is described in the following equation. The household recovery level (*RECH*) is computed in the same manner as for businesses.

$$PT(t) = \begin{cases} 0 & , \; if \quad LEAV = 1 \quad and \quad REC_h > 0 \\ 0.333 * (CRIT(t) + HLTH(t) + SHEL(t)) & , \; if \quad REC_h = 0 \\ 0.2\sqrt{EMPL(t)}(\sqrt{BL(t)} + LL(t) + CRIT(t) + & , \; if \quad REC_h = 0.25 \\ \quad + HLTH(t) + SHEL(t)) & \\ 0.25EMPL(t)(BL(t) + LL(t) + HLTH(t) + (1 - DEBT(t))) & , \; if \quad REC_h = 0.5 \\ 0.333EMPL^2(t)(BL^2(t) + LL^2(t) + (1 - DEBT(t))^2 & , \; if \quad REC_h = 0.75 \\ 1 & , \; if \quad REC_h = 1.0 \end{cases} \tag{10.3}$$

10.4.2 Prototype Implementation using Simulink

A prototype application was conducted in the Simulink modeling environment for Mathwork's MATLAB. Simulink is specifically designed for implementing complex, time-based simulations using a graphical language consisting of operators (Simulink blocks) and data flows, which connect the blocks. This framework makes implementation relatively simple after constructing the data flow diagrams of the function model. Simulink makes simulation implementation relatively simple (as opposed to coding the simulation using a programming language). Simulink affords significant advantages in terms of model building and execution, and robust capabilities for building graphical user interface elements. The greatest advantage of Simulink is probably the modularity it provides, which is very compatible with an object-oriented design. Simulink allows models to be built in both a hierarchical and encapsulated manner, which greatly increases the organization and understandability of the simulation components. Once simulation components, such as a household, are designed and implemented with the specific attributes and operations, they can be saved in a library and duplicated to create large simulations. The simulation can be executed within the MATLAB command-line environment or within MATLAB programming scripts, which facilitates stochastic simulations or detailed sensitivity analyses. The disadvantages of Simulink include the tediousness of assembling a large number of simulation components, such as households and businesses, and, to a lesser extent, the initially cumbersome way in which flow control (*e.g.*, IF-THEN) is implemented.

A fixed-step, discrete solver is used to run the simulation. The simulation time interval was chosen as a half-week to allow sampling (reporting) of the simulation at weekly intervals or coarser. A one-week resolution is typical of lifeline restoration timeframes in earthquake disasters; for example, in the 1995 Kobe (Japan) earthquake, electric power was fully restored within one week and water within 10 weeks. Investigating the effect of the selected time-step on model behavior remains a topic for further research.

10.4.3 Initial Simulation Scenarios

The prototype simulation was run to evaluate the conceptual framework and model. A simulated community was designed and parameterized to reflect a high level of community planning and preparedness. As such, the community has a disaster response plan ($PLAN = 1$), mutual aid agreements in place ($MUT = 1$), and a reliance on short term housing ($STH = 1$). The community is made up of three neighborhoods. The neighborhoods are located similarly with respect to the seismic hazard (*e.g.*, fault line or rupture area), and thus experience the same earthquake shaking intensity. In this case, the earthquake severity is set to a maximum value for each neighborhood ($EQ = 1$). The neighborhoods were also assigned equal priority for emergency response and post-disaster restoration ($PRTY = 1$). In this prototype, each neighborhood contains 12 households and 12 businesses. The demographics of the three neighborhoods are given in figure 10.6. The demographics were chosen to generally represent "old core", "new core" (downtown), and suburb neighborhoods. These are distinguished by the age mix of buildings, income distribution of households, and sectoral and size distribution of businesses.

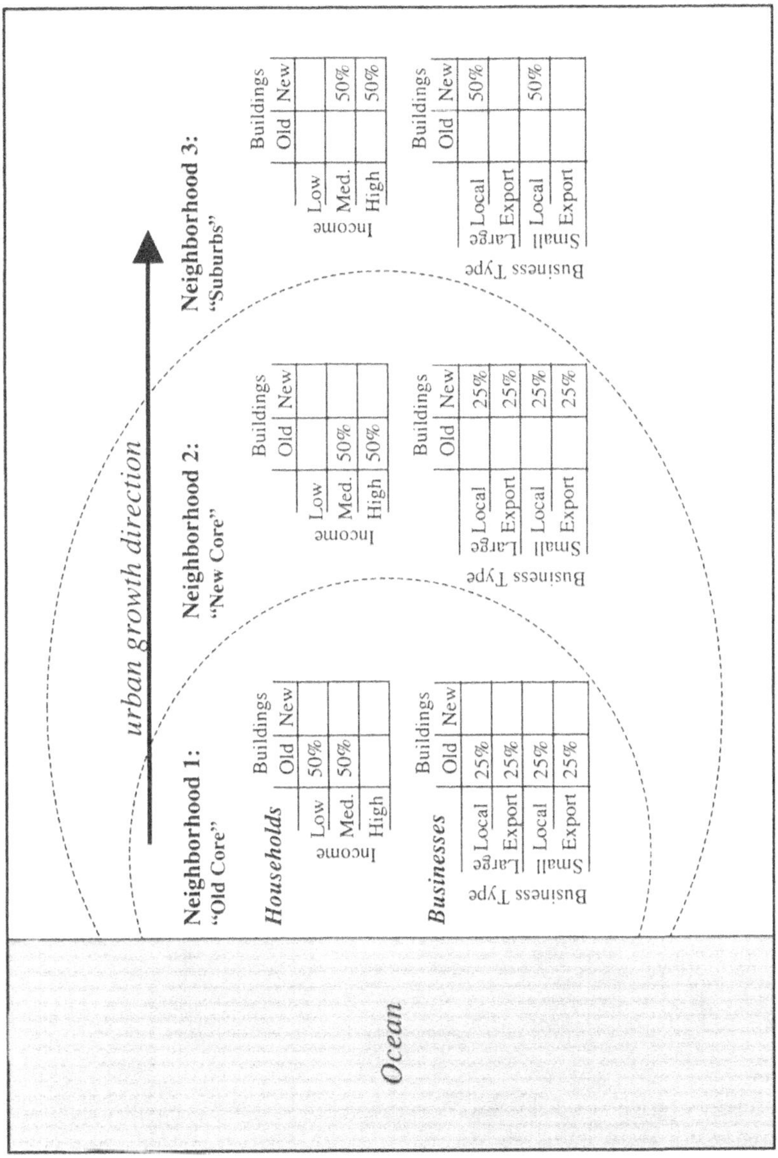

Figure 10.6 Household and Business Characteristics of Neighborhoods in Simulated City

The model parameters described above were held constant to demonstrate the effects of changing the values of lifeline mitigation variables. The simulation was run for two simple scenarios: (1) mitigation measures taken for all lifelines (*TMIT*=1, *EMIT*=1, *CMIT*=1, *WMIT*=1, *WALT*=1 for all neighborhoods) and (2) no mitigation measures taken for any lifelines (same variables set to zero for all neighborhoods). (Figure 10.5, which describes the functional relationships for computing business recovery, demonstrates the effects of changing the lifeline mitigation variables.) For these scenarios, all randomness was taken out to better understand the sensitivity of changing the lifeline variables.

10.4.4 Initial Simulation Results

The simulation described in the previous subsection was parameterized and run using a programming script. The simulation takes 1 to 2 minutes, depending on available memory, to run on an 850Mhz processor. Thus, it is likely that much larger simulations will have reasonable run times, especially if the storage input-output is minimized in the simulation code. The outputs of the two simulation scenarios described above are quite reasonable with respect to the influence of input variable values. Very little time was spent calibrating the default (restoration) values that drive the simulation, so the absolute performance of the model can be evaluated and improved in the future. Nevertheless, this initial application demonstrates advantages of the conceptual model and representing recovery at the business-household level.

The output of the simulated scenarios is summarized by the average business and household recovery level for each neighborhood, together with a listing of how many businesses failed or households left in each neighborhood. Figures 10.7 and 10.8 describe the recovery of businesses and households, respectively, with no lifeline mitigation. Table 10.1 lists how many businesses failed and households left in each neighborhood.

Table 10.1 Results of Simulation, No Lifeline Mitigation Scenario

Neighborhood / sector	Result
Neighborhood 1	
Businesses:	6 failed
Households:	6 left
Neighborhood 2	
Businesses:	0 failed
Households:	0 left
Neighborhood 3	
Businesses:	0 failed
Households:	0 left

It is easy to see how the different demographics of the three neighborhoods affect the business recovery levels (figure 10.7). First, there is a slight lag between Neighborhoods 2 and 3. This reflects the fact that Neighborhood 3 has only businesses with local markets, while Neighborhood 2 contains businesses for a mix of sectors. In Neighborhoods 2 and 3, all of the businesses occupy new or retrofitted buildings. As a result, Neighborhood 2 and 3 reached a higher average business recovery level than Neighborhood 1, which has old or unretrofitted buildings that are more likely to be damaged. For similar reasons, no businesses failed in Neighborhood 2 and 3, whereas the six small businesses failed in Neighborhood 1 (table 10.1). Business failures affect the average recovery level and prevent the level from reaching the maximum value of one in neighborhood 1. In the case of Neighborhoods 2 and 3, the reason that the average recovery levels do not reach one is not entirely clear. It may be because that Neighborhoods 2 and 3 contains small, local businesses that, because of the model algorithm, do not completely recover.

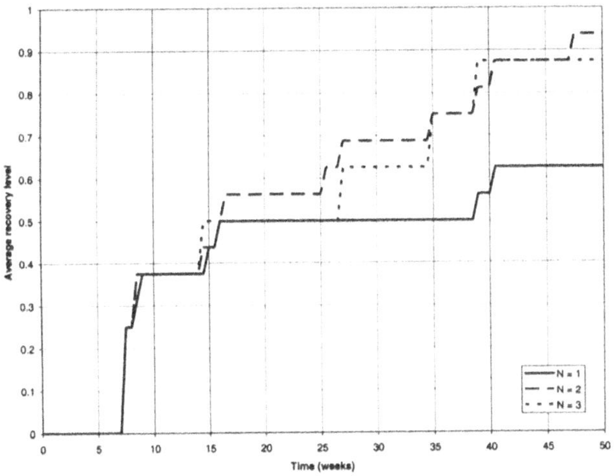

Figure 10.7 Average Business Recovery by Neighborhood: "No Mitigation" Case

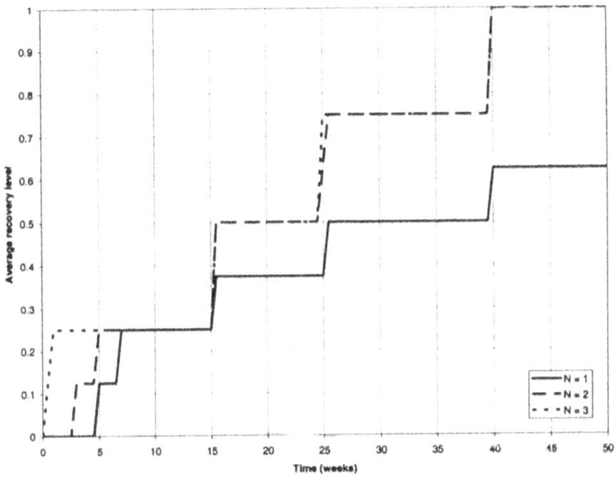

Figure 10.8 Average Household Recovery by Neighborhood: "No Mitigation" Case

The household recovery output for the scenario of "no lifeline mitigation" is similar to that of the business output. The slight lag, early on, between Neighborhood 2 and 3 is due to the difference in residential building characteristics. (The small lag might suggest that the model may not be sensitive enough to building attribute in some contexts.) In contrast, Neighborhood 1, which contains all old or unretrofitted houses, did not completely recover and the six low-income households left (table 10.1). It is interesting to note that all three neighborhoods took several weeks to reach the first level of recovery. Likelihood of reaching the first level of recovery (*PT*) is a function of critical facilities recovery (*CRIT*), household health (*HLTH*), and whether or not shelter is available (or needed) (*SHEL*).

Looking at the contrasting mitigation scenario, the effect of lifeline mitigation on the model output is readily apparent. Figures 10.9 and 10.10 describe the recovery of businesses and households, respectively, in the case with lifeline mitigation. For both the businesses and households, recovery time is significantly shorter. One of the most significant aspects of this is how much more quickly all businesses and households reached the first level of recovery (0.25). In addition, no businesses failed, nor did any households leave the area. For businesses recovery, the difference between the three neighborhoods is less obvious because of the overall speed of recovery. The effect of building type and mitigation can be seen in the slight lag in recovery of Neighborhood 1, which has only old or unretrofitted buildings for its businesses, behind Neighborhood 2. The behavior of Neighborhood 3 is more difficult to interpret. The only indication that the neighborhood has only local business is the fact it does not reach the same maximum recovery level as the other neighborhoods. In the case of household recovery, the difference between neighborhoods is easily distinguished. This reflects the fact that each neighborhood is increasingly, from Neighborhood 1 to 3, less vulnerable. Interestingly, for this scenario, the time to the first recovery level is similar for each neighborhood, but quite different for the remaining three levels (0.5, 0.75, 1).

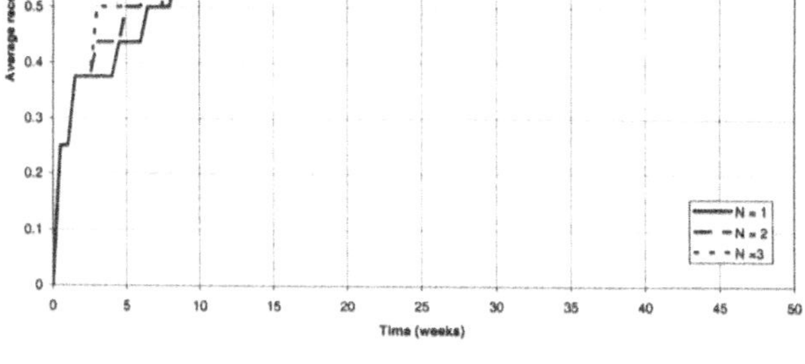

Figure 10.9 Average Business Recovery by Neighborhood: Lifeline Mitigation Case

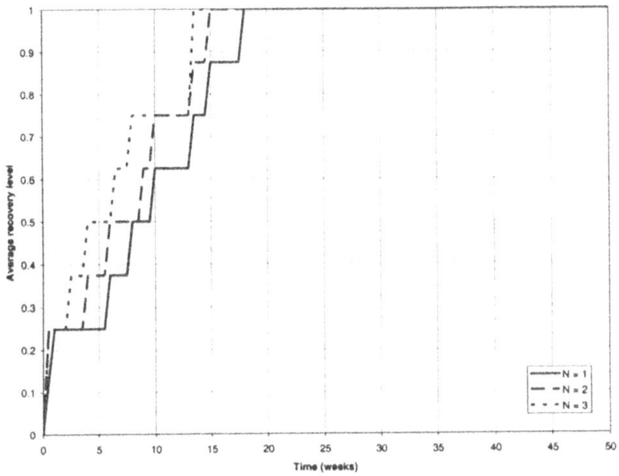

Figure 10.10 Average Household Recovery by Neighborhood: Lifeline Mitigation Case

The general result that lifeline failures play a critical role in community recovery is consistent with observations in actual disasters. The prototype application illustrates how investing in pre-disaster mitigation of lifelines can help reduce losses to the community and hasten recovery, as well as diminish the socio-economic and spatial disparities in disaster impacts. Clearly, a range of preparedness, mitigation, and planning scenarios could be investigated using the model. This could provide a very useful, multi-faceted understanding of the implications and benefits of such decisions.

10.5 Conclusions

This study has developed and implemented a rudimentary model of the recovery process for a disaster-stricken community. It has emphasized how spatial disparities in disaster recovery may arise from differences in the recovery timepaths of individual agents within the community. Capturing interactions between sectors – lifeline infrastructure systems, households, businesses, and community planning – in even a rudimentary fashion reveals enormous complexities in disaster recovery. These interactions may magnify and complicate the impacts of public sector decisions regarding mitigation investment, response, and recovery.

The framework and model can be improved in many ways to approach a more realistic and implementable decision-support system. Conceptually, many refinements are warranted; for example, agents should be allowed to relocate within the region. Modeling detail should be added with respect to the number of agents and neighborhoods, earthquake effects and damage (*e.g.*, distinguishing between ground shaking and fire damage), representing certain objects such as lifelines (*e.g.*, modeling individual substations rather than the electric power system as a single object) and disaster plans, and characterizing the attributes of agents. In terms of implementation, it will be important to integrate the model with GIS and spatial analysis and to develop a graphical user interface to facilitate exploring alternative plans and decisions. Model outputs should be evaluated in the context of sensitivity analysis and parameterized case studies.

Perhaps most critical of all, however, is the need to verify and quantify key relationships in the conceptual model with empirical data. While case studies of recovery processes in actual disasters have provided valuable guidance in designing the recovery model, conversely, the model's structure and outcomes can suggest priorities for data collection and analysis in future disasters. For example, greater attention should be paid to simply measuring recovery at the individual, neighborhood, and community levels. Another priority would be to systematically identify the key decisions made by public authorities with regard to disaster preparedness, response, and recovery planning. Data should be collected more systematically across disaster events to allow comparative study. Clearly, developing these understandings will require substantial collaboration across the many social sciences – between economists and sociologists, urban planners and decision scientists, geographers and policy analysts.

References

Alesch, D.J. and J.N. Holly. 1998. "Small Business Failure, Survival, and Recovery: Lessons from the January 1994 Northridge Earthquake," *NEHRP Conference and Workshop on Research on the Northridge, California Earthquake of January 17, 1994*. Richmond, CA; CUREe.

Berke, P.R., J. Kartez, and D. Wenger. 1993. "Recovery after Disaster: Achieving Sustainable Development, Mitigation and Equity," *Disasters*, 17 (2): 93-109.

Blaikie, P., T. Cannon, I. Davis, and B. Wisner. 1994. *At Risk: Natural Hazards, People's Vulnerability, and Disasters*. New York, NY; Routledge.

Bolin, R. 1993. *Household and Community Recovery After Earthquakes*. Boulder, CO; Institute of Behavioral Science, University of Colorado.

Bolin, R. and P. Bolton. 1986. *Race, Religion, and Ethnicity in Disaster Recovery*. Boulder, CO; Institute of Behavioral Science, University of Colorado.

Bolin, R. and L. Stanford. 1991. "Shelter, Housing and Recovery: A Comparison of U.S. Disasters," *Disasters*, 15 (1): 24-34.

Brookshire, D.S., S.E. Chang, H. Cochrane, R.A. Olson, A. Rose, and J. Steenson. 1997. "Direct and Indirect Economic Losses from Earthquake Damage," *Earthquake Spectra*, 13 (4): 683-701.

Chang, S.E. 2001. "Structural Change in Urban Economies: Recovery and Long-Term Impacts in the 1995 Kobe Earthquake," *The Kokumin Keizai Zasshi (Journal of Economics & Business Administration)*, 183 (1): 47-66.

Chang, S.E., W.D. Svekla, and M. Shinozuka. 2002. "Linking Infrastructure and Urban Economy: Simulation of Water Disruption Impacts in Earthquakes," *Environment and Planning B*, 29 (2): 281-301.

Cho, S., P. Gordon, J.E. Moore II, H.W. Richardson, M. Shinozuka, and S.E. Chang. 2001. "Integrating Transportation Network and Regional Economic Models to Estimate the Costs of a Large Urban Earthquake," *Journal of Regional Science*, 41 (1): 39-65.

Cochrane, H.C. 1974. "Predicting the Economic Impact of Earthquakes," Boulder, CO; Institute of Behavioral Science, University of Colorado, Working Paper No.15.

Durkin, M.E. 1984. "The Economic Recovery of Small Businesses after Earthquakes: The Coalinga Experience," paper presented at the International Conference on Natural Hazards Mitigation Research and Practice, New Delhi, India.

Gordon, P., H.W. Richardson, and B. Davis. 1998. "Transport-Related Impacts of the Northridge Earthquake," *Journal of Transportation and Statistics*, 1 (2): 21-36.

Haas, J.E., R.W. Kates, and M. Bowden eds. 1977. *Reconstruction Following Disaster*. Cambridge, MA; MIT Press.

Hewitt, K. 1997. *Regions of Risk: A Geographical Introduction to Disasters*. Essex, UK; Addison Wesley Longman Limited.

Hirayama, Y. 2000. "Collapse and Reconstruction: Housing Recovery Policy in Kobe after the Hanshin Great Earthquake," *Housing Studies*, 15 (1): 111-128.

Hogg, S. J. 1980. "Reconstruction Following Seismic Disaster in Venzone, Friuli," *Disasters*, 4 (2): 173-185.

Kawashima, K. and T. Kanoh. 1990. "Evaluation of Indirect Economic Effects Caused by the 1983 Nihonkai-Chubu, Japan Earthquake," *Earthquake Spectra*, 6 (4): 739-55.

Kroll, C.A., J.D. Landis, Q. Shen, and S. Stryker. 1991. "Economic Impacts of the Loma Prieta Earthquake: A Focus on Small Businesses," Berkeley, CA; University of California at Berkeley, U.C. Transportation Center and the Center for Real Estate and Urban Economics, working paper #91-187.

Miles, S.B. and S.E. Chang. 2003. *Urban Disaster Recovery: A Framework and Simulation Mode,. Technical Report MCEER-03-0005*. Buffalo, NY; Multidisciplinary Center for Earthquake Engineering Research, University at Buffalo, the State University of New York.

Okuyama, Y., G.J.D. Hewings, and M. Sonis. 2000. "Sequential Interindustry Model (SIM) and Impact Analysis: Application for Measuring Economic Impact of Unscheduled Events," paper presented at the 47th North American Meetings of the Regional Science Association International, Chicago, IL.

Rose, A., J. Benavides, S.E. Chang, P. Szczesniak, and D. Lim. 1997. "The Regional Economic Impact of an Earthquake: Direct and Indirect Effects of Electricity Lifeline Disruptions," *Journal of Regional Science*, 37 (3): 437-458.

Rose, A. and G. Guha. 1999. "Computable General Equilibrium Modeling of Electric Utility Lifeline Losses from Earthquakes," paper presented at the 46th North American Meetings of the Regional Science Association International, Montreal, Canada.

Rubin, C. 1991. "Recovery from Disaster," in T.E. Drabek and G.J. Hoetmer eds. *Emergency Management: Principles and Practice for Local Government*. Washington, DC; ICMA: 224-259.

Rubin, C.B. and R. Popkin. 1990. "Disaster Recovery after Hurricane Hugo in South Carolina," Boulder, CO; Institute of Behavioral Science, University of Colorado.

Rumbaugh, J., M. Blaha, W. Premerlani, F. Eddy, and W. Lorenson. 1991. *Object-Oriented Modeling and Design*. Upper Saddle River, NJ; Prentice-Hall.

Tierney, K.J. and J. M. Dahlhamer. 1998. "Business Disruption, Preparedness and Recovery: Lessons from the Northridge Earthquake," *NEHRP Conference and Workshop on Research on the Northridge, California Earthquake of January 17, 1994*. Richmond, CA; CUREe.

West, C.T. and D.G. Lenze. 1994. "Modeling the Regional Impact of Natural Disaster and Recovery: A General Framework and an Application to Hurricane Andrew," *International Regional Science Review*, 17 (2): 121-150.

Appendix

10.A.1 Definition of Variables

AID	= availability of reconstruction aid in community
BL	= availability of building for use
BL_h	= availability of housing
BMIT	= pre-earthquake structural mitigation
BYR	= year building built
CAP_c	= recovery capacity of community (proxy for integration, consensus)
CDMG	= extent of damage to critical facilities
CMIT	= pre-earthquake mitigation to critical facilities
CODE	= compliance of building with seismic code (Y/N)
CRIT	= availability of critical facilities
CYR	= year seismic code effective
DAID	= default aid availability status
DBL	= default building repair status
DCR	= default critical facility availability (restoration curve)
DEBT	= extent of indebtedness
DEL	= default electricity availability (restoration curve)
DEM	= demand for product post-earthquake
DHSE	= default availability of own housing (restoration curve)
DINJ	= default health restoration curve
DINS	= default availability of insurance
DMG	= damage state of building (extensive, moderate, slight, none)
DTRN	= default transport accessibility (restoration curve)
DWAT	= default water availability (restoration curve)
EDMG	= extent of damage to electricity network
ELEC	= availability of electricity
EMIT	= pre-earthquake mitigation to electric power system
EMPL	= availability of employment/income
EQ	= severity of earthquake's physical effects (gr.shaking, gr.failure)
FAIL	= occurrence of business failure
HSE	= availability of own home
INC	= income group of household
INJ	= debilitating effect of injury in household
INSP	= speed of safety inspections
LEAV	= status of household leaving region
LL	= overall lifeline availability status
LOAN	= amount of reconstruction loan taken out
MARG	= pre-disaster financial marginality (Y/N), proxy for building condition
MUT	= provision for mutual aid in restoration plan
$PLAN_c$	= availability of restoration and recovery plan
$PRTY_n$	= restoration priority accorded to neighborhood
PT	= probability of transition to next higher recovery level
RBL	= overall physical reconstruction
REC_b	= business economic recovery level
REC_h	= household economic recovery level
$RECB_c$	= overall business recovery in community (proxy for suppliers)
$RECH_c$	= overall household recovery in community (proxy for labor)
RES	= financial resources for recovery
RN	= a random number (for simulation)
SECT	= type of sector (local, export, construction)
SHEL	= availability of temporary shelter

SIZE = business size (large, small)
STH = reliance on short-term housing provision in recovery plan
TDMG = extent of damage to transport network
TMIT = pre-earthquake mitigation to transportation system
TRNS = transport accessibility
WALT = provision for alternate water sources (water trucks) in plan
WAT = availability of water
WDMG = extent of damage to water network
WMIT = pre-earthquake mitigation to water system
WNET = water availability from network (Y/N)

10.A.2 Recovery Model Structure

Notes: (1) Exogenous variables in **bold**. Policy variables in **bold underline**. Recovery indicators in *italic underline*; (2) Positive and negative signs indicate direction of influence; (3) Subscript indices: b=business, h=household, n=neighborhood, c=community or city.

Businesses Variables

$REC_b(t) = f_1(+REC_b(t\text{-}1), +PT_b(t), \mathbf{RN})$ (10.A.1)
$PT_b(t) = f_2(+BL_b(t), +LL_n(t), +CRIT_c(t), +DEM_b(t), +RECB_c(t\text{-}1),$
$+RECH_c(t\text{-}1), -FAIL_b(t))$ (10.A.2)
$BL_b(t) = f_3(\mathbf{+DBL_b(t)}, -DMG_b, +RES_b(t\text{-}1), +INSP_c)$ (10.A.3)
$DMG_b = f_4(\mathbf{+EQ_n}, -CODE_b, \mathbf{-BMIT_b}, +MARG_b)$ (10.A.4)
$CODE_b = f_5(\mathbf{BYR_b}, \underline{\mathbf{CYR_c}})$ (10.A.5)
$MARG_b = f_6(\mathbf{-SIZE_b}, \mathbf{RN})$ (10.A.6)
$RES_b(t\text{-}1) = f_7(\mathbf{+SIZE_b}, \mathbf{+DINS_b(t\text{-}1)}, +LOAN_b(t\text{-}1))$ (10.A.7)
$LOAN_b(t\text{-}1) = f_8(+LOAN_b(t\text{-}2), \mathbf{-SIZE_b}, +DMG_b, +AID_c(t\text{-}1))$ (10.A.8)
$DEM_b(t) = f_9(\mathbf{SECT_b}, \mathbf{+SIZE_b}, +RECH_n(t\text{-}1), +RECH_c(t\text{-}1), +TRNS_n(t))$ (10.A.9)
$FAIL_b(t) = f_{10}(+FAIL_b(t\text{-}1)^*, -REC_b(t\text{-}1), -REC_b(t\text{-}2), -REC_b(t\text{-}3),$
$+LOAN_b(t\text{-}1), +MARG_b)$ **if FAIL(t-1)=1, then FAIL(t)=1.* (10.A.10)

Household Variables

$REC_h(t) = f_{11}(+REC_h(t\text{-}1), +PT_h(t), \mathbf{RN})$ (10.A.11)
$PT_h(t) = f_{12}(+BL_h(t), +LL_n(t), +EMPL_n(t), -INJ_h(t), -DEBT_h(t), -LEAV_h(t))$ (10.A.12)
$BL_h(t) = f_{13}(+HSE_h(t), +SHEL_c(t))$ (10.A.13)
$HSE_h(t) = f_{14}(\mathbf{+DHSE_h(t)}, -DMG_h, +RES_h(t\text{-}1), +INSP_c, \mathbf{+INC_h})$ (10.A.14)
$DMG_h = f_{15}(\mathbf{+EQ_n}, -CODE_h, \mathbf{-BMIT_h}, +MARG_h)$ (10.A.15)
$CODE_h = f_{16}(\mathbf{BYR_h}, \underline{\mathbf{CYR_c}})$ (10.A.16)
$MARG_h = f_{17}(\mathbf{-INC_h}, \mathbf{RN})$ (10.A.17)
$RES_h(t\text{-}1) = f_{18}(\mathbf{+INC_h}, \mathbf{+DINS_h(t\text{-}1)}, +LOAN_h(t\text{-}1))$ (10.A.18)
$LOAN_h(t\text{-}1) = f_{19}(+LOAN_h(t\text{-}2), \mathbf{-INC_h}, +DMG_h, +AID_c(t\text{-}1))$ (10.A.19)
$INJ_h(t) = f_{20}(\mathbf{+DINJ_h(t)}, +DMG_h)$ (10.A.20)
$DEBT_h(t) = f_{21}(+DEBT_h(t\text{-}1), +DMG_h, -EMPL_n(t\text{-}1), +INJ_h(t\text{-}1),$
$+LOAN_h(t\text{-}1))$ (10.A.21)
$LEAV_h(t) = f_{22}(+LEAV_h(t\text{-}1), -REC_h(t\text{-}1), -REC_h(t\text{-}2), -REC_h(t\text{-}3),$
$+DEBT_h(t\text{-}1), \mathbf{-INC_h})$ **if LEAV(t-1)=1, then LEAV(t)=1.* (10.A22)

Neighborhood & Community Variables

$$LL_n(t) = f_{23}(\, +WAT_n(t),\ +ELEC_n(t),\ +TRNS_n(t),\ +CRIT_c(t)\,) \qquad (10.A.23)$$

$$WAT_n(t) = f_{24}(\, +WNET_n(t),\ +\underline{\mathbf{WALT}_c}\,) \qquad (10.A.24)$$

$$WNET_n(t) = f_{25}(\, -WDMG_c,\ +\mathbf{DWAT}_c\mathbf{(t)},\ +\underline{\mathbf{MUT}_c},\ +ELEC_n(t),\ +TRNS_n(t),\ +\underline{\mathbf{PRTY}_n}\,) \qquad (10.A.25)$$

$$WDMG_c = f_{26}(\, +\mathbf{EQ}_{n=1\ldots N},\ -\underline{\mathbf{WMIT}_c}\,) \qquad (10.A.26)$$

$$ELEC_n(t) = f_{27}(\, -EDMG_c,\ +\mathbf{DEL}_c\mathbf{(t)},\ +\underline{\mathbf{MUT}_c},\ +TRNS_n(t),\ +\underline{\mathbf{PRTY}_n}\,) \qquad (10.A.27)$$

$$EDMG_c = f_{28}(\, +\mathbf{EQ}_{n=1\ldots N},\ -\underline{\mathbf{EMIT}_c}\,) \qquad (10.A.28)$$

$$TRNS_n(t) = f_{29}(\, -TDMG_c,\ +\mathbf{DTRN}_c\mathbf{(t)},\ +\underline{\mathbf{MUT}_c},\ +\underline{\mathbf{PRTY}_n}\,) \qquad (10.A.29)$$

$$TDMG_c = f_{30}(\, +\mathbf{EQ}_{n=1\ldots N},\ -\underline{\mathbf{TMIT}_c}\,) \qquad (10.A.30)$$

$$CRIT_c(t) = f_{31}(\, -CDMG_c,\ +\mathbf{DCR}_c\mathbf{(t)}\,) \qquad (10.A.31)$$

$$CDMG_c = f_{32}(\, +\mathbf{EQ}_{n=1\ldots N},\ -\underline{\mathbf{CMIT}_c}\,) \qquad (10.A.32)$$

$$AID_c(t\text{-}1) = f_{33}(\, +\mathbf{DAID}_c\mathbf{(t\text{-}1)},\ +\underline{\mathbf{CAP}_c}\,) \qquad (10.A.33)$$

$$INSP_c = f_{34}(\, +\underline{\mathbf{CAP}_c},\ +\underline{\mathbf{PLAN}_c}\,) \qquad (10.A.34)$$

$$SHEL_c(t) = f_{35}(\, -RBL_c(t\text{-}1),\ +\underline{\mathbf{STH}_c}\,) \qquad (10.A.35)$$

$$EMPL_n(t) = f_{36}(\, +RECB_c(t\text{-}1),\ TRNS_n(t)\,) \qquad (10.A.36)$$

Recovery Indicators

$$\underline{\mathit{RECB}_n}(t\text{-}1) = f_{37}(\, +REC_{b=1,\ldots Bn}(t\text{-}1)\,) \qquad (10.A.37)$$

$$\underline{\mathit{RECB}_c}(t\text{-}1) = f_{38}(\, +RECB_{n=1,\ldots N}(t\text{-}1)\,) \qquad (10.A.38)$$

$$\underline{\mathit{RECH}_n}(t\text{-}1) = f_{39}(\, +REC_{h=1,\ldots Hn}(t\text{-}1)\,) \qquad (10.A.39)$$

$$\underline{\mathit{RECH}_c}(t\text{-}1) = f_{40}(\, +RECH_{n=1,\ldots N}(t\text{-}1)\,) \qquad (10.A.40)$$

$$\underline{\mathit{RBL}_n}(t) = f_{41}(\, +BL_{b=1,\ldots Bn}(t),\ +HSE_{h=1,\ldots Hn}(t)\,) \qquad (10.A.41)$$

$$\underline{\mathit{RBL}_c}(t) = f_{42}(\, +RBL_{n=1,\ldots N}(t)\,) \qquad (10.A.42)$$

11 Earthquake Disaster Mitigation for Urban Transportation Systems: An Integrated Methodology that Builds on the Kobe and Northridge Experiences*

Peter Gordon,[1] James E. Moore II,[2] Harry W. Richardson,[1] Masanobu Shinozuka,[3] Donghwan An[4], and Sungbin Cho[5]

[1] School of Policy, Planning, & Development and Department of Economics, University of Southern California, USA
e-mail: pgordon@usc.edu, hrichard@usc.edu
[2] Department of Industrial and Systems Engineering and School of Policy, Planning, & Development, University of Southern California, USA
e-mail: jmoore@usc.edu
[3] Department of Civil and Environmental Engineering, University of California, Irvine, USA
e-mail: shino@uci.edu
[4] Department of Agricultural Economics and Rural Development, Seoul National University, Republic of Korea
e-mail: dha@snu.ac.kr
[5] School of Policy, Planning, & Development, University of Southern California, and Imagecat, USA
e-mail: sungbinc@usc.edu and sc@imagecatinc.com

11.1 Introduction

11.1.1 Applied Economics and Policy Analysis

Around the world, natural disasters kill thousands each year and inflict billions of dollars in damage. Better analysis has the potential to save lives and resources on a large scale. One of the most important applications of economic analysis is to the evaluation of proposed projects and policy measures, usually benefit-cost analysis. A related but different approach involves regional economic impact analysis. Whereas benefit-cost analysis can be used to rank policy measures in terms of their efficient use of resources, impact analysis offers a reading of how far these measures deviate the local economy from current performance levels. The simplest examples are the widely reported multiplier analyses wherein proponents of certain projects (sports stadia, convention centers, etc.) claim that some multiple of annual expenditures will enhance

* This research was supported by NSF Award Number: CMS 9812503 (NSF Program 1499: Earthquake Systems Integration, US-Japan Cooperative Research in Urban Earthquake Disaster Mitigation), and in part by NSF Award No. CMS 9728805 to the Institute for Civil Infrastructure Systems (ICIS) at New York University (in partnership with Cornell University, Polytechnic University of New York, and the University of Southern California). Any opinions, findings, and conclusions or recommendations expressed in this document are those of the authors and do not necessarily reflect the views of the National Science Foundation. This chapter was substantially improved by changes requested by the editors of this volume. The faculty authors remain responsible for any errors or omissions.

the regional economy because of various ripple effects. Our claim in this research is that some available economic impact models when properly modified and elaborated lend themselves to the problem of determining plausible evaluations of earthquake mitigation and reconstruction policies for metropolitan areas.

A review of the literature (Cho *et al.*, 2001) shows that there has been limited attention given to the socioeconomic impacts of earthquakes. Progress in economic impact research is recent. Most of the research on earthquakes has been in the engineering and geological fields. Earthquake engineering is an established field, but integrating the economic impacts of earthquakes with engineering models remains a challenge.

The most widely used models of regional economic impacts are versions of inter-industry models. These trace intra- and inter-regional shipments at a high level of industrial disaggregation. They only account for losses via backward linkages, because they are demand driven.

11.1.2 Integrating Regional Economic Modeling and Urban Earthquake Policy Analysis

A considerable earthquake engineering literature devotes itself to the estimation of "direct" damages from a past or expected natural disaster. These estimates become benchmarks for calculating possible loss reductions (benefits), which can be weighed against the costs of achieving these reductions. It is now widely recognized that this approach is inadequate because these losses also have a time dimension: for how long will the services of the facility be diminished? The latter are often labeled as "indirect" effects, a possibly misleading descriptor because indirect has a slightly different meaning in the regional economic impact assessment literature. Researchers in this field have recently adopted "business interruption" or "loss of functionality" as descriptors of many of these effects.

Yet, there are other effects to consider. It is important to be comprehensive because policy analysis should begin with a full accounting of losses before any plausible policy recommendations are made. A full accounting supposes the ability to trace the full effects of the losses of any facility through the regional economy. Interindustry economic analysis, usually input-output (IO) analysis, has been applied to aspects of this problem for many years. The appeal of this approach is that the interdependence of shipments is depicted in considerable detail. Yet, it may seem ironic that the medium over which shipping occurs is usually absent from these models. Our conjecture is that, for the case of highways, the fact that highway services are not obtained on a contractual basis, with payment for these services normally treated as an indirect cost to firm operations (taxes and license fees) rather than being associated with individual shipments, explains the paradox. This institutional fact of life, however, does not absolve modelers and policy makers from attempting to integrate models that determine shipping costs with models that estimate the associated production costs. In terms of operational models, this compels us to fuse regional input-output models with regional highway network models. That effort also opens the possibility of merging earthquake engineering models of seismic activity and structures into the framework.

The following sections of this report provide the details of aspects of model integration that were accomplished. We also discuss the applications of the resulting model to: i) the simulation of the full costs of a hypothetical earthquake; ii) the determination of bridge reconstruction costs; iii) the calculation of plausible bridge reconstruction budgets, determined in light of expected production shortages and bottlenecks.

11.2 Southern California Planning Model 2 (SCPM2)

11.2.1 Background

Regional economists have invested much time examining interindustry models. The details of intersectoral linkages in these models are useful for exploring regional economic structure. However, this approach has not permitted an adequate treatment of transportation costs, not all of which are transacted because most roads are publicly provided. This problem has recently been addressed at the national level by the Bureau of Transportation Statistics effort to create Transportation Satellite Accounts (Han and Fang, 2000).

Spatial elaborations of input-output and related approaches require explicit treatment of the resources consumed by flows between origin-destination pairs (Moses 1960, Okuyama *et al.* 1997). Explicit representation of the transportation network is usually not necessary in multiregional approaches. It is another matter at the intrametropolitan level, because congestion dominates line-haul costs.

Southern California Planning Model 1 (SCPM1) by Richardson *et al.* (1993) combined a metropolitan level input-output model with a Garin-Lowry model to spatially allocate induced economic impacts. This operationalized spatial input-output analysis at the intrametropolitan level. That model did not treat the transportation network explicitly. Congestion effects were ignored, and transportation flows were exogenous.

Integrating a transportation network into SCPM1 provides important opportunities. Distance decay relationships (destination choice) can be endogenized, permitting an improved spatial allocation of indirect and induced economic impacts. Also, this integration makes it possible to better account for the economic consequences of changes in transportation network capacity.

Our interest is in the regional economic consequences of earthquakes, which result in some of the most dramatic changes in regional economic and infrastructure capacity. The costs-of-earthquakes literature emphasizes the measurement of structure and contents losses. More recently, social-science-based research on earthquakes has addressed the measurement of business interruption costs (Gordon, Richardson, and Davis, 1998; Rose and Benavides, 1998; Boarnet, 1998). Yet, there are still few studies that examine the role of infrastructure and its interactions with the metropolitan economy.

Several research questions motivated this work. First, we wanted to integrate regional economic, transportation, bridge performance, and other structural response models in a way that respects feedback relationships between land use and transportation. Second, we sought to apply such integrated, operational models to the problem of estimating the costs of a large earthquake. Third, we wanted to account for the costs of damage to infrastructure, with special attention to bottlenecks and shortages that are created in the course of large-scale reconstruction. Further, because "all politics are local," we wanted to describe these costs and benefits at the submetropolitan level.

To accomplish this, we integrated a) bridge and other structure performance models, b) transportation network models, c) spatial allocation models, and d) inter-industry (input-output) models. We then used the integrated model to begin analysis of various bridge reconstruction scenarios.

The Southern California Planning Model (SCPM1) was developed for the Los Angeles metropolitan region. The study area includes Los Angeles, Orange, Riverside, San Bernardino and Ventura counties. The area covers more than 35,000 square miles. The 2000 population of the five-county area was over 16.3 million. At this time, data for the urbanized portions of the metropolitan area were not yet available. In 1990, the urbanized portions extended to 1,966 square miles; population density in the urbanized area was about 5,801 people per square mile, the highest in the U.S. The urbanized area is described in terms of SCAG's 1527 disaggregate traffic analysis zones (TAZs). The regional highway network includes 22,244 links.

Table 11.1 provides additional recent aggregate data describing the study area. The total households in the area were 5.4 million in 1998 (US Bureau of Census). The non-farm employment in the SCAG region was over 5.8 million in 1997. Personal income in the area was $329.6 billion and per capita personal income was $21,542 in 1994. 21.6 percent of income was goods-related and 78 percent of income was service related. The employment distribution across industry sectors is: 34.3 percent in services, 16.4 percent in manufacturing, 13.3 percent in government, 9.6 percent in retail, and 7.3 percent in FIRE. International exports from the five-county area have been reported to be $35.7 billion in 1996 (Exporter Location Series, US Bureau of the Census). Our analysis suggests, however, that this is a significant underestimate (Gordon and Richardson, 1999).

Table 11.1 Socio-Economic Profile, SCAG Five-County Area

County	Population (persons)	Households (1,000)	Employment (paid employees)	Total Personal Income ($1,000)	Per Capita Income ($)	Land Area (square miles)
Year	2000	1998	1997	1994	1994	1990
Los Angeles	9,519,338	3,136.6	3,693,537	197,289,098	21,562	4,060
Orange	2,846,289	941.0	1,212,689	64,892,666	25,516	790
Riverside	1,545,387	1,037.9[a]	319,904	25,086,809	18,543	7,208
San Bernardino	1,709,434	[a]	406,859	26,477,943	17,043	20,062
Ventura	753,197	239.9	211,591	15,899,444	22,625	1,846
Five-County	16,373,645	5,355.4	5,844,580	329,645,960	21,542	33,966

Source: U.S. Census Bureau's state and county quick facts (http://quickfacts.census.gov/qfd/)
Notes: a. Data for Riverside-San Bernardino PMSA

b. Values of employment (private nonfarm employment) and land area from People QuickFacts for each individual county. Values of population come from USA Counties General Profile for each individual county. Values of total personal income and per capita personal income are from Local Area Personal Income data of Bureau of Economic Analysis (BEA).

The model allocates impacts in terms of jobs or the dollar value of output to 308 sub-regional (municipal) zones. Analysis of Northridge earthquake business interruption effects utilized SCPM1 (Gordon, Richardson, and Davis, 1998). That model was driven by reduced demands on the part of damaged businesses, as ascertained from survey results.

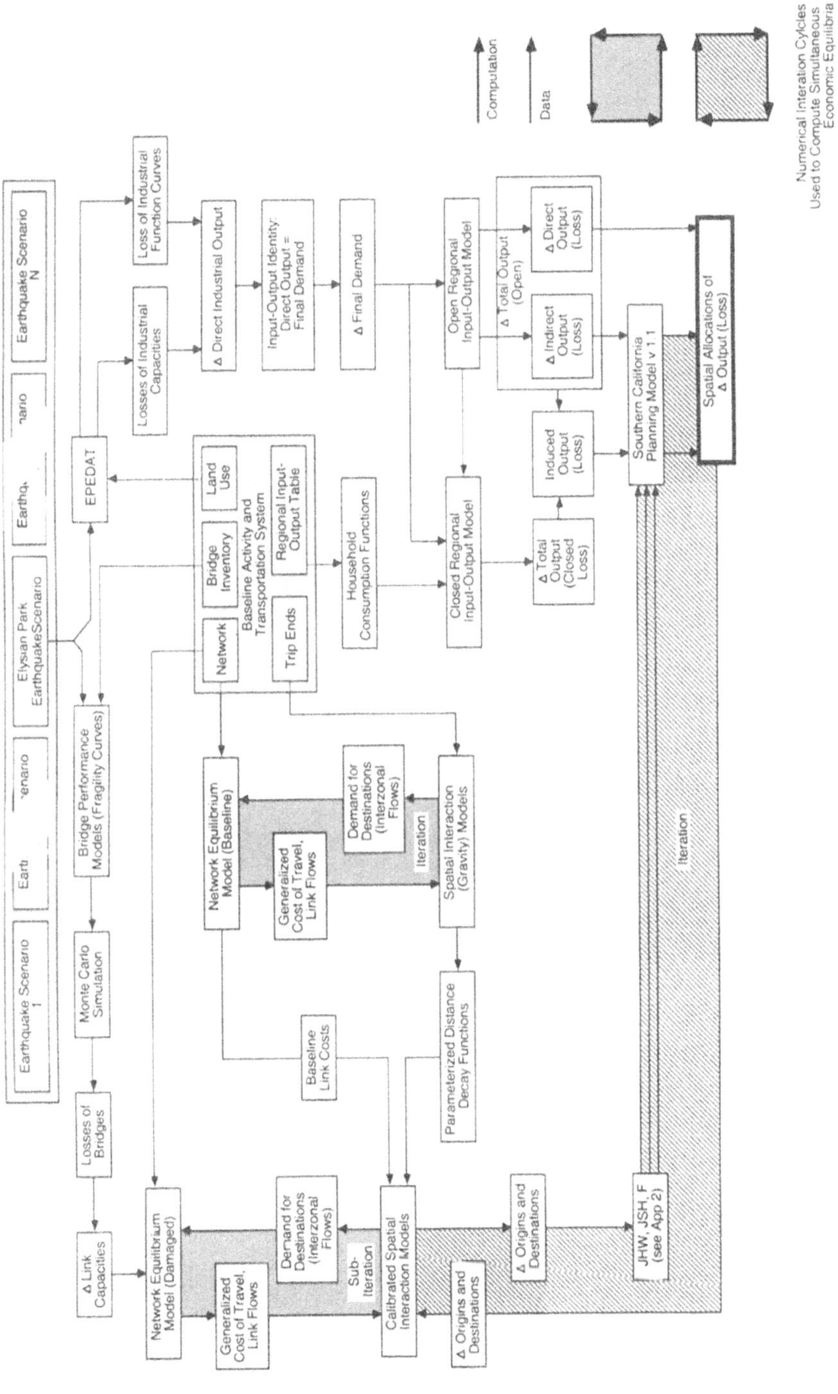

Figure 11.1 Summary of the Southern California Planning Model 2 (SCPM2)

In this exercise, we focused on a hypothetical earthquake, a magnitude 7.1 maximum credible earthquake (MCE) event on the Elysian Park blind thrust fault. In this case, results of structure damage to businesses, as developed by EQE International's Early Post-Earthquake Damage Assessment Tool (EPEDAT), were used to drive a new version of SCPM, SCPM2, that has been improved to include the regional transportation network. EQE's EPEDAT is a GIS-based earthquake loss estimation program that estimates ground motion, structural damage, and direct business interruption losses associated with a specific earthquake (Eguchi *et al.*, 1997; Campbell, 1997). EPEDAT predicts, among other values, the lengths of time for which firms throughout the region will be non-operational. This allows the calculation of exogenously prompted reductions in demand by these businesses. These are introduced into the inter-industry model as reductions in final demand (Isard and Kuenne, 1953). Explicit treatment of the transportation network made it possible to model the concurrent impact of transportation cost changes on the activity system, including reductions in regional network capacity resulting from large numbers of bridge failures.

Figure 11.1 summarizes our approach. Engineering models predict damage to transportation structures by location for the Elysian Park scenario. EPEDAT predicts spatial loss of industrial function. The I/O model translates this production shock into direct, indirect, and induced costs, and the indirect and induced costs are spatially allocated in terms consistent with the endogenous transportation behaviors of firms and household. See Cho *et al.* (2001) for a detailed description of SCPM2.

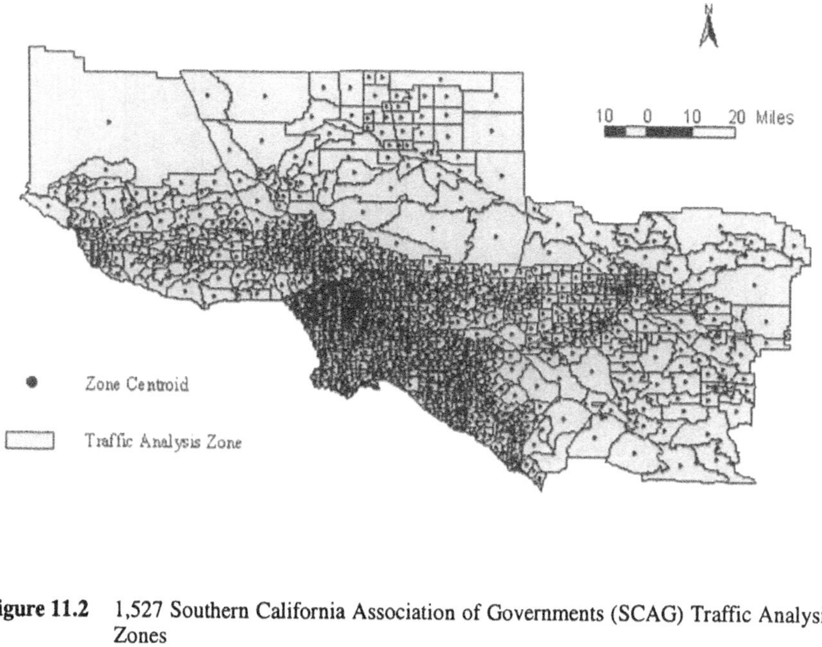

Figure 11.2 1,527 Southern California Association of Governments (SCAG) Traffic Analysis Zones

Implementing this approach is a data intensive effort. SCPM2 aggregates the Southern of California Association of Governments (SCAG) 1,527 traffic analysis zones (see figure 11.2) into 308 political jurisdictions, and aggregates to 17 the 515 sectors represented in the Regional Science Research Corporation's PC I-O model Version 7 (Stevens, 1996) based on the work of Stevens, Treyz, and Lahr (1983). SCPM2 treats the transportation network explicitly, endogenizing otherwise exogenous matrices describing the travel behavior of households, achieving consistency across network costs and origin-destination requirements, and better allocating indirect and induced economic losses over zones in response to direct earthquake losses to industrial and transportation capacity. Making distance decay relationships and congestion endogenous also endogenizes the spatial allocation of indirect and induced economic losses.

Our goal was to model the effects of earthquakes on industrial capacity and system-wide transportation demand and supply. We also wanted to measure as fully as possible the economic impacts associated with both of these effects. Our first step was to compute a pre-earthquake baseline that is consistent with respect to equilibrium network costs, network flows, and inter-zonal flows and origin-destination requirements. The information needed to model an internally consistent baseline is also sufficient to treat changes in configuration of the network and the activity system. Following an earthquake, there will be losses of network capacity and simultaneous losses of industrial capacity. The former reduces transportation capacity and raises costs. The latter will reduce demands imposed on the network. The bridge performance models and building fragility curve analysis (EPEDAT) ascribe consistent losses of both types to particular earthquake scenarios. The spatial interaction elements of our approach made it possible to capture the changes in transportation requirements associated with changes in network performance. These changes and changes resulting from earthquake damage to industrial facilities were treated simultaneously and consistently.

11.2.2 Application: Towards Determining the Full Costs an Elysian Park 7.1 Maximum Credible Event

SCPM2 was applied to the Los Angeles metropolitan area for the scenario defined by a maximum credible earthquake (magnitude 7.1) on the Elysian Park thrust ramp. This Elysian Park scenario was selected on the basis of its potential to produce major damage and casualties; despite this, the fault has received little systematic attention. Also, the annual probability of an earthquake associated with this fault is relatively high within the Southern California fault system. Like the 1994 Northridge earthquake, the Elysian Park scenario occurs on a blind thrust fault. While the maximum size earthquakes that seismologists believe are possible on the blind thrust faults are lower than those on, for example, the San Andreas Fault, they are expected to have the potential to result in more severe damage because of their proximity to metropolitan Los Angeles. The planar earthquake source representation for the Elysian Park event varies in depth from 11.0 to 16.0 km below the surface. The surface projection of this source includes a broad, densely populated area of central Los Angeles County, including downtown Los Angeles. Figure 11.3 provides a map of predicted peak ground accelerations by 1990 census tract for this event.

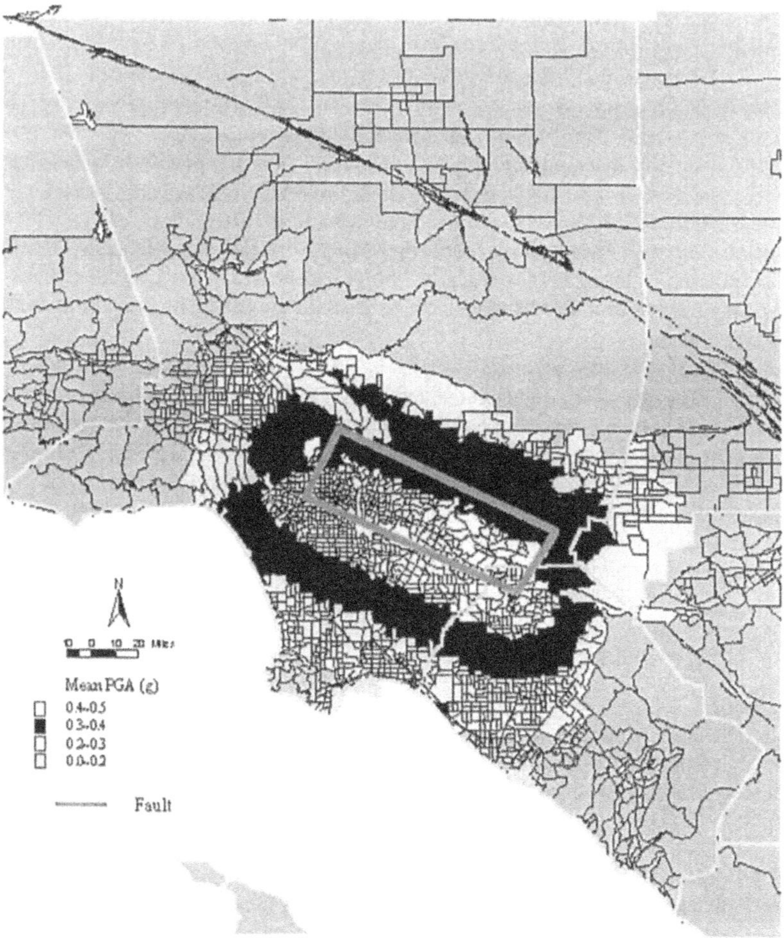

Figure 11.3 Predicted Peak Ground Accelerations for a Magnitude 7.1 Earthquake on the Elysian Park Blind Thrust Fault

Bridge fragility curves give the probability distribution of bridge damage states conditioned by bridge type and earthquake event, in this case the Elysian Park scenario. These damage states were defined in terms of a bridge damage index (BDI) ranging from 0 (no damage) to 1 (collapse). BDI intervals are mapped to each of four damage states as shown in table 11.2.

Empirical fragility curves (Shinozuka, 1998 and 1999) are developed on the basis of bridge damage records made after the 1994 Northridge Earthquake. These curves are expressed in the form of two-parameter lognormal distribution functions. The two parameters (median and log-standard deviation) are estimated via a maximum likelihood method. The Peak Ground Acceleration (PGA) is used to characterize the intensity of the seismic ground motion, although use of other intensity measures such as Peak Ground Velocity (PGV), Spectral Acceleration (SA), Spectral Intensity (SI), and Modified Mercalli Intensity (MMI) are reasonable.

$$L = \prod_{i=1}^{N} [F(a_i)]^{x_i} [1 - F(a_i)]^{1-x_i} \tag{11.1}$$

where L = likelihood;
$F(a_i)$ = the fragility curve for a specific bridge damage state;
a_i = is the PGA value to which bridge i is subjected,
x_i = 1 or 0, depending on whether or not the bridge achieves damage state $F(a_i)$ under PGA = a_i; and
N = the total number of bridges inspected after the earthquake.

Under the lognormal assumption, $F(a)$ takes the analytical form

$$F(a) = \Phi\left[\frac{\ln\left(\frac{a}{c}\right)}{\xi}\right] \tag{11.2}$$

where a = PGA value,
c = median,
ζ = logstandard deviation, and
$\Phi[\cdot]$ = is the standardized normal distribution function.

The two parameters c and ζ in equation (11.2) are computed as c_0 and ζ_0 maximizing the log of the likelihood function, ln$L$, and hence L.

$$\frac{d\ln L}{dc} = \frac{d\ln L}{d\zeta} = 0 \tag{11.3}$$

This optimization is straightforward. This procedure produces fragility curves classified by bridge damage state.

Table 11.2 Bridge Damage States, Bridge Damage Index, and Fragility Curve Parameters

Bridge Damage State / Fragility Curve	> BDI Lower Bound	≤ BDI Upper Bound	Median	Log Standard Deviation
Minor Damage	0.050	0.200	0.83	0.82
Moderate Damage	0.200	0.525	1.07	0.82
Major Damage	0.525	0.850	1.76	0.82
Collapse	0.850	1.000	3.96	0.82

A family of four separate fragility curves for (1) at least minor damage, (2) at least moderate damage, (3) at least major damage, and (4) collapse states is estimated simultaneously on the basis of the PGA values and damage states reported by California Department of Transportation engineers for 1,998 bridges damaged by the Northridge Earthquake. Each Fragility curve describes the cumulative probability of achieving a given damage state as a function of PGA, consequently the fragility curves for the four damage states never intersect (see figure 11.4). The medians and log-standard deviations of these fragility curves are given in table 11.2.

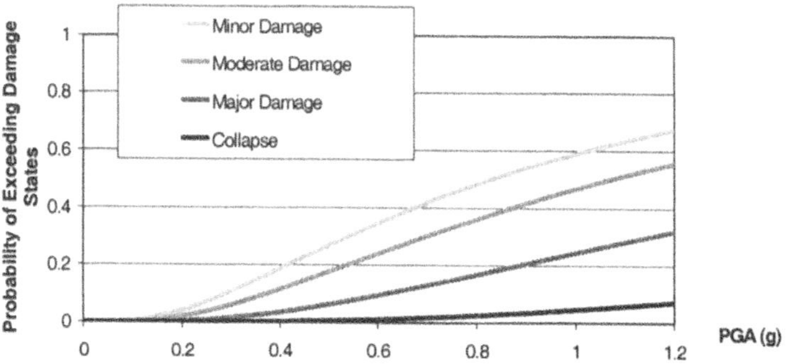

Figure 11.4 Bridge Fragility Curves Estimated from Caltrans Northridge Earthquake Data

From a network management perspective, the key operational question is "At what bridge damage index value should the bridge be closed?" Our approach made it possible to systematically investigate the cost implications of alternative bridge closure criteria.

The approximate midpoints of the bridge damage index intervals associated with moderate and severe damage states are 0.3 and 0.75, respectively. We treated these values as the most conservative and riskiest BDI thresholds that transportation authorities are likely to accept as bridge closure criteria. A conservative, safety oriented policy would close damaged structures to traffic, including bridges with a damage index ≥ 0.30. This would increase delay and other transportation costs. A less risk averse policy intended to emphasize an emergency focus on maintaining regional economic function would leave moderately damaged structures open, closing only bridges with a damage index ≥ 0.75. No authority would open the most dangerous structures.

Earthquakes induce changes in industrial production due to effects on building stocks, particularly factories, warehouses, and office buildings. Damage to production facilities was translated into an exogenous change in final demand. Building damage causes direct loss in industrial production. EPEDAT's loss-of-function curves convert damage to building stocks to loss of production by zone and sector. The loss-of-function curves structural damage states to business closure times and direct business interruption (production) losses. Inputs are commercial and industrial building damage estimates from EPEDAT, expressed as the percent of structures in each of four damage states by use class and by each of the 308 SCPM zones. Outputs are estimates of direct business interruption loss for the region by industry, month (over the first year following the earthquake), and SCPM zone.

EPEDAT projects structure losses in the five-county Los Angeles metropolitan region of between \$21.7 billion and \$36.2 billion for the Elysian Park event. If building contents are included, property damage is estimated at \$33.9 to \$56.6 billion. Residential damage accounts for approximately two-thirds of the total. These estimates do not include damage to bridges or other infrastructure. About 72 percent of the structural damage to buildings is estimated to occur in Los Angeles County.

A corresponding change in final demand drives SCPM2, ultimately providing changes in output and employment for 17 sectors across 308 zones. This is an iterative calculation. Direct changes are exogenous, and already spatially identified. SCPM2 allocates indirect and induced changes in a way that respects both observed travel behavior and new network costs. A core contribution of this research is the ability to more completely endogenize submetropolitan freight and passengers flows and destinations. In this case, nine classes of passenger flows were combined with four classes of freight and loaded on a common network.

Bridge damage results were generated for 200 Monte Carlo simulations of the Elysian Park scenario earthquake. The bridge damage index achieved by any specific structure varies across each simulation, but each outcome is drawn from the fixed stochastic process corresponding to the Elysian Park scenario. Collectively, these simulations correspond to a distribution of damaged transportation networks. Each network is characterized (in part) by a vector of 2,810 bridges, each assigned a BDI value. The alternative bridge closure criteria (BDI > 0.30, BDI > 0.75) are applied to every bridge in every network in this set, producing two new distributions. The transportation networks in these distributions are still characterized by a vector of 2,810 bridges, but each bridge is now open (1), or closed (0).

Our model of the Los Angeles economy is convergent, but it is computationally infeasible to exhaustively investigate each network state represented in these distributions of damaged networks. Instead, we selected representative members of each distribution. The 200 simulations were rank ordered in terms of the baseline vehicle-miles that would otherwise be traveled across the damaged links. This rank ordering made it possible to identify those simulations that are:

- maximally disruptive with respect to baseline transportation flows; and
- representative in a median sense.

An example of preliminary simulation results describing the full costs of a magnitude 7.1 Elysian Park event are summarized in table 11.3. Row A reflects the midpoint of the range of structure damage predicted by EPEDAT, $45.25 billion, including $29 billion in structure loss. Row B is the sum of direct, indirect, and induced losses computed by the input-output model of the five-county, Los Angeles metropolitan area. This sum is $46.7 billion. These aggregate values are identical across all other simulations (Cho *et al.*, 1999). Row C summarizes the post earthquake network equilibrium transportation costs in light of reduced production and reduced network capacity. These values do vary across all simulations. Table 11.3 corresponds to median simulated disruption of baseline transportation combined with a risk tolerant bridge closure criteria that leaves moderately damaged structures open to normal traffic. In this case 122 of 16,946 network links are closed due to bridge damage. This loss of almost 480 lane miles results in a substantial retention of transportation network capacity, and a relatively small increase in transportation costs of almost $1.5 billion.

11.3 Bridge Reconstruction

The previous discussion extended our abilities to account for both the levels and spatially disaggregated nature of earthquake losses. The objective of our efforts is to support and improve pre- and post- earthquake policy decisions. Identification of efficient reconstruction strategies is an obvious post-earthquake objective.

11.3.1 Application of SCPM2 to the Evaluation of Bridge Reconstruction Strategies

There are numerous decisions that affect bridge repair costs. These include how damaged bridges are grouped to define repair projects, and associated equipment management, traffic diversion, and network delay costs.

Table 11.3 Total Loss ($Billions): Elysian Park Magnitude 7.1 Earthquake, Maximum Simulated Disruption to Baseline Transportation (Closure at Bridge Damage Index ≥ 0.75)

Loss Type	Baseline		Elysian Park Scenario: Risk Tolerant Bridge Closure Criterion	
A Structure Loss[a]			$ 45.250 billion (48.35% of total)	
Business Loss				
Direct Loss[b]			28.155	
Indirect Loss[c]			9.627	
Induced Loss[d]			8.955	
B Business Loss Subtotal			46.737 billion (49.95% of total)	
Network Costs[e]	PCU Minutes	$Billions	PCU Minutes	$Billions
Personal Travel Cost	85,396,813	21.290	89,945,131.	22.424
Freight Cost	10,298,781	4.550	10,966,123.	4.844
Total Travel Cost	95,695,594	25.839	100,911,255.	27.268
Network Loss = Δ Network Costs			PCU Minutes	$Billions
Δ Personal Travel Cost			4,548,318.	1.134
Δ Freight Cost			667,343.	0.295
C Δ Total Travel Cost			5,215,661.	1.429 (1.5% of total)
D Bridge Repair Cost (Excludes Delay Cost)			Median $Billions	Mean $Billions
			0.071	0.219
Loss Total = A + B + C +D			$ 93.487	$ 93.635

Notes: a. Midpoint EPEDAT outputs, EQE International.
b. EPEDAT, EQE International.
c. Regional Science Research Institute PC-IO (RSRI) Model for the Los Angeles Metropolitan Area.
d. Difference between the RSRI solution with the processing sector closed with respect to labor and the RSRI solution with the processing sector open with respect to labor.
e. Network cost is the generalized total transportation cost associated with a simultaneous equilibrium across choice of destinations and routes. These estimates reflect 365 travel days per year, an average vehicle occupancy of 1.42 for passenger cars, 2.14 passenger car units per truck, a value of time for individuals of $6.5/hour, and $35/hr for freight.

Row D in table 11.3 includes preliminary bridge repair cost estimates based on a discriminant analysis of Loma Prieta and Northridge Earthquake bridge damage states and estimated repair costs. Mean and median repair costs are reported. The full costs of the earthquake are estimated to be almost $93.5 billion, close to 14 percent of the

SCAG area's 1990 GRP, although direct (business interruption) costs account for about seven percent. In this case, transportation costs account for a small share of the full cost of the earthquake. However, these costs include an optimistic assumption: None of the damaged bridges left open to traffic ever collapse.

The loss-of-function curves utilized in this research describe production capacity over a one-year period following the earthquake. Production capacity was predicted to approach pre-earthquake levels within six months. Restoration of transportation network capacity is less rapid.[1] Bridges were assumed to remain closed for one year following the earthquake. During this period they are repaired or replaced. Other assumptions or empirical relationships can certainly be accommodated to further refine these preliminary results. State DOT officials provided very different expert estimates of the time required for repair following extensive damage.

SCPM2 provides unprecedented disaggregation of economic impacts over metropolitan space. Employment and production losses are calculated at the level of 1,527 traffic analysis zones, and the aggregated for 310 cities and communities, which together exhaust the urbanized portions of the Los Angeles five-county metropolitan area. More complete tabular results, maps, and narrative summaries for this element of the research are available on our website (www.usc.edu/schools/sppd).

Corresponding results were calculated for other representative bridge-closure simulations. All of these results included the change in network costs associated with reductions in supply of transportation services. The resulting redistribution of economic activities are just one source of local (city level) losses. Increases in network transportation costs are another significant source of local impacts. These costs are more difficult to disaggregate. There is insufficient information to reliably allocate these transportation costs to economic sectors, but these costs can be geographically distributed to traffic origins and destinations.

These new network costs may also influence the distribution of indirect and induced economic losses via the distance decay relationship between travel cost and destination choice. But in all our simulations, the overall GRP changes associated with indirect and induced economic losses remain modest. Differences in spatially distributed impacts are also modest.

The Southern California region has a highly redundant road and highway system, and these findings corroborate the economic importance of the regional transportation network's high levels of redundancy. The high level of travel endogeneity associated with the travel choices represented in SCPM2 is explained by the redundancy of the Los Angeles regional transportation network. The various bridge closure simulations affect between 84 and 326 directional network links, including freeway and arterial links. The representation of the network contained in SCPM2 includes 16,946 links. Bridge closures do impact total travel cost and route choice. A comparison of our simulations indicated that the cumulative value of increased network cost can be significant, but the day-to-day increase does not induce profound changes in destination choice, and thus does not have a pronounced impact on the spatial distribution of economic losses.

These results suggest several hypotheses relating to the relationships accounted for by SCPM2 and the way these relationships are parameterized.

- This application of SCPM2 remains incomplete. The loss-of-function curves apply only to production activities. The impact on households, *i.e.*, on the production of labor, has not yet been accounted for, and changes in the spatial

[1] All SCPM2 outputs are for one-year periods. Outputs for shorter intervals are scaled proportionately.

distribution of activities and losses do not reflect the impact of changes in household consumption.

- Destination choice may be more sensitive to post-earthquake travel costs than to pre-earthquake costs. The distance decay functions in SCPM2 are estimated with pre-earthquake data. Post earthquake responses to travel cost may be different. Travelers may be more risk averse than the distance decay functions in SCPM2 imply.
- Travelers may also diminish trip frequencies in response to the cost of travel. In SCPM2, demand for freight transportation changes as a result of the earthquake, but passenger trip generation rates remain unchanged. If trip generation rates are endogenized, some longer passenger trips would be removed from these results, and this would intensify changes in the geographic distribution of activities and losses. However, these two latter limitations of earlier versions of SCPM2 have been remedied in Section IV below.

We can execute this procedure for any relevant earthquake, mitigation, or reconstruction scenario. The baseline exercise describes pre-earthquake conditions. The simulations described above summarize post-earthquake outcomes conditioned on present levels of mitigation. These results can be contrasted with results that include mitigation measures. The difference between these full-cost results measures the benefits of the mitigation, to be compared against the costs of implementing the mitigation. Importantly, the benefits measured in this manner are provided at the local submetropolitan level. This includes municipalities, and in the case of the City of Los Angeles, Council districts. If all politics are indeed local, then results like this are critically important to policy implementation.

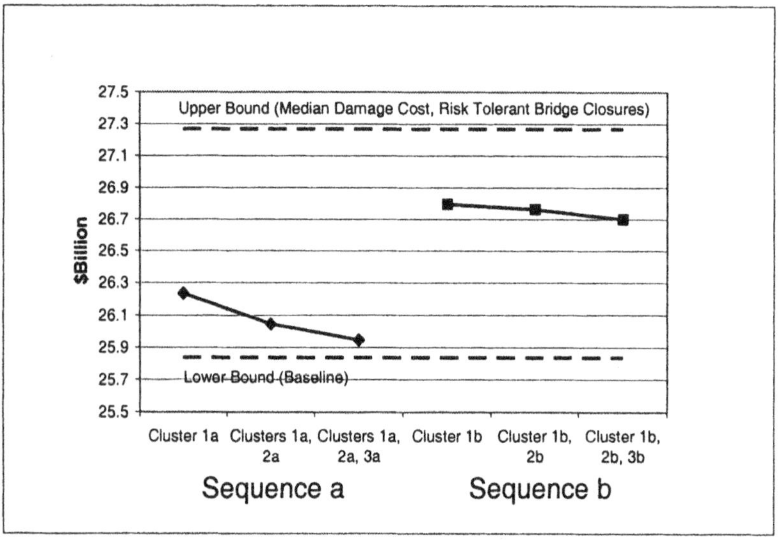

Figure 11.5 Transportation Network Costs of Two Alternative Cluster Repair Sequences (Cho *et al.*, 2000)

There is considerable interest in efficient bridge reconstruction approaches. SCPM2 is well suited to comparing the economic benefits of alternative schemes. Figure 11.5 summarizes the results of some preliminary simulations. These include the following sequence of steps:

- Identify spatial clusters of bridges. The highway agency is likely to want to repair bridges in groups that are spatially proximate. This reduces equipment staging and project set-up costs. Our procedure used a traveling salesman algorithm to identify eight spatial clusters of damaged links. This is one of several alternative clustering algorithms that can be applied. Our preliminary effort does not address improving these initial clusters, but this is a reasonable extension.
- Calculate the total pre-event traffic link volumes associated with each cluster. This is a simple measure of the importance of the facilities in each cluster. More sophisticated alternatives that account more carefully for changes in post event flows are available.
- Select an efficient sequence of bridge repairs. This selection algorithm may be either heuristic or an optimum-seeking dynamic program (Kiyota, Vandebona, and Tauoue, 1999).
- Estimate network cost improvements as cluster repair benefits associated with the repair sequence.

In an optimization exercise, the last two steps would likely be combined. These steps could be separate in some heuristic procedures. In preliminary work, we selected a cluster repair sequence based on the pre-event traffic volumes for the cluster. The highest-volume cluster was repaired first, followed by the next-highest volume cluster, etc. The magnitudes of network cost reductions are plotted in figure 11.5. System improvements are measured in terms of post-repair network flows. These benefits should be compared to the bridge reconstruction costs in row D of table 11.3.

The lower bound in figure 11.5 describes the network user costs on an undamaged network, $25.839 billion (see table 11.3). The upper bound is network costs given median damage (200 simulations) associated with the Elysian Park scenario, and a risk-tolerant bridge closure policy that leaves moderately damaged bridges open to traffic.

The lower left-hand curve shows the network cost improvements associated with repairing the clusters in order of their total pre-event link volumes, starting with the highest volume cluster. The upper right-hand curve reverses this sequence, repairing the lowest volume cluster first. The plots have the expected shapes. In the lower-left, benefits from repairing additional clusters of bridges become available in ever smaller increments. In the case of the upper right curve, benefits become available in ever larger increments.

11.3.2 Discussion

Our integration of seismic, transportation network, spatial allocation, and input-output models permits the study of how the economic impacts of industrial and transportation structure loss are distributed over metropolitan space. Some of this loss is produced directly by the earthquake, which destroys industrial capacity. The procedure accounts for the impact of industrial structure losses and resulting direct production losses. The model computes further indirect and induced losses, and makes the spatial distribution of these losses sensitive to increases in network costs resulting from transportation structure losses.

These preliminary research results permit us to address the problem of bridge reconstruction prioritization. To do so, we first assess the earthquake risk to the transportation system and the urban economy by accounting for a wide range of outcomes associated with damage to bridges and production facilities. The costs of efficient bridge reconstruction improve the accounting of the costs of the earthquake. This approach has four elements, specifying an integrated model, assembling data from disparate sources, achieving computability, and identifying bridge reconstruction strategies.

While these results are preliminary, they demonstrate the way SCPM2 can be applied. We are currently testing alternate bridge repair sequences and plan to compare these with actual experience from the Kobe and Northridge bridge repair efforts.

11.4 SCPM2 Extensions: Endogenizing Work and Shopping Trip Attractions and Productions

In SCPM2, all distance-decay functions were endogenously determined (Cho *et al.*, 2001). In light of severe shocks to network capacity and also to the demand for network services, this seemed to be a necessary step in developing an integrated model.

Yet, it can also be argued that passenger trip generation characteristics would change in response to shocks of this nature. In the applications that follow, we modified SCPM2 to account for some of these effects. We endogenized trip attractions and productions for work trips and for shopping trips. The following four extensions implement these changes. The resulting elaboration of SCPM2 is convergent.

Zone specific economic impacts will produce changes in work trips attracted to each zone. Direct job impacts V_d^z are given by zone. However, total impacts also include indirect impacts V_n^z and induced impacts V_u^z. These indirect and induced estimates are model outputs. The requirement that trips attracted match trips produced implies a corresponding change in work trip productions. The objective is to predict mutually consistent changes in work trip ends and economic impacts. Define the change in work trips attracted to zone z to be:

$$dA_{w,i+1}^z = \rho \cdot \sum_k \left(V_{d,k,i}^z + V_{n,k,i}^z + V_{u,k,i}^z \right) \tag{11.4}$$

where ρ = a constant that converts economic impacts to trip changes, *i.e.*, a ratio of the total number of trips / total regional output,
k = an industrial sector index, and
i = an iteration counter.

Changes in work trips produced in and attracted to each zone should be consistent with SCPM2's journey-from-home-to-work (*JHW*) matrix. This *JHW* matrix gives the proportion of work trips terminating in each zone for each origin. Balancing work trip productions and attractions requires that:

$$dP^z_{w,i} = P^z_{w,i} - JHW \cdot \left(A^z_{w,i} + dA^z_{w,i}\right) = \\ = P^z_{w,i} - JHW \cdot \left(A^z_{w,i+1}\right) \tag{11.5}$$

Reductions in work trips and employment reduce aggregate household income in each zone. This reduces shopping trips produced in each zone. Define the change in shopping trips produced in zone z to be dP^z_s. Fewer shopping trips will occur, but assume expenditures per shopping trip do not change. If the income-driven change in shopping trip productions is proportional to the change in work trip productions, then:

$$P^z_{s,i+1} = P^z_{s,i} \cdot \left[\frac{P^z_{w,i} + dP^z_{w,i}}{P^z_{w,i}}\right] = P^z_{s,i} \cdot \left[\frac{P^z_{w,i+1}}{P^z_{w,i}}\right] \tag{11.6}$$

Reduced zonal retail and service sector activity reduces the number of shopping trips attracted to each zone. Define the change in shopping trips attracted to zone z to be dA^z_s. $X^z_{retail,i}$ and $X^z_{service,i}$ are total retail and service activity calculated at iteration i, respectively. Total baseline retail and service output are the initial values. Changes in retail and service activities result from direct, indirect, and induced impacts on those economic sectors. That is:

$$dX^z_{retail,i} = V^z_{d,retail,i} + V^z_{n,retail,i} + V^z_{u,retail,i} \tag{11.7.a}$$

And:

$$dX^z_{service,i} = V^z_{d,service,i} + V^z_{n,service,i} + V^z_{u,service,i} \tag{11.7.b}$$

These zone specific changes in retail and service activities produce changes in the shopping trips attracted to each zone:

$$A^z_{s,i+1} = A^z_{s,i} \cdot \left[\frac{\left(X^z_{retail,i} + X^z_{service,i}\right) - \left(dX^z_{retail,i} + dX^z_{service,i}\right)}{X^z_{retail,i} + X^z_{service,i}}\right] = \\ = A^z_{s,i} \cdot \left[\frac{X^z_{retail,i+1} + X^z_{service,i+1}}{X^z_{retail,i} + X^z_{service,i}}\right] \tag{11.8}$$

This procedure modifies the outer iteration loop identified in the lower portion of figure 11.1. This extension is convergent. This iterative procedure terminates when the changes in total trip productions and attractions become relatively small. As generalized travel costs stabilize across SCPM2 iterations, total trip productions and attractions stabilize. This simultaneously defines the trip matrices (OD requirements) generated in SCPM2's inner loop, and zone specific economic impacts defined in the outer loop of figure 11.1.

We have applied the approach described in equations (11.4) through (11.8) to the problem of endogenizing shopping trip ends. Results follow below. Reductions were calculated for both working and shopping trips, but only the changes in shopping trip requirements were imposed on the network equilibrium model and generalized cost calculations. An argument can be mounted for reducing both shopping and work trips, but shopping trips are certainly the more discretionary of the two.

Table 11.4 summarizes transportation network cost results for several representative bridge damage and policy scenarios. As before, scenarios are defined in terms of bridge damage, bridge closure policies, and the pre-earthquake traffic volumes on links affected by bridge closures. In the maximum scenario, the sum of affected baseline link traffic volumes is a maximum from a set 200 Monte Carlo realizations of earthquake damage. The median scenario is the scenario that provides the median affected total traffic volume. Recall that bridge damage is expressed in terms of a bridge damage index that ranges from 0 (no damage) to 1 (collapse). Closing only severely damaged bridges, *i.e.*, BDI ≥ 0.75, is a risk tolerant policy. Row 5 in each table modifies the results from row 1 by adding the effects of shopping trip end adjustments. These adjustments greatly reduce the demand for transportation, and thus associated flows and delays. This more meaningfully models the expected level of service on the network, but no attempt has been made to account for the economic value of these missing trips. The nature of earthquake-induced adjustments in work trip ends remains an important research question.

11.5 Application of SCPM2 to the Determination of Bridge Reconstruction Costs

11.5.1 Price Effects, Bottlenecks, and Budget Forecasts

Large-scale reconstruction efforts such as those identified above produce a variety of economic impacts. These include substantial indirect and induced production activities prompted throughout the metropolitan economy, as well as related price effects. It is important to anticipate these price effects as best we can because price increases intensify pressure on reconstruction budgets. Most important, local increases in wages augment reconstruction budget requirements. Additional system-wide price effects follow from the additional earnings accruing to households. SCPM2 is applied iteratively in the manner shown below to determine the size and location of all reconstruction employment and income effects. We do this for bridge repair budgets determined above, which we now augment by the price effects that the model computes.

11.5.2 Determining Reconstruction Budgets with Endogenous Price Adjustments

Linear interindustry models have been elaborated in many ways (Miller and Blair, 1985). In what follows, we describe how we utilized some of the elaborated models to endogenize price effects. Some of the critical baseline data are from the 1993 PC-IO package from the Regional Science Research Institute. These data includes X_0, a vector of baseline total outputs for the region; and A, a matrix of technical coefficients for the regional economy. Line D in table 11.3 provides an initial estimate of bridge reconstruction and repair costs. These are merely direct costs, and are likely to cause substantial multiplier effects.

Table 11.4 Total Travel Cost (before repairs) by Earthquake Scenario (Passenger Car Units × Minutes)

Total Travel Cost (before bridge repairs) by Earthquake Scenario (Passenger Car Units × Minutes)			
	Driver Delay	Freight Delay	Total Delay
Baseline	85,396,813	10,298,781	95,695,594
Close if BDI ≥ 0.30, Maximum Scenario	225,830,486	28,285,954	254,116,440
Close if BDI ≥ 0.30, Median Scenario	117,493,842	15,602,872	133,096,713
Close if BDI ≥ 0.75, Maximum Scenario	94,349,424	11,581,677	105,931,101
Close if BDI ≥ 0.75, Median Scenario	89,945,131	10,966,123	100,911,255
Close if BDI ≥ 0.30, Maximum Scenario with Trip End Adjustments	90,175,132	10,483,089	100,658,221
Total Travel Cost (before bridge repairs) by Earthquake Scenario ($Billions)			
	Driver Delay	Freight Delay	Total Delay
Baseline	21.290	4.550	25.839
Close if BDI ≥ 0.30, Maximum Scenario	56.300	12.495	68.795
Close if BDI ≥ 0.30, Median Scenario	29.291	6.893	36.184
Close if BDI ≥ 0.75, Maximum Scenario	23.522	5.116	28.638
Close if BDI ≥ 0.75, Median Scenario	22.424	4.844	27.268
Close if BDI ≥ 0.30, Maximum Scenario with Trip End Adjustments	22.481	4.631	27.112
Changes in Total Travel Cost Relative to Baseline (before bridge repairs) by Earthquake Scenarios (PCU × Minutes)			
	Driver Delay	Freight Delay	Total Delay
Close if BDI ≥ 0.30, Maximum Scenario	140,433,673	17,987,173	158,420,846
Close if BDI ≥ 0.30, Median Scenario	32,097,029	5,304,091	37,401,119
Close if BDI ≥ 0.75, Maximum Scenario	8,952,611	1,282,896	10,235,507
Close if BDI ≥ 0.75, Median Scenario	4,548,318	667,343	5,215,661
Close if BDI ≥ 0.30, Maximum Scenario with Trip End Adjustments	4,778,318	184,309	4,962,627
Changes in Total Travel Cost Relative to Baseline (before repairs) by Earthquake Scenarios ($Billions)			
	Driver Delay	Freight Delay	Total Delay
Close if BDI ≥ 0.30, Maximum Scenario	35.010	7.946	42.956
Close if BDI ≥ 0.30, Median Scenario	8.002	2.343	10.345
Close if BDI ≥ 0.75, Maximum Scenario	2.232	0.567	2.799
Close if BDI ≥ 0.75, Median Scenario	1.134	0.295	1.429
Close if BDI ≥ 0.30, Maximum Scenario with Trip End Adjustments	1.191	0.081	1.273

Notes: BDI = Bridge Damage Index. Closing a link for which BDI ≥ 0.30 is risk averse. Closing a link for which BDI ≥ 0.75 is risk tolerant.

Various economic translators are published by the Regional Science Research Institute. These translators are generic final demand vectors calibrated to unit expenditures on various kinds of projects. Each kind of project is defined by the mix of inputs required. RSRI translator #37 specifies plausible proportions of expenditures corresponding to the final demand sectors involved with bridge and highway construction. Application of these allows any project budget to be decomposed into a vector of final demands. Analysts can then use the IO model to calculate the full economic impacts of any major construction project. Our extension of this standard procedure is to calculate price effects that can be expected when project budgets become large.

Using standard input-output notation, the calculation procedure is as follows. Run the version of the I-O model that is closed with respect to the household sector. This gives:

$$dX_{i=0} = (I-A)^{-1} \cdot dY_{i=0} \tag{11.9}$$

where dY_0 = change in direct demand associated with bridge construction;
dX_0 = the associated change in total output; and
i = an iteration counter.

Concurrent with this result, there is a vector of household income changes, dW_0, as a result of the reconstruction program. We address the difficult problem of evaluating labor supply elasticity effects by adoption of an overtime rate. The burden of bridge reconstruction would require additional labor inputs to expedite rebuilding the bridges within a reasonable time period. If the construction sector were close to full employment, this could require attracting construction workers from outside the region. Such workers would have to be offered higher wages to pull them in and to affect what in some cases would be short-term (*e.g.* less than a year) local subsistence costs. Laub (1993) uses the example of Florida's Hurricane Andrew in 1992, to make the case that price increases, particularly with respect to labor, are highly likely following a natural disaster; and in fact, necessary to attract the materials and services needed for rapid recovery. But would any of these wage premia spillover into the wages received by local workers? To avoid having to address this difficult question, an alternative solution was adopted. Instead of attracting more workers from outside, the existing construction labor force could be used more intensively by allowing enough overtime to accommodate the additional labor requirements for bridge construction. The additional labor costs and associated price effects would not necessarily be the same as those resulting from offering higher wages to workers from outside the region, but they offer an acceptable alternative estimate.

Assuming a 24-hour, seven-day weekly work schedule, and a 50 percent wage premium for all overtime work, the average wage premium is 38 percent. Applying this assumed overtime rate to the construction sector:

$$dW_{i=0}^{c} = dW^{c} \cdot 0.38 \tag{11.10}$$

where dW^c is the change in labor cost in construction sectors. Thus the revised budget, including labor overtime costs, is:

$$dY_{i=1} = dY_{i=0} + dW_{i=0}^{c} \tag{11.11}$$

In addition we account for the impact of price effects on final demand. These price effects are calculated via the following iterative procedure. Given the revised budget from equation (11.11), run the closed IO model to obtain

$$dX_{i=1} = \left(I - A\right)^{-1} \cdot dY_{i=1} \tag{11.12}$$

Equation (11.12) updates the values of $dX_{i=0}$ obtained in equation (11.9). The household row of the transactions table gives $dW_{i=1}$, an initial vector of wage changes in all sectors. Higher wages in the system mean higher prices. Initial changes in wages are used to generate an initial estimate of changes in prices:

$$dP_{i=1} = \left(I - A^{T}\right)^{-1} \cdot \frac{dW_{i=1}}{X_0} \tag{11.13}$$

These higher prices further increase the required reconstruction budget:

$$dY_{i+1} = dY_i \cdot \left(1 - dP_i\right)^{-1} \tag{11.14}$$

The feedback between reconstruction budget requirements and prices described by equations (11.12) through (11.14) eventually attenuates, when the quantity $dP_{i+1} - dP_i$ is sufficiently small. A summary of this iterative procedure appears in figure 11.4.

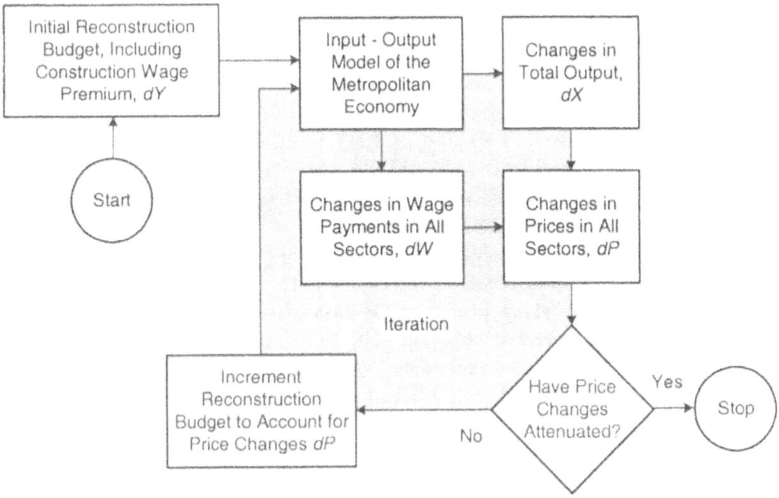

Figure 11.4 Computing a Reconstruction Budget that Accounts for Exogenous Price Effects

11.5.3 Simulation Results for Bridge Reconstruction with Endogenous Price Effects

We apply mean and median reconstruction cost alternatives to one of the risk tolerant bridge closure scenario (DS = 0.75, median traffic disruption). Table 11.5 summarizes the damage associated with this scenario, and associated mean and median facility repair costs. The cost distribution is skewed, and mean costs are more than twice median costs. As noted above, bridge repair cost estimates are based on a discriminant analysis of Loma Prieta and Northridge Earthquake bridge damage states and estimated repair costs.

Table 11.5 Bridge Damage Data

Bridge Damage State	Collapsed	Major	Moderate	Minor	Total
Number of Bridges	3	43	120	67	233
Median Repair Cost/Bridge ($1,000)	17,260	362	28	11	
Mean Repair Cost/Bridge ($1,000)	47,274	1,278	138	89	
Total Repair Cost ($1,000)					
Median Scenario	51,780	15,555	3,337	707	71,379
Mean Scenario	141,823	54,962	16,541	5,992	219,317

Applying this procedure for the median and mean bridge repair costs generates the results in tables 11.6 and 11.7, respectively. Each of the two tables show model IO results for the various modeling steps described immediately above. The first column describes how the Regional Science Research Institute's bridge and highway construction translator #37 allocates expenditures throughout the construction and other sectors. Rather than $71 million or $219 million spent within the region, the model determines how much there is in leakages (expenditures that accrue to firms and workers residing outside the metropolitan area). In table 11.6, leakages are $9.62 million. In table 11.7, they are $29.67 million.

We have disaggregated the construction sector appearing in the 17-sector model because the translator provides extra levels of detail. The second columns in tables 11.6 and 11.7 show the same programs of expenditure if the contracting agency absorbs the 38 percent overtime charges. The third columns show the same budget after completion of the iterative procedure. This is the new vector of direct effects (column 4), used to calculate indirect, induced and total effects (columns 5, 6, and 7).

Table 11.8 details the additional transportation costs incurred in the process of reconstruction. Table 11.9 contrasts the mean and median reconstruction cost outcomes relative to the baseline described in table 11.3. Accounting for endogenous price effects increases mean reconstruction budget requirements by approximately $42 million. This assumes bridges are reconstructed at the mean cost observed in the data.

Table 11.6 Reconstruction Budget and Interindustry Economic Effects of Median Cost[a] Reconstruction Activity

Sector	Reconstruction Budget[b]			Economic Effects of Reconstruction[b]			
	Initial Budget	Budget With Overtime[c]	New Budget Considering Price Effects	Direct	Indirect	Induced	Total
1. Agriculture	0.00	0.00	0.00	0.00	0.08	0.29	0.37
2. Mining	0.46	0.46	0.46	0.46	0.21	0.30	0.97
3. Construction	38.01	49.00	49.28	49.28	0.50	0.62	50.40
32. general const. contractors	4.39	5.65	5.66	5.66	0.07	0.09	5.82
33. highway & street construction	3.14	3.92	3.95	3.95	0.00	0.00	3.95
34. other heavy const. contractors	9.41	11.96	12.07	12.07	0.02	0.03	12.11
35. plumb/heat/air cond. contrctrs	1.66	2.17	2.17	2.17	0.03	0.04	2.23
36. painting, papering, decorating	0.00	0.01	0.01	0.01	0.01	0.01	0.03
37. electrical const. contractors	6.10	8.02	8.05	8.05	0.03	0.05	8.13
38. masonry, drywall & plastering	2.22	2.88	2.89	2.89	0.03	0.03	2.95
39. carpentering & flooring	0.00	0.01	0.01	0.01	0.01	0.01	0.03
40. roofing & sheet metal work	0.00	0.01	0.01	0.01	0.01	0.02	0.03
41. concrete work	1.11	1.44	1.44	1.44	0.01	0.02	1.47
42. water well drilling	0.00	0.00	0.00	0.00	0.00	0.00	0.00
43. special trade contractors, nec	9.98	12.95	13.04	13.04	0.02	0.03	13.09
44. maint & rep: residential bldgs	0.00	0.00	0.00	0.00	0.02	0.10	0.12
45. maint & rep: non-res. bldgs.	0.00	0.00	0.00	0.00	0.23	0.16	0.39
46. maint & rep: farm residences	0.00	0.00	0.00	0.00	0.00	0.00	0.00
47. maint & rep: other farm bldgs.	0.00	0.00	0.00	0.00	0.00	0.00	0.00
58. maint & rep: streets & h'ways	0.00	0.00	0.00	0.00	0.00	0.00	0.00
59. maint & rep: petr. & gas wells	0.00	0.00	0.00	0.00	0.01	0.01	0.02
60. maint & rep: other nonbldg fac	0.00	0.00	0.00	0.00	0.01	0.01	0.02
4. Manufacturing (nondurable)	1.71	1.71	1.71	1.71	1.82	5.94	9.47
5. Manufacturing (durable)	14.57	14.57	14.57	14.57	3.73	2.21	20.50
6. Transportation	0.87	0.87	0.87	0.87	0.54	0.47	1.87
7. Communications and utilities	0.22	0.22	0.22	0.22	0.71	2.15	3.09
8. Wholesale trade	2.02	2.02	2.02	2.02	1.90	1.77	5.68
9. Retail	0.00	0.00	0.00	0.00	1.44	6.48	7.92
10. F.I.R.E.	0.00	0.00	0.00	0.00	1.76	9.69	11.45
11. Business services	0.00	0.00	0.00	0.00	2.71	2.19	4.90
12. Personal services	0.00	0.00	0.00	0.00	0.19	0.80	0.99
13. Entertainment and recreation	0.00	0.00	0.00	0.00	0.05	0.86	0.91
14. Health	0.00	0.00	0.00	0.00	0.00	0.80	0.80
15. Educational services	0.00	0.00	0.00	0.00	0.01	0.61	0.62
16. Professional and related	3.52	3.52	3.52	3.52	1.13	2.59	7.23
17. Government	0.00	0.00	0.00	0.00	0.26	1.04	1.30
Sum	61.38	72.37	72.65	72.65	17.03	38.80	128.48
Leakage	9.62		11.39				
Total	71.00		84.04				

Notes: a. Median total bridge repair/replacement cost = $71 Million.
b. $Millions.
c. Prime Rate for Overtime-Labor in Construction Sector = $[40+(128\times 1.5)]/168=1.38$.

Table 11.7 Reconstruction Budget and Interindustry Economic Effects of Mean[a] Cost Reconstruction Activity

Sector	Reconstruction Budget[b]			Economic Effects of Reconstruction[b]			
	Initial Budget	Budget With Overtime[c]	New Budget Considering Price Effects	Direct	Indirect	Induced	Total
1. Agriculture	0.00	0.00	0.00	0.00	0.25	0.90	1.15
2. Mining	1.43	1.43	1.43	1.43	0.63	0.95	3.01
3. Construction	117.25	151.14	153.88	153.88	1.56	1.92	157.36
32. general const. contractors	13.54	17.42	17.49	17.49	0.23	0.28	18.00
33. highway & street construction	9.67	12.08	12.39	12.39	0.00	0.01	12.41
34. other heavy const. contractors	29.02	36.90	37.93	37.93	0.05	0.08	38.06
35. plumb/heat/air cond. contrctrs	5.12	6.68	6.71	6.71	0.08	0.12	6.91
36. painting, papering, decorating	0.00	0.02	0.02	0.02	0.03	0.04	0.08
37. electrical const. contractors	18.82	24.72	25.07	25.07	0.09	0.15	25.31
38. masonry, drywall & plastering	6.85	8.89	8.96	8.96	0.08	0.11	9.14
39. carpentering & flooring	0.00	0.02	0.02	0.02	0.03	0.04	0.09
40. roofing & sheet metal work	0.00	0.02	0.02	0.02	0.03	0.05	0.10
41. concrete work	3.42	4.44	4.47	4.47	0.03	0.05	4.45
42. water well drilling	0.00	0.00	0.00	0.00	0.00	0.01	0.01
43. special trade contractors, nec	30.80	39.93	40.80	40.80	0.05	0.11	40.97
44. maint & rep: residential bldgs	0.00	0.00	0.00	0.00	0.06	0.32	0.38
45. maint & rep: non-res. bldgs.	0.00	0.00	0.00	0.00	0.72	0.49	1.21
46. maint & rep: farm residences	0.00	0.00	0.00	0.00	0.00	0.00	0.00
47. maint & rep: other farm bldgs.	0.00	0.00	0.00	0.00	0.00	0.01	0.01
58. maint & rep: streets & h'ways	0.00	0.00	0.00	0.00	0.00	0.01	0.01
59. maint & rep: petr. & gas wells	0.00	0.00	0.00	0.00	0.04	0.02	0.06
60. maint & rep: other nonbldg fac	0.00	0.00	0.00	0.00	0.02	0.03	0.05
4. Manufacturing (nondurable)	5.28	5.28	5.28	5.28	5.63	18.51	29.42
5. Manufacturing (durable)	44.93	44.93	44.95	44.95	11.52	6.88	63.35
6. Transportation	2.68	2.68	2.68	2.68	1.66	1.45	5.79
7. Communications and utilities	0.69	0.69	0.69	0.69	2.21	6.70	9.60
8. Wholesale trade	6.22	6.22	6.22	6.22	5.89	5.50	17.60
9. Retail	0.00	0.00	0.00	0.00	4.47	20.19	24.66
10. F.I.R.E.	0.00	0.00	0.00	0.00	5.46	30.18	35.64
11. Business services	0.00	0.00	0.00	0.00	8.43	6.81	15.24
12. Personal services	0.00	0.00	0.00	0.00	0.58	2.50	3.08
13. Entertainment and recreation	0.00	0.00	0.00	0.00	0.16	2.69	2.84
14. Health	0.00	0.00	0.00	0.00	0.01	2.48	2.49
15. Educational services	0.00	0.00	0.00	0.00	0.03	1.90	1.93
16. Professional and related	10.86	10.86	10.86	10.86	3.49	8.06	22.41
17. Government	0.00	0.00	0.00	0.00	0.82	3.23	4.06
Sum	189.33	223.22	225.99	225.99	52.80	120.86	399.65
Leakage	29.67		35.42				
Total	219.00		261.41				

Notes: a. Mean total bridge repair/replacement cost = $219 Million.
b. $Millions.
c. Prime Rate for Overtime-Labor in Construction Sector = [40+(128 × 1.5)]/168=1.38.

Table 11.8 Additional Travel Cost Associated with Mean and Median Reconstruction Activities

Median Reconstruction Cost Scenario	Passenger Delay	Freight Delay	Total Delay
Total Travel Cost (Baseline Flows + Reconstruction Flows)			
(PCU × Minutes)	91,702,850	11,070,634	102,773,483
($Billion)	22.862	4.890	27.752
Change Due to Reconstruction Flows			
(PCU × Minutes)	1,757,719	104,511	1,862,228
($Billion)	0.438	0.046	0.484

Mean Reconstruction Cost Scenario	Passenger Delay	Freight Delay	Total Delay
Total Travel Cost (Baseline Flows + Reconstruction Flows)			
(PCU × Minutes)	92,349,189	11,110,640	103,459,830
($Billion)	23.023	4.908	27.931
Change Due to Reconstruction Flows			
(PCU × Minutes)	2,404,058	144,517	2,548,575
($Billion)	0.599	0.064	0.663

Table 11.9 Bridge Reconstruction Cost Summary ($Billions)

	Bridge Repair Costs		Additional Travel Cost Associated with Reconstruction[b]
	Ignoring Endogenous Price Effects[a]	Accounting for Endogenous Price Effects	
Median Reconstruction Cost Scenario	$ 0.071	$ 0.084	0.484
Mean Reconstruction Cost Scenario	$ 0.219	$ 0.261[b]	0.663

Notes: a. Table 11.3.
b. Table 11.8.

11.6 Conclusions

This chapter addresses some often ignored repercussions of a major earthquake by supplementing the standard structural damage impacts with business loss effects (although these have been a few studies of this issue) and disruptions to the transportation network. These include increased freight and travel costs resulting from damage to the network, and the supply inelasticities associated with a major bridge reconstruction effort. The goal of this effort is a full-cost approach to measuring earthquake impacts, but in fact the estimates derived here ($94.34 billion) are loose lower bounds. There are several reasons. First, the increase in travel costs in the post-earthquake situation is minimized by adopting a risk tolerant response to bridge damage. Only those facilities for which the bridge damage index exceeds 0.75 are closed. If the more conservative threshold (BDI $\geq$ 0.30) had been adopted, the median case travel costs increase would have surged to $10 billion. Second, the research does not include all costs inflicted on households. This study accounts only for residential structure damage and increased personal travel costs because of the damaged network. Dislocation costs if households have to move to temporary accommodation are ignored, as are possible reductions in consumption because the difficulty of getting to destinations. Third, there is no attention given in this research to the costs of possible deaths and injuries associated with an earthquake of this magnitude. Of course, such estimates would be contingent upon a number of unknowns, such as the time of day when the earthquake occurred. When all these considerations are taken into account, the full-costs of the earthquake would certainly be much higher.

References

Arnott, R., A. Palma, and R. Lindsey. 1993. "Properties of Dynamic Traffic Equilibrium Involving Bottlenecks, Including a Paradox and Metering," *Transportation Science,* 27 (2): 148-160.

Blunden, W.R. 1971. *The Land-Use/Transportation System Analysis and Synthesis.* Oxford, UK; Pergamon Press.

Boarnet, M. 1998. "Business Losses, Transportation Damage and the Northridge Earthquake," *Journal of Transportation and Statistics,* 1 (2): 49-64.

Bureau of Transportation Statistics. 1990. *Census Transportation Planning Package.* US DOT.

Campbell, K.W. 1997. "Empirical Near Source Attenuation Relationships for Horizontal and Vertical Components of Peak Ground Acceleration, Peak Ground Velocity, and Pseudo-Absolute Acceleration Response Spectra," *Seismological Research Letters,* 68 (1), January/February.

Chang, S.E. 1995. "Urban Economic Impact of Transportation Disruption: Evidence from the January 17, 1995 Hyogo-ken Nambu Earthquake," presented at the 14th Conference of the Pacific Regional Science Association (PRSA, Taipei, July 25).

Charney, A. and J. Leones. 1997. "Implan's Induced Effects Identified Through Multiplier Decomposition," *Journal of Regional Science,* 37 (3): 503-517.

Cho, S., P. Gordon, J.E. Moore II, H.W. Richardson, M. Shinozuka, and S.E. Chang. 2001. "Integrating Transportation Network and Regional Economic Models to Estimate the Costs of a Large Urban Earthquake," *Journal of Regional Science,* 41 (1): 39-65.

Cho, S., P. Gordon, J.E. Moore II, H.W. Richardson, and M. Shinozuka. 2000. "Analyzing Transportation Reconstruction Network Strategies: A Full Cost Approach," *Review of Urban & Regional Development Studies,* 12 (3): 212-227.

Cho, S., P. Gordon, J.E. Moore II, H.W. Richardson, M. Shinozuka, and S.E. Chang. 1999. "Effect of Earthquakes on Urban Highway Infrastructure Productivity, Volume 2: Integrating Transportation Network and Regional Economic Models to Estimate the Costs of A Large Earthquake," Report to the National Science Foundation.

Cochrane, H.C. 1997. "Forecasting the Economic Impact of a Midwest Earthquake," in B.G. Jones ed. *Economic Consequences of Earthquakes: Preparing for the Unexpected.* Buffalo, NY; NCEER.

Conrad, K. 1997. "Traffic, Transportation, Infrastructure and Externalities," *The Annals of Regional Science,* 31: 369-389.

Dietzenbacher, E. 1997. "In Vindication of the Ghosh Model: A Reinterpretation as a Price Model," *Journal of Regional Science,* 37 (4): 629-651.

DRI-McGraw Hill. 1994. *Interregional Goods Movement Database Final Report.* Los Angeles, CA; Southern California Association of Governments.

Eguchi, R.T., J.D. Goltz, C.E. Taylor, S.E. Chang, P.J. Flores, L.A. Johnson, H.A. Seligson, and N.C. Blais. 1998. "Direct Economic Losses in the Northridge Earthquake: A Three-Year Post-Event Perspective," *Earthquake Spectra,* 14 (2): 245-264.

Eguchi, R.T., J.D. Goltz, H.A. Seligson, P.J. Flores, N.C. Blais, T.H. Heaton, and E. Bortugno. 1997. "Real-Time Loss Estimation as an Emergency Response Decision Support System: The Early Post-Earthquake Damage Assessment Tool," *Earthquake Spectra,* 13 (4): 815-832.

Evernden, J.F. and J.M. Thomson. 1988. "Predictive Model for Important Ground Motion Parameters Associated with Large and Great Earthquakes," United States Geological Survey, Bulletin 1838.

Florian, M. 1984. *An Introduction to Network Models Used in Transportation Planning, Transportation Planning Models*. Amsterdam, The Netherlands; Elsevier Science.

Gordon, P. and H.W. Richardson. 1999. "Review Essay: The City: Los Angeles and Urban Theory at the End of the Twentieth Century," *Urban Studies,* 36 (3): 575-591.

Gordon, P., H.W. Richardson, and B. Davis. 1996. "The Business Interruption Effects of the Northridge Earthquake," Report to the National Science Foundation, Grant No. CMS-9416758, the University of Southern California, Los Angeles, CA.

Gordon, P., H.W. Richardson, and B. Davis. 1998. "Transport-Related Impacts of the Northridge Earthquake," *Journal of Transportation and Statistics,* 1 (2): 21-36.

Han, X., and B. Fang. 2000. "Four Measures of Transportation's Economic Importance," *Journal of Transportation Statistics,* 3 (1): 15-30.

Isard, W. and R.E. Kuenne. 1953. "The Impact of Steel upon the Greater New York-Philadelphia Industrial Region," *Review of Economics and Statistics,* 35 (4): 289-301.

Israilevich, P., G.J.D. Hewings, M. Sonis, and G. Schindler. 1997. "Forecasting Structural Change with a Regional Econometric Input-Output Model," *Journal of Regional Science,* 37 (4): 565-590.

Jackson, R., M. Madden, and H. Bowman. 1997. "Closure in Cole's Reformulated Leontief Model," *Papers in Regional Science,* 76 (1): 21-28.

Kimbell, L.J. and N. Bolton. 1994. "The Impact of the Northridge Earthquake on the Economies of Los Angeles and California," presented to the Seismic Safety Commission of the State of California (Burbank, CA, March 2).

Kiyota, M., U. Vandebona, and H. Tauoue. 1999. "Multistage Optimization of Reconstruction Sequence of Highways," *ASCE Journal of Transportation Engineering,* 125 (5): 456-462.

Laub, P.M. 1993. "Insurance Companies, Banks, and Economic Recovery in South Florida in the Wake of Hurricane Andrew," Report to FEMA, Development Technologies, Inc. Washington, DC; US Government Printing Office.

Leontief, W. 1951. *The Structure of the American Economy, 1919-1939: 2nd Edition.* New York, NY; Oxford University Press.

de Mesnard, L. 1997. "A Biproportional Filter to Compare Technical and Allocation Coefficient Variations," *Journal of Regional Science,* 37 (4): 541-564.

Miller, R.E. and P.D. Blair. 1985. *Input-Output Analysis: Foundations and Extensions.* New Jersey; Prentice-Hall.

Moses, L.N. 1960. "A General Equilibrium Model of Production, Interregional Trade, and Location of Industry," *Review of Economics and Statistics,* 42: 373-399.

Okuyama, Y., G.J.D. Hewings, and M. Sonis. 1997. "Interregional Analysis of an Unscheduled Event," presented at the 44th North American Meetings of the Regional Science Association International (RSAI, Buffalo, October 31).

Putnam, S.H. 1983. *Integrated Urban Models: Policy Analysis of Transportation and Land Use.* London; Pion Ltd.

Richardson, H.W., P. Gordon, M.J. Jun, and M.H. Kim. 1993. "PRIDE and Prejudice: The Economic Impacts of Growth Controls in Pasadena," *Environment and Planning A,* 25 (A): 987-1002.

Rose, A., J. Benavides, S.E. Chang, P. Szczesniak, and D. Lim. 1997. "The Regional Economic Impact of an Earthquake: Direct and Indirect Effects of Electricity Lifeline Disruptions," *Journal of Regional Science,* 37 (3): 437-458.

Rose, A. and J. Benavides. 1998. "Regional Economic Impacts," in M. Shinozuka *et al.* eds. *Engineering and Socioeconomic Impacts of Earthquakes: An Analysis of Electricity Lifeline Disruptions in the New Madrid Area.* Buffalo, NY; MCEER.

Rose, A. and D. Lim. 1997. "The Economic Impact of Electricity Lifeline Disruptions Stemming from the Northridge Earthquake," Final Report to the National Science Foundation, Department of Energy, Environmental, and Mineral Economics, The Pennsylvania State University, University Park, PA.

Sheffi, Y. 1985. *Urban Transportation Networks-Equilibrium Analysis with Mathematical Programming Methods*. New Jersey: Prentice-Hall.

Shinozuka, M. 1998. "Development of Bridge Fragility Curves," Proceedings of the Joint US-Italy Workshop on Seismic Protective Systems for Bridges, New York, NY; Columbia University, April; 26-28.

Shinozuka, M. 1999. "Effect of Earthquakes on Urban Highway Infrastructure Productivity, Volume I," Report to the National Science Foundation.

Shinozuka, M., S.E. Chang, R.T. Eguchi, D.P. Abrams, H.H.M. Hwang, and A. Rose. 1997. "Advances in Earthquake Loss Estimation and Application to Memphis, Tennessee," *Earthquake Spectra*, 13 (4): 739-748.

Shoven, J. and J. Whalley. 1992. *Applying General Equilibrium*. Cambridge, UK; Cambridge University Press.

Southern California Association of Governments. 1993. *1991 Origin-Destination Survey*. Los Angeles, CA.

Stevens, B.H. 1996. *PC I-O Version 7*. Hightstown, NJ; Regional Science Research Corporation.

Stevens, B.H., G. Treyz, and M. Lahr. 1983. "A New Technique for the Construction of Non-Survey Regional Input-Output Models," *International Regional Science Review*, 8 (3): 271-286.

Stopher, P.R. 1975. *Urban Transportation Modeling and Planning, D.C. Health and Company*. Lexington; Lexington Books.

Thomas, R. 1991. *Traffic Assignment Techniques*. UK; Avebury Technical.

U.S. Bureau of the Census. 1993a. *1992 Economic Census CD-ROM*. Washington, DC; Economics and Statistics Administration, Bureau of the Census.

U.S. Bureau of the Census. 1993b. *1992 Economic Census: Census of Transportation, Communication, and Utilities--1993 Commodity Flow Survey: TC92-CF-52*. Washington, DC; Economics and Statistics Administration, Bureau of the Census.

U.S. Department of Transportation, Bureau of Transportation Statistics. 1997. *1993 Commodity Flow Survey: U.S. Highlights*. Washington, DC.

U.S. Department of Transportation. 1990. *National Transportation Statistics Annual Report*. Report DOT-TSC-RSPA-90-2, Research and Special Programs Administration. Washington, DC; U.S. Government Printing Office.

Wilson, A.G. 1970. *Entropy in Urban and Regional Modeling*. London, UK; Pion.

12 Analysis of Economic Impacts of an Earthquake on Transportation Network*

Jungyul Sohn,[1] Geoffrey J.D. Hewings,[2] Tschangho John Kim,[3] Jong Sung Lee,[3] and Sung-Gheel Jang[3]

[1] National Center for Smart Growth Research and Education, University of Maryland, and Regional Economics Applications Laboratory, University of Illinois, USA
e-mail: jsohn@umd.edu
[2] Regional Economics Applications Laboratory, University of Illinois, USA
e-mail: hewings@uiuc.edu
[3] Department of Urban and Regional Planning, University of Illinois, USA
e-mail: tjohnkim@uiuc.edu, jonglee1@uiuc.edu, sjang3@uiuc.edu

12.1 Introduction

Prior to the 1990s, natural disasters and their economic impacts were not a major field of study for regional economic analysts even though there was a sizeable literature based on structural engineering and geotechnical approaches. The latter approaches attempted to understand the behavior of earthquakes and to explore ways to prevent or minimize damage from the disaster should it occur. However, when decisions needs to be made on the retrofit of existing facilities as a prevention or the restoration of disrupted facilities after damages, economic considerations related to budgeting priorities have not been prominently featured. As a consequence, decisions about retrofit strategies tend to focus on engineering-based criteria (for example, bridge 21 on route 50 should be retrofitted because it presents the greatest probability of collapsing given an earthquake of magnitude x) rather than on economic criteria (for example, a 10% loss of capacity on bridge 10 on route 60 would create the greatest economic disruption under a similar earthquake scenario and hence would have the highest priority for retrofit). Hence, there is a clear need to provide some interface to explore the ways in which engineering-based assessments can be compared with those based on economic analysis tools. The current research described in this chapter provides such an interdisciplinary research effort.[1]

The purpose of the present research is to explore various economic impacts of the earthquake especially on the transportation network and to provide essential information for a rational retrofit strategy of this transportation network. The analysis is timely in that most other analyses have focused on disruption of economic facilities such as factories and offices, but have tended to ignore a strategic approach to transportation facilities. For the purpose of the research, a 25-year span of the final demand vectors are estimated as the reference values starting from the year 1993, the

* This work was supported primarily by the Earthquake Engineering Research Centers Program of the National Science Foundation under Award Number EEC-9701785 to the Mid American Earthquake Center at the University of Illinois at Urbana-Champaign.

[1] For example, some of the functions in the scenario analysis come from the civil engineering research groups in the Mid-America Earthquake Center (MAEC).

base year of the integrated commodity flow model (ICFM).[2] The framework of the scenario analysis is set up based on the combination of the final demand loss function (FDLF) as well as the ICFM. A hypothetical scenario is set up and examined to check the performance of the framework of the analysis.

In the next section, a brief review of approaches to the economic analysis of earthquakes will be provided. Section three describes the analysis framework, while the data preparation for the final demand estimation is discussed in section four. The final demand loss function is introduced in section five and the transportation network model in section six. The scenario analysis is described in section seven with some concluding comments completing the final section.

12.2 Economic Impacts of Earthquakes: A Brief Review

Two aspects of economic impacts led by earthquakes are of a major concern of this chapter: transportation cost increase and economic damage (defined as final demand loss). Assessing economic damage resulted from earthquakes, or more generally from natural disasters, has been a popular topic among those who are interested in natural disasters from a socioeconomic viewpoint. There have been various measures for economic assessment introduced and used in association with many different tools and methods. Chang (1998) may be classified in the group of research to identify the most direct economic impact by examining physical damages and the consequent impacts in production system. Her study is also unique in that she tried to make a stochastic approach on the loss function. Davis and Salkin (1984) used output change as the measurement of economic impact. They tried to estimate the changes in the outputs of unconstrained sectors given the reduced outputs of the supply-constrained (by the earthquake) sectors. Instead of estimating the change of outputs, Cochrane (1974) focused on estimating the change of the sectoral employment based on an intensity damage relation curve showing the relationship between the modified Mercalli intensity (MMI) and mean damage ratio. Quigley (1998) tried to link fatalities with their monetary cost and used them as a measure of economic impact. It was one of the few that focused on a benefit-cost analysis of the retrofit policies. The author compared the annualized costs incurred by three different policy options with their expected annual savings in terms of damage reduced and fatalities averted.

There has been relatively more number of studies that examined final demand change as a measure of economic impact along with input-output techniques. Rose *et al.* (1997) and Rose and Benavides (1998) devised an impact analysis model to examine final demand change led by the disruption of electricity lifeline. The model is a revised version of the standard input-output impact formula, specified to measure the disruption impact on the economy. As an extension of the input-output technique, Cole (1998) tried to combine the Social Accounting Matrix (SAM) approach with the GIS in his exploratory approach to explore the spatial as well as the economic impact of the earthquake. HAZUS (a natural hazard loss estimation methodology) may be more comprehensive extension of the economic impact analysis tool that combines input-output technique. As noted in Brookshire *et al.* (1997) and Cochrane (1997), this

[2] This research is the second phase of a project on the economic impact of earthquakes co-funded by the NSF and MAEC; the first phase focused on the construction of the ICFM.

model handles both direct and indirect economic damages for multiple years and impacts are measured in different ways: output, income and employment.

Some researchers have been more interested in efficiencies of management and planning of economic system during the post-disaster period, in addition to focusing on measuring damages themselves. Several studies have adopted linear programming techniques to derive the optimal distribution pattern after the damage occurs (see, for example, Rose *et al.*, 1997; Cochrane, 1974; Rose, 1981; and Rose and Benavides, 1998). The results from such applications can be used as benchmarks against which more pessimistic assumptions could be compared.

Impact assessment on transportation network performance led by natural disasters has been more recent addition to the school of studies in the impact analysis of disasters. A series of works by Kim *et al.* (1999), Ham (2001), and Kim *et al.* (2002) have been devoted to developing the integrated commodity flow model that optimizes network flows under a given condition. The model runs with assumptions on link damages and the resulting transportation cost of the model is compared with the one under the normal condition. The model is flexible in that it adopts different levels of disruption ratio (partial or complete damages) on network links before the model runs. Even if they have not explicitly tried to link transportation network performance and the resulting cost with economic damage, final demand vector should be defined before the model runs and this enables an extension of the model that explicitly combines transportation and economic impacts.

Combining transportation cost with economic damage provides a comprehensive analysis framework on impact studies. Not many attempts, however, have been made to link those two aspects of impact. Okuyama *et al.* (1998, and 1999) tried not only to extend the economic impact analysis model of electricity lifeline disruption into the one for transportation networks, but also to set up a foundation to link economic damage with transportation network performance at a conceptual level. Cho *et al.* (2000) developed an integrated transportation network and regional economic model. It integrated four sub-models: bridge and other structure performance model, transportation network model, spatial allocation model, inter-industry model. The model was devoted to measuring transportation and economic cost led by an earthquake.

This chapter belongs to the last group of research that explicitly incorporates transportation and economic cost. Some of the improvements of the chapter compared to earlier works in the group include: (1) real highway/railway network considered; (2) partial closure of the bridges and links allowed; and (3) reduced final demands due to the disruption of the transportation network accounted for. Also, the model in this chapter is able to provide damage cost by an individual economic sector by each zone, which has not been achieved by other economic impact analyses.

12.3 Methodological Approach

This section discusses the framework to estimate the economic impacts of the disruption on the transportation network generated by an earthquake and to calibrate system-wide transportation cost increases and final demand loss. The sum of these costs is considered as the potential benefit, since this is the opportunity cost (or shadow price) of the retrofit on the highway bridges. Figure 12.1 shows the framework of the scenario analysis.

First of all, the transportation network model (ICFM) is run to estimate the system transportation cost under normal conditions, which will be used as the reference cost. An appropriate earthquake scenario requires three elements: (1) Location of the epicenter; (2) Moment magnitude; and (3) Year.

The location of the epicenter and the earthquake magnitude are critical components in the estimation of the damage on the transportation network. Once the year of the earthquake is determined, then the appropriate final demand vector is applied to the transportation network model and the final demand loss function. In addition, the present value of costs and benefits should be generated by discounting future values using a proper discount rate in order for the values from different years to be comparable. While it may be arbitrary to determine the most appropriate discount rate, Quigley (1998) suggested using higher interest rates for discounting future values to be conservative in evaluating retrofit policy.

Once the earthquake occurs according to the scenario, the network loss function needs to be derived to estimate the damage of the earthquake on the transportation network. The resulting loss on the transportation network can be scaled as the ratio of the total capacity that spans from zero (no damage) to one (complete destruction).

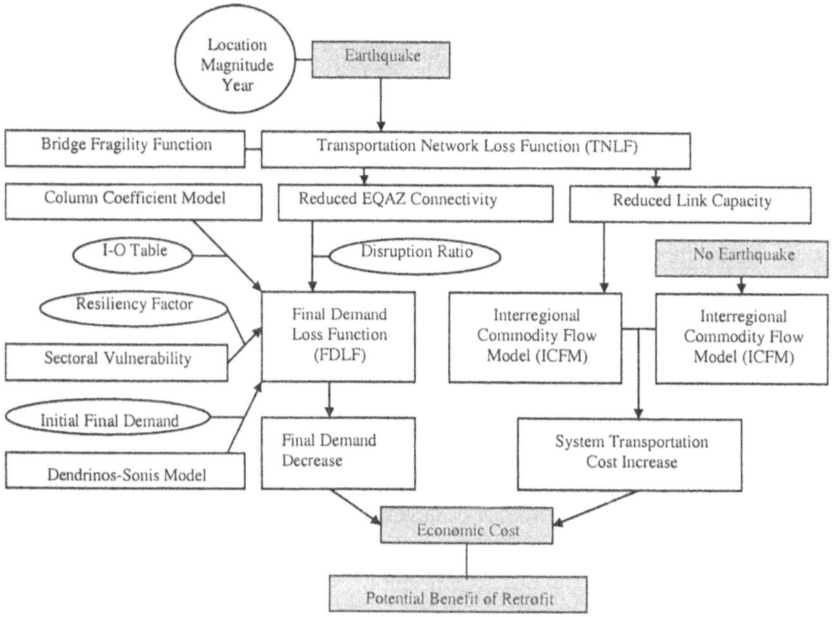

Figure 12.1 Analysis Framework

While the loss function is related to the direct impact, the indirect economic impact can be estimated through the interindustry relationships. The damage or the strength of the earthquake weighted by the distance from the origin of the earthquake can be translated into the adjusted final demands in each zone through the process incorporating the input-output table, the initial final demands, the zone-based network disruption ratio and the resiliency factor.

Once the network capacity in the transportation network model is adjusted according to a scenario, the ICFM runs again. Now, the model provides the overall transport cost for the optimized freight shipment under the changed network capacity. This value can be compared to the one obtained without any disruption of the network and the difference provides the increased system cost.

12.4 Data Preparation: Final Demand Estimation

While there are several measures to reflect the economic impact, there are certain advantages to using final demands as a measure of the economic impact. First, if we are more interested in indirect economic impact, final demand is an appropriate variable to initiate the generation of the disruptive impacts. As mentioned in the previous section, the direct economic impact is related to the disruption of buildings, facilities or factories. If this is the case, total output may be a more appropriate variable in that the level of production is directly related to the level of loss. However, the impact of transportation network disruption on economic activities is indirect in the sense that it affects production (level) not by damaging production facilities themselves, but by restricting the mobility of goods and adjusting the demand and supply mechanism. In other words, even if disruption of the network does not affect production facilities, it is still influential on regional demand in such a way that less supply due to network disruption may cause demand setback. Secondly, if we are also interested in the interindustry effects of certain damages, working through changes in final demands offers an appropriate vehicle.

There are 36 EarthQuake Analysis Zones (EQAZs) and 13 economic sectors in this analysis. The time period of the research starts from the base year, 1993, and extends to 2017 completing 25-year period that is typical to economic forecasting. Figure 12.2 shows the 36 EQAZs in the continental U.S. Since the main focus of the analysis lies in the Midwest, it is disaggregated into 29 smaller zones while the rest of the U.S. is divided into 7 macro regions. There are 9 states in the Midwest in the analysis: Illinois, Indiana, Iowa, Michigan, Missouri, Kentucky, Ohio, Tennessee and West Virginia. 29 EQAZs are created first by dividing all the zones based on the state and NTAR (National Transportation Analysis Region) boundaries[3] and then by aggregating some smaller zones with neighbors so that the average number of EQAZs is between 2 to 5 per state.

Table 12.1 lists the 13 economic sectors that are used in the analysis. Emphasizing the characteristics of the Midwestern economy, seven of the sectors are manufacturing (for more detail, see Okuyama *et al.*, 1998, and 1999).

[3] In a practical sense, it provides an easier way to redistribute the relevant economic data since the Commodity Flow Survey (CFS) published by the Bureau of Transportation Statistics gives both sets of information: the state and the NTAR.

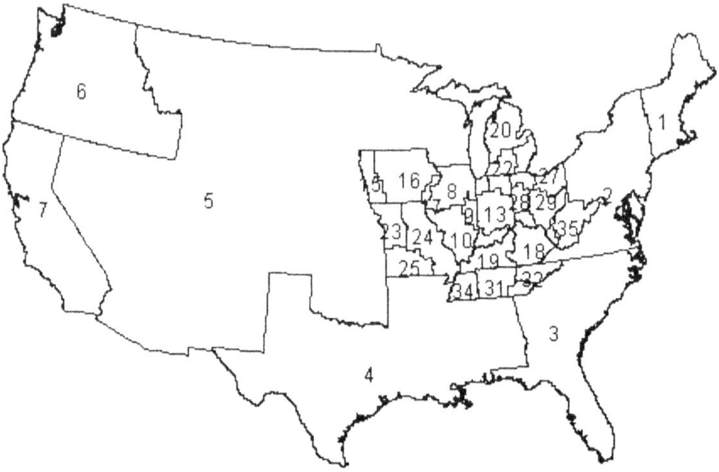

Figure 12.2 36 EQAZs in the Continental U.S.

Table 12.1 Sector Classification

Sector	Description
1	Agriculture, forestry and fisheries
2	Mining
3	Construction
4	Food and kindred products
5	Chemicals and allied products
6	Primary metals industries
7	Fabricated metal products
8	Industrial machinery and equipment
9	Electronic and electric equipment
10	Transportation equipment
11	Other non-durable manufacturing
12	Other durable manufacturing
13	TCU*, services and government enterprises

Note: * Transportation, Communication and Utilities.

Three steps are involved in estimating the 25 years of final demands for each of the 13 sectors by the 36 EQAZs. These are (1) obtaining available time series regional shares of total employment, population and government employment, (2) forecasting time series regional shares of total employment, population and government employment up to 2017, and (3) redistributing final demands obtained from Regional Econometric Input-Output Model (REIM) into regions using forecast regional shares of total employment, population and government employment.

The first step is to extract and rearrange the time-series data on total employment, population and government employment by each zone and then to obtain the zonal shares of them in the state in which it is located. For forecasting, the time series of those three data series have been gathered from the Bureau of Census: the total population from 1970 to 1998, the total employment from 1988 to 1997, and the government employment from 1969 to 1994.[4]

The second step is to forecast the shares of those two types of employment and the population for 25-year span starting from 1993 up to 2017. Since some portion of the data already includes the time series after 1993, those are used as they are.[5] The primary methodology used in this step is the Dendrinos-Sonis (D-S) model. It was first introduced in Sonis and Dendrinos (1987) for modeling the relative dynamics between two locations for one stock using a nonlinear method. It was later extended by using numerical computations in Dendrinos and Sonis (1987) and the first empirical application of the model with multiple locations and one stock was developed by Dendrinos and Sonis (1988). Hewings *et al.* (1996) were the first to adopt economic variables in the model and Magalhães *et al.* (1999) further applied the model to examine the regional competition and complementarity in the U.S. and Brazil. The model has been shown to be useful in understanding and forecasting growth dynamics in a multiregional system (Jackson and Sonis, 2001). One of the advantages of the model is that it is not necessary to assume a linear trend of change as most time-series models do. However, since the model is relative dynamics, the system-wide forecast (growth or decline) of total regions has to be provided exogenously to estimate individual region's forecast.

The model is an iterative mapping of relative discrete dynamics. It is noted that since relative, rather than absolute, growth is modeled, the result always reflects a zero-sum relationship. In other words, the expected growth in one region should be compensated by the decline of another.

Now consider a set of arbitrary positive real-valued functions, $F_{jt} = (F_{1t}, F_{2t}, \ldots, F_{St})$, such that each F_{jt} is defined at each time period t by a subset of Y_{st}. The general discrete nonlinear process can then be defined as:

$$Y_{s,t+1} = \frac{F_{st}}{\sum_j F_{jt}} \qquad j = 1,2,3,\ldots,S \tag{12.1}$$

where $Y_{s,t+1}$ = observation (employment, population or government employee) in region s at time $t+1$; and
F_{st} = arbitrary positive real-valued function in region s at time t.

If the first region is taken to be the numéraire (reference) region, *i.e.*, if the following relation is used:

$$F_{ojt} = \frac{F_{jt}}{F_1} \qquad j = 1,2,3,\ldots,S \tag{12.2}$$

where F_{ojt} = relative arbitrary positive real-valued function in region j at time t to the numéraire region; and
F_1 = arbitrary positive real-valued function in the numéraire region.

[4] Population between 1990 and 1998 is from http://eire.census.gov/popest/data/counties/tables/CO-EST2001-12.php, between 1980 and 1990 from http://eire.census.gov/popest/archives/1980.php, between 1970 and 1980 from http://eire.census.gov/popest/archives/pre1980.php, total employment is from http://www.census.gov/pub/epcd/cbp/download/cbpdownload.html, and government employment is from http://govinfo.kerr.orst.edu/reis-stateis.html.

[5] Those are 1993-1998 population, 1993-1997 total employment, and 1993-1994 government employment.

Then, the process defined in (12.1) can also be represented by:

$$Y_{1,t+1} = \frac{1}{1+\sum_j F_{ojt}} \qquad j = 1,2,3,\ldots,S \tag{12.3}$$

$$Y_{s,t+1} = \frac{F_{ost}}{1+\sum_j F_{ojt}} \qquad j = 1,2,3,\ldots,S \tag{12.4}$$

where $\sum_s Y_{st} = 1$ $\qquad s = 1,2,3,\ldots,S$; and

$$\frac{Y_{s,t+1}}{Y_{1,t+1}} = F_{ost} \qquad j = 1,2,3,\ldots,S.$$

F_{ojt} is interpreted as a relative term; it is the temporal comparative advantages of location s relative to the numéraire location (Dendrinos and Sonis, 1990). According to Dendrinos and Sonis (1988), F_{ojt} can also be expressed as:

$$F_{ost} = A_s \prod_{k=1}^{S} Y_{kt}^{a_{sk}} \qquad [F_{st}>0;\ s = 2,\ldots,S] \tag{12.5}$$

where A_s = coefficient representing the location advantages of all regions $s \in S$ (when $A_s > 0$); and

$a_{sk} = \dfrac{\partial \ln F_{ost}}{\partial \ln Y_{kt}}$, $[s = 2,3,\ldots,S;\ k = 1,2,\ldots,S]$; regional growth elasticities $[-\infty < a_{sk} < \infty]$.

Log-linearization of equation (12.5) results in:

$$\ln Y_{s,t+1} - \ln Y_{1,t+1} = \ln A_s + \sum_{k=1}^{S} a_{sk} \ln Y_{kt} \tag{12.6}$$

As Nazara *et al.* (2001) noted, the coefficient a_{sk} as well as A_s are central to the analysis, both in terms of sign and magnitude. A positive a_{sk} would indicate complementary growth in shares between the two regions s and k. That is, every one percent population (or employment) growth in share in region k would correspond to an a_{sk} percent population (or employment) growth in share in region s. On the other hand, a negative value of a_{sk} would indicate a competitive relationship between the two regions. If the share in one region grows, the other's share will decline. A_s is the mean locational advantage of region s. If region s has experienced a relative growth over the period of analysis, it will show a positive value and vice versa. In practice, the model can be estimated by a least square estimator since the specific form in equation (12.6) is linear, in which case, a_{sk} is obtained as the coefficients of each region and A_s as the intercept.

A total of ten such models are needed to complete the forecasting of one economic variable since there are 9 Midwestern states and the rest of the U.S. (RUS). Each of them is considered as an individual independent economy in the D-S model and the relative shares are calculated based on the state total, except for the RUS case. In addition, since there are three economic variables to be estimated (total population, total and government employment), the total number of such models to be estimated is 30.

There are several cases in which one region absorbs all the economic activities and the rest of the regions have nothing in the future according to the estimates of the model (a mutual exclusion problem). Since this is not probable in the real world, some adjustments are needed in the model to prevent such extreme estimates. One adjustment made in the D-S model is to set the maximum rate of growth per each year in estimating forecasts. In other words, if the estimated growth of a certain year is greater than the maximum rate obtained from the historical trends[6], the adjusted estimated rate is set equal to the maximum rate (equation 12.7). Otherwise, the model runs as defined (above):

$$\textit{Adjusted Estimated Share} = \\ = \min \{\textit{regional share estimated by the D-S model}, \max (r_t - r_{t-1})\} \quad (12.7)$$

where r_t = regional share at time t; and
t = 1970-1994 (government sector), 1989-1997 (employment), and 1971-1998 (population).

The last step is to redistribute the final demand obtained from the REIM based on five Midwestern states and the RUS into nine Midwestern states and the RUS and finally into 36 EQAZs for the corresponding years. The forecast ratios obtained from the previous stage are used to perform this step of work. The REIM has ten final demand components in the model and the three different ratio forecasts are used in the redistribution process of the relevant components as seen in table 12.2.

Table 12.2 Final Demand Components and Redistribution Method

Component	Description	Redistribution Method
CA	Consumption, autos and parts	Population
COD	Consumption, other durables	
CND	Consumption, nondurables	
CS	Consumption, services	
IRES+INRES	Investment, residential structures + investment, non-residential structures	Employment
IE	Investment, equipment	
OFGC	Output, federal government, civilian expenditures	Government employment
GCON	Federal government, construction	
OSLG	Other state and local government expenditures	
SLCON	State and local government construction	

There are four different types of states that require different treatments for redistributing final demands: (1) states that belong to both the Midwestern five states of the REIM model estimate and the Midwestern nine states in this analysis (IL, IN, OH and MI); (2) the state that belongs to the five states, but not to the nine states (WI); (3) states that do not belong to the five Midwestern states, but belong to the nine Midwestern states (IA, KY, MO, TN and WV); and (4) the RUS. For the first case states, the ten components of the final demand of a state in any given year are

[6] Availability of time series data from the Bureau of Census online resource limits the span of years of each variable.

redistributed into the ones of individual EQAZs in each of the specific state using population, employment and government share correspondingly. The ten final demand components of Wisconsin are added to the ones for the RUS in this analysis. For the third case, two steps are involved. First, each of the final demand components of five states is prorated from the final demand components of the RUS (obtained from REIM model) using population, employment and government employment shares. The estimated final demand components of each of the five states are then redistributed into the corresponding EQAZs using three types of shares in table 12.2. The adjusted final demand components of the RUS (in the sense that it includes WI, but does not include IA, KY, MO, TN and WI after processing the treatments for the three types of states just described) are redistributed into seven RUS EQAZs using the three types of shares in table 12.2.

12.5 Final Demand Loss Function

It is not surprising that there are only a few studies that have estimated the change of the final demand resulting from an earthquake, considering the linkage between the earthquake and the indirect economic impact has received attention only recently. Among those, Rose and Benavides (1998) and Rose *et al.* (1997) provided a more relevant type of function and process to the research of interest in this chapter. Equation (12.8) is a modified version of the final demand loss function proposed in those two earlier papers. Since they focused on the earthquake damage on the electricity network, the function is modified accordingly to fit the transportation network model. Note that the equation is in the matrix terms.[7]

$$\Delta f = (I - A)\{[D \otimes (1_{13} - R)] \circ [(I - A)^{-1} f]\} \tag{12.8}$$

where Δf = change of the final demand by sector by zone (468×1);
A = 13 sector direct input coefficient matrix by zone (468×468);
D = network disruption ratio by zone (36×1);
$1_{13} - R$ = one minus sectoral resiliency factor vector (13×1); and
f = final demand by sector by zone before the earthquake (468×1).

$\otimes$ is the tensor defined as $B \otimes G = \begin{pmatrix} b_{11}G & \cdots & b_{1j}G & \cdots & b_{1n}G \\ \vdots & \ddots & \vdots & \ddots & \vdots \\ b_{i1}G & \cdots & b_{ij}G & \cdots & b_{in}G \\ \vdots & \ddots & \vdots & \ddots & \vdots \\ b_{m1}G & \cdots & b_{mi}G & \cdots & b_{mn}G \end{pmatrix}_{mp \times nq}$, where $\mathrm{B} = (b_{ij})_{m \times n}$ and $\mathrm{G} = (g_{ij})_{p \times q}$ are arbitrary matrices. And, $\circ$ is defined as

[7] The number 468 in the equation is the product of 13 sectors and 36 zones.

$$B \circ G = \begin{pmatrix} b_{11}g_{11} & \cdots & b_{1j}g_{1j} & \cdots & b_{1n}g_{1n} \\ \vdots & \ddots & \vdots & \ddots & \vdots \\ b_{i1}g_{i1} & \cdots & b_{ij}g_{ij} & \cdots & b_{in}g_{in} \\ \vdots & \ddots & \vdots & \ddots & \vdots \\ b_{m1}g_{m1} & \cdots & b_{mi}g_{mi} & \cdots & b_{mn}g_{mn} \end{pmatrix}$$, where $B = (b_{ij})_{m \times n}$ and $G = (g_{ij})_{m \times n}$ are arbitrary matrices.

The network disruption ratio needs to be calculated for individual link of the network initially. More specially, the disruption ratios for individual bridges need to be calculated at the beginning of the model.[8] A series of studies by Hwang *et al.* (2001) and Hwang *et al.* (2000) have been devoted to deriving fragility curves and estimating disruption ratios for different types of bridges. Different types of fragility curves derived from those works are used in this analysis for obtaining a better estimate of disruption ratio for different types of bridges. Three steps are required to obtain the bridge disruption ratio. The first step is to calculate the peak ground acceleration (PGA) for the location of each bridge in the area following a set of equations below. It is noted that equation (12.9) and the supplementary equations below are drawn from Hwang *et al.* (2000).[9]

$$PGA = F_s * PBA \tag{12.9}$$

where $\ln(F_s) = -2.705 * PBA + 1.001$

$$\ln(PBA) = -2.904 + 0.926M -$$
$$-1.271\ln\left[\sqrt{R^2 + H^2} + 0.06e^{0.7M}\right] - 0.00302\sqrt{R^2 + H^2}$$

PGA = peak ground acceleration;
F_s = site coefficient;
PBA = peak bedrock acceleration;
M = moment magnitude;
R = epicentral distance (km); and
H = focal depth in the study region (10km for the New Madrid Seismic zone).

There are two types of bridge fragility curves associated with three types of damage status: no/minor, repairable and significant damage. Hwang *et al.* (2000) also provided different types of fragility curves for eight bridge groups. Original scattergrams for the fragility curves are regressed based on a nonlinear specification.[10] The second step calibrates the probability of damage for three damage statuses using an appropriate type of the fragility curves, while the last step calculates the expected

[8] In the model, it is assumed that the disruption of the bridges on the highway provides more serious damages on the commodity flow since bridges are hard to be restored in a short period of time. Considering the economic and transportation cost calculation in this analysis is yearly based, road disruption might not be as influential as the bridges in that it can be recovered more quickly. It is also noted that the model does not consider railroad disruption due to the lack of proper network loss functions studied, even if railroad network is combined in the commodity flow optimization process.

[9] It is noted that Hwang *et al.* (2000) distinguished two types of ground surface categories and the corresponding equations in calculating the site coefficient, F_s. In this analysis, only the equation for the worse scenario category (suggesting more damage when there is an earthquake) is used.

[10] For more information on bridge types classified, refer to Hwang *et al.* (2000).

damage ratio on a bridge considering those three possible damage statuses. For each group of the three statuses (no/minor, repairable and significant damages), the maximum disruption ratios are set to 0.1, 0.5 and 1 respectively (see equation (12.10)).

$$E(d) = 1.0P(\textit{significant damage}) + 0.5P(\textit{repairable damage}) + \\ + 0.1P(\textit{no / minor damage}) \qquad (12.10)$$

where $E(d)$ = expected disruption ratio; and
$P(status)$ = probability with which a certain status occurs.

The resulting ratio ranges from zero to one, representing no damage to complete loss respectively.[11] The disruptions on the bridges are aggregated for the highway links on the network. Link is defined as a line segment between two nodes on the highway network. In case that there is more than one bridge on a single link, the worst case is adopted. For example, if one of two bridges on a link is disrupted as much as 50% and the other by 30%, it is decided such that the capacity of the link is disrupted by 50%. Once all the disruption ratios are determined for the links on the network, they are aggregated to the zonal level to be used as an input to the final demand loss function.[12] In other words, if there are two links between two zones and one of which is damaged 25% out of total capacity of 200 and the other 50% out of 100, the total disruption ratio for the links between these two zones is (200*.25+100*.5)/300 =.333... Equation (12.11) summarizes the relationship.

$$d_i = \frac{\sum_{j=1}^{n} c_{ij}^D}{\sum_{j=1}^{n} c_{ij}^T} \qquad (12.11)$$

where d_i = transportation network disruption ratio for link(s) to/from zone i and element of the D matrix;
c_{ij}^D = disrupted link(s) capacity between zone i and all adjacent zones (j's);
c_{ij}^T = total link(s) capacity between zone i and all adjacent zones (j's); and
n = total number of adjacent zones.

Since the model is, by nature, the interregional flow model, intrazonal flows as well as intrazonal disruption ratios are not considered in the analysis.

Resiliency of economic sectors on the transportation network is defined as the share of production remaining after the complete disruption of the transportation network. For example, the resiliency factor 0.1 for the manufacturing means 10% of the manufacturing still remains even after the 100% disruption of the transportation network. Chang (1998) compared two sets of resiliency factors of economic sectors on electricity surveyed in different ways: ATC (1991) and Tierney and Dahlhamer (1998). Those are based on the actual surveys either of an expert group or the

---

[11] By taking expected values on the disruption of individual bridges, the approach does not consider the joint probabilities of damage for the entire set of bridges. For an example of the model that approaches this issue, refer to Chang *et al.* (2000).

[12] But the ICFM in the later section uses link-based disruption ratio rather than more aggregated zone-based ratio.

entrepreneurs in the relevant economic sectors. A similar comparison is also found in Chang *et al.* (1995) for the resiliency on water and natural gas lifelines.

For the transportation network resiliency, three factors are considered. The first factor is the level of intrazonal flow: what proportion of the commodity flow is intrazonal? Based on the 36x36 zone/13 sector commodity flow derived from the 1993 CFS data, the proportion of the intrazonal flow is obtained for each economic sector. Commodity flow in terms of tonnage is used rather than the flow based on dollar unit because the analysis deals with the resiliency on the transportation network related to some physical movement that may be reflected better in the ton-based flows. It is thought that any sector with a higher rate of intrazonal flow has higher resiliency on the interregional network disruption. The second factor is the modal share of highways: the proportion of the highway flows among various transportation modes. It is evident that higher usage of and as a result, higher dependence on the interzonal highway networks imply lower resiliency to the disruption of the highway networks. Similarly, higher usage of other modes than highway (*i.e.*, rail, air, water and so forth) results in higher resiliency on the highway networks. The third factor is the average shipment distance on the highway. Generally, commodities with a longer shipping distance are more vulnerable to the highway disruption than the ones with a shorter travel distance. Unlike previous two issues, however, this is not measured as the ratio or proportion. As a result, the values are standardized around the average distance of all commodities. Here, the signs are switched, so that higher values represent higher resiliency. The first factor is multiplied by the second one to obtain the resiliency ratio in the worst case (in the sense that if any of the two ratios is 0, the resulting resiliency goes to 0), rather than taking the (arithmetic or geometric) mean of the two. The third factor, then, is added as a deviation of the resiliency of a given sector.[13] Equation (12.12) summarizes this process.

$$r_k = r_k^{IZ}\left(1 - r_k^{HW}\right) + r_k^{AD} + \varepsilon_k \tag{12.12}$$

where r_k = resiliency factor of sector k;

$r_k^{IZ} = \dfrac{\sum_{m=1}^{36} f_{mm}^k}{f^k}$ = relative ratio of intrazonal flow of sector k;

f_{mm}^k = commodity flow of sector k from m to m (intrazonal flow);

f^k = total commodity flow of sector k;

r_k^{HW} = modal share of highway in sector k;

r_k^{AD} = standardized inversed average shipment distance of sector k on highway; and

ε_k = a random variable that satisfy $N\left(0,\sigma^2\right)$.

As a result, the term, $D \otimes \left(1_{13} - R\right)$, represents the ratio of the substantial damage on the flows of each zone by each sector. ε_k is a random variable that enables the equation to incorporate a stochastic process, so that the resiliency might have variable

[13] This is used to function as a deviation because it is not as explicitly related to the highway resiliency and it is not the relative ratio between zero and one as the other two.

factor values combining random numbers.[14] Table 12.3 summarizes the resiliency factors for 13 economic sectors. The sectoral resiliency factors range from .0152 (Electronic and electric equipment) to .4246 (Agriculture, forestry and fisheries) excluding construction sector. The resiliency for all economic sectors is .2397. It should also be noted that most manufacturing sectors show relatively lower level of resiliency.

Table 12.3 Resiliency Factors for 13 Economic Sectors

	Description	r^{IZ}	r^{HW}	r^{AD}	r
Sector 1	Agriculture, forestry and fisheries	.7666	.4662	.0154	.4246
Sector 2	Mining	.7146	.5548	.0500	.3682
Sector 3	Construction	N/A	N/A	N/A	1.0000
Sector 4	Food and kindred products	.5980	.8640	.0131	.0944
Sector 5	Chemicals and allied products	.5908	.5870	.0006	.2446
Sector 6	Primary metals industries	.4708	.8002	-.0175	.0766
Sector 7	Fabricated metal products	.5340	.9126	-.0299	.0168
Sector 8	Industrial machinery and equipment	.4978	.8532	-.0457	.0273
Sector 9	Electronic and electric equipment	.4542	.8569	-.0498	.0152
Sector 10	Transportation equipment	.4615	.6857	-.0361	.1090
Sector 11	Other non-durable manufacturing	.7172	.5021	.0336	.3907
Sector 12	Other durable manufacturing	.7465	.8591	.0223	.1275
Sector 13	TCU*, services and government enterprises	.4666	.6694	.0281	.1824
Total		.6513	.6320	.0000	.2397

Note: No data are available on the construction sector (sector 3) from the CFS and it is assumed that there is no interregional flow in sector 3 and consequently it is not affected by any disruption of the interzonal highway systems.

The multiregional input-output coefficient matrix, A, was estimated by the column coefficient model (see Moses (1955) and Okuyama (1997) for more information). The column coefficient model assumes an identical regional supply pattern, in which regional composition of any given input is identical among all purchasers in a region. Some advantages for this method include: (1) Less data are required since trade coefficients are derived from the regional totals and (2) Separate treatment of technology and trade can be accommodated. Equation (12.13) shows how to obtain the multiregional input-output coefficients using the regional and the trade coefficient.

$$b_{ij}^{pq} = a_{ij}^{q} t_{i}^{pq} \tag{12.13}$$

where b_{ij}^{pq} = technical coefficient between sector i in region p and sector j in region q;

a_{ij}^{q} = technical coefficient between sector i and sector j in region q; and

t_{i}^{pq} = trade coefficient of output i entering from region p to q per unit of output of i in q.

[14] In this scenario run, however, an expected value of r_k is calculated, so that ε_k is not counted in the calculation since $E[\varepsilon_k]=0$.

The 13 by 13 sector national input-output table is used for each of the 36 regional input-output tables by assuming identical interindustry structure among regions.[15] Trade coefficients are obtained from the estimated commodity flows among 36 EQAZs for 12 economic sectors.[16] The result is the 468 (36 EQAZs by 13 sectors) by 468 multiregional input-output matrix. $(I-A)^{-1}$ is a Leontief inversed input-output coefficient matrix and f is the final demand before the earthquake. Alternatively, X, the total output before the earthquake, can be used instead of $(I-A)^{-1}f$.

12.6 Transportation Network Model

For the calculation of the transportation system network cost, the integrated commodity flow model developed by Kim *et al.* (1999) and later adjusted by Ham (2001) and Kim *et al.* (2002) is used in the scenario analysis. The model consists of interregional commodity flows to meet each region's final and intermediate demand and assesses the direct and indirect damages that an earthquake could cause. The model presumes that the transportation network is represented by routes and links connecting the US highway and the primary railway systems (see figures 12.3 and 12.4).

Figure 12.3 U.S. Interstate Highway System

[15] Identical input-output table for all regions implies that all regions have the same (composition of) production technology.
[16] It is noted that the construction sector has the identity matrix (intrazonal flow only).

Figure 12.4 U.S. Primary Railway Systems

The model assumes that exports and imports by commodity for each subregion are given by the regional model, and allocated to the subregional level for the period of analysis of one year. Therefore, the problem is to estimate the flows by sector/commodity between each pair of subregions and allocate these flows to the transportation network. The integrated commodity flow model is defined as in equations (12.14) to (12.16).

$$\min_{\bar{h},\bar{x}} \quad Z(\bar{h},\bar{x}) = \sum_a \int_0^{f_a} d_a(\omega)d\omega + \sum_{mj} c_{jj} \frac{x_{jj}^m}{g^m} + \sum_m \frac{1}{\beta^m g^m} \sum_{ij} x_{ij}^m \ln\left(x_{ij}^m\right)$$

$$\text{s.t.} \quad \sum_i x_{ij}^m = \sum_n a^{mn} \sum_k x_{jk}^n + y_j^m \qquad \text{for } m, j \quad \text{Material balance} \qquad (12.14)$$

$$\sum_r h_{ijr}^m = \frac{x_{ij}^m}{g^m} \qquad \text{for } m, i, j \quad \text{Conservation of flow} \qquad (12.15)$$

$$h_{ijr}^m \geq 0 \qquad \text{for } m, r, i, j \quad \text{Non-negativity} \qquad (12.16)$$

where the exogenous variables are:

a^{mn} = technical input-output coefficient for input of sector m required to make one unit of sector n;

β^m = parameter for each sector m;

g^m = conversion factor of sector m from dollar to ton ($/ton);

y_j^m = final demand (consumption, investment, government, etc) of sector m in zone j ($);

the endogenous variables are:

c_{jj} = intrazonal travel cost within zone j (mile);

$d_a(f_a)$ = length function of total flow on link a (mile);

f_a = total flow on link a (ton) = $\sum_m \sum_{ijr} h_{ijr}^m \delta_{ijr}^a$;

δ_{ijr}^a = 1 if link a belongs to route r from i to j, and 0 otherwise;

h_{ijr}^m = flow of output of sector m from i to zone j on route r (ton); and

x_{ij}^m = flow of the output of sector m from zone i to zone j (\$).

The objective function to be minimized for the entire United States is the sum of the total costs in the system. This is obtained by summing the network assignment costs, the intrazonal travel costs, and the OD flow distribution costs. Then the commodity flows of links, routes, and origin-destination for each sector are estimated as a result of the dispersion of commodity flows among zones and the choices of minimum cost travel routes. As mentioned earlier, when the earthquake occurs, adjusted link capacities (c_a in the function $d_a(f_a)$) replace the initial value in the model and it is run again to obtain the adjusted transportation cost. Once the value of the objective function is obtained, a sectoral system-wide transportation cost is calculated in the following way:

$$\Delta TC_i = \Delta TC * \frac{DM_1^i - DM_0^i}{\sum_{i=1}^{13}(DM_1^i - DM_0^i)} \qquad (12.17)$$

where $\Delta TC = 6{,}943{,}988 * \frac{Z_1 - Z_0}{Z_0}$ = system transportation cost change (million \$);

6,943,988 = total commodity flow in the U.S. in 1997 based on 1997 CFS (million \$);

Z_1 = objective function value of ICFM after the earthquake;

Z_0 = objective function value of ICFM before the earthquake;

ΔTC_i = sectoral system transportation cost change;

DM_1^i = total dollar-mile for sector i commodity after the earthquake; and

DM_0^i = total dollar-mile for sector i commodity before the earthquake.

According to Kim *et al.* (2002), the integrated commodity flow model with Evans algorithm converges quickly until a near optimal solution (within 10 iterations), but slowly at a near optimal solution. Wilson's iterative balancing method, a sub-problem in the Evans algorithm, also shows that the model converges within 50 iterations with an acceptable stopping error, confirming that the model converges with stability.

12.7 Scenario Analysis

It is advantageous to make use of the prior (1812) earthquake as the one in the scenario analysis. Most of all, the large volume of previous studies either related to earthquake engineering or economic impact provides detailed information and knowledge for the scenario analysis. This earthquake recorded a value of 8.7 on the Richter scale and 8 moment magnitude. The epicenter is at New Madrid, TN. The New Madrid Seismic zone lies within the central Mississippi Valley, extending from northeast Arkansas, through southeast Missouri, western Tennessee, western Kentucky to southern Illinois. In the scenario analysis, the year is set to 2001 while all the other earthquake-related settings remain as same as in the 1812 earthquake. Figure 12.5 shows the software interface of the customized program developed to perform a scenario analysis. In the map window, especially around the Midwestern states, several important elements are shown: the epicenter (triangle), bridges on the highway (circle), highway network and EQAZ boundary.

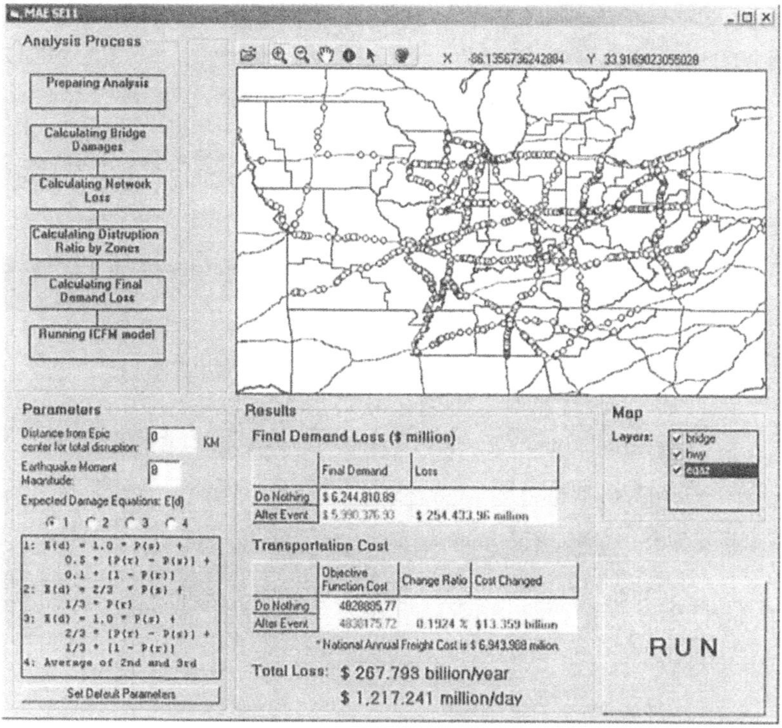

Figure 12.5 Customized Software to Perform the Scenario Analysis

The cost is calculated in two parts – final demand loss and transportation cost increase. For the estimation of the sectoral transport cost increase, the sectoral share of the total dollar-mile cost is calculated both for highway and railway.[17] In table 12.4, the column labeled "before" shows the total sectoral dollar-mile before the earthquake and "after" presents the one after the earthquake under the scenario. The differences of both highway and railway are summed to obtain the relative ratio of the total transport cost increase.

Table 12.4 Change of Sectoral Total Dollar-Miles Before and After the Earthquake (billion$-miles)*

Sector	Highway			Railway			Total	
	before	after	diff	before	after	diff	Sum	share**
1	32976	33160	184	56911	57416	505	689	.022
2	434	465	31	199	202	2	33	.001
3	0	0	0	0	0	0	0	.000
4	275585	278120	2534	130904	132939	2035	4569	.146
5	169976	171092	1116	175213	177317	2104	3220	.103
6	69823	70758	936	43597	44235	638	1574	.050
7	117264	118382	1118	58169	59046	877	1995	.064
8	146198	146838	640	197282	198880	1598	2237	.072
9	353037	354847	1810	271422	274511	3089	4898	.157
10	151465	152034	568	218952	220991	2039	2607	.083
11	286961	289331	2370	207829	210945	3116	5486	.176
12	259909	260313	404	414065	417596	3531	3935	.126
13	0	0	0	0	0	0	0	.000
Total	1863628	1875339	11711	1774544	1794077	19533	31243	1.000

Notes: * Calculated over the entire U.S.
** Calculated by dividing sectoral sum by sum total (31243).

Table 12.5 summarizes the sectoral economic and transport cost increase as a result of the New Madrid earthquake. The largest cost increase is seen in sector 13 (TCU, services and government enterprises). However, because of the aggregation scheme (in the sense that this sector is the rest of the economic activity other than primary and manufacturing) and the interests in commodity flow in the model, it is not of great interest in this analysis. Some of the sectors with the major economic cost include sectors 4 (food and kindred products), 9 (electronic and electric equipment) and 12 (other durable manufacturing). These are sectors that (1) have a relatively lower level of resiliency against the disruption of the transportation network and (2) are largely distributed over the Midwestern states relatively near the epicenter. There are also the sectors with higher final demands (negative values) than before the earthquake, leading to a higher level of production as a consequence: sectors 1 (agriculture, forestry and fishery), 2 (mining), 3 (construction) and 6 (primary metal industries). Sectors 1, 2, and 3 have relatively higher resiliency factors than others while the sector 6 activities might be distributed relatively farther from the significant earthquake damage zone considering the lower level of the resiliency factor. It is probable that those sectors with less damage might be more demanded as a compensation to greater cutback of the final demands (due to greater damage) in other sectors through the input-output relationship.

[17] Sectoral total dollar-miles are obtained along with the value of the objective function after the model runs. Since the value of the objective function does not provide sectoral costs, the sectoral total dollar-miles are used to prorate the total cost increase into individual sectors.

Table 12.5 Economic and Transport Cost Increase Led by the Earthquake (million dollars)*

Sector	Economic Cost	Transport Cost**	Sectoral Total
1	-2,101	295	-1,806
2	-935	14	-921
3	-6,693	0	-6,693
4	21,260	1,954	23,214
5	10,163	1,377	11,540
6	-856	673	-183
7	3,891	853	4,744
8	8,134	957	9,091
9	17,679	2,094	19,773
10	11,536	1,115	12,651
11	13,346	2,346	15,692
12	15,897	1,682	17,579
13	163,114	0	163,114
Total	254,434	13,359	267,793

Notes: * Calculated over the entire U.S.
** Calculated from equation (12.17)

The transport cost increase pattern also shows a similar pattern as in the economic loss case. Sectors 4, 9, 11 (other non-durable manufacturing) and 12 are big losers under the earthquake while sectors 1, 2 and 6 operate relatively well. Recall that sectors 3 and 13 are assumed to have no interregional commodity flows in the ICFM, so that the transportation costs of both sectors have not changed under the scenario. The last column of the table shows the sectoral total cost increase under the scenario.

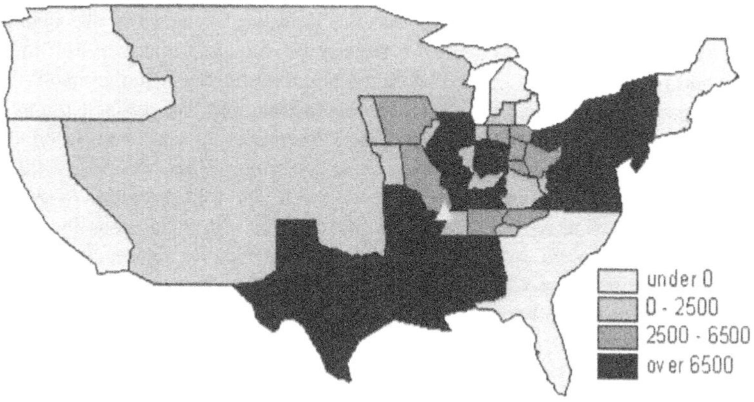

Figure 12.6 Spatial Distribution of Final Demand Loss (million dollars)

Figure 12.6 represents the spatial distribution pattern of the final demand loss by 36 EQAZs. The white triangle is the epicenter of the earthquake under the scenario. One important finding is that some of the largest losses are observed not only around the Midwestern states but the Northeast and the Southern states, reflecting the strong economic linkage between regions: Northeast and Midwest and South and Midwest. Another point to be noted is that the economic loss is not totally dependent on the distance from the epicenter and the level of physical disruption as well.[18] For example, even if the network links are significantly damaged, the resulting damage to the whole economy might not be so critical either if the zone of interest contains few economic activities or if the zonal economies are more independent and self-sufficient. In this respect, it is rather important to identify such important spatial economic links between zones as well as some key zones in the spatial economy.

12.8 Conclusions

The chapter focused on developing the scenario analysis framework on the economic assessment for the retrofit of transportation network in relation to the earthquake. For this purpose, final demands for 25 years starting from the base year, 1993, were estimated using the Dendrinos-Sonis model. The overall framework for the scenario analysis has been established. Several key models and functions including the final demand loss function and the transportation network model were incorporated in the system. A hypothetical scenario has also been built to test the framework based on the 1812 New Madrid earthquake with the moment magnitude of 8. Two types of cost were estimated for the economic damage from the earthquake: final demand loss and transport cost increase. The estimation result of the scenario analysis was summarized by individual sectors as well as EQAZs.

One of the achievements in this analysis is that the model is specified to be able to deal with real world situations more precisely. For example, compared to other impact models, the model is able to (1) consider real highway/railway networks, (2) allow partial closure of bridges and links, and (3) account for reduced final demands due to the disruption of transportation network. The model, in the form of customized software, can be used by policy makers and planners before they make an investment decision on retrofit of transportation network facilities to examine the significance of facilities of interest and the consequent priority of retrofit related to a certain earthquake scenario. The model can be easily extended to incorporate various types of engineering models. For example, a road (or railway) fragility function, in addition to bridge fragility functions, can be easily updated in the network loss function to calculate disruption ratio of the network once it is available.

A few limitations should be noted related to the extension of the model towards benefit-cost analysis. First, the model only considers impact differentials that contribute to benefits, so that other types of benefits as well as costs are not discussed. Also, the analysis discusses how impact differentials can be evaluated for a particular scenario earthquake, but does not tackle how the results of multiple scenarios can be probabilistically combined, which is essential in benefit-cost analysis. Similarly, since the model is designed to calibrate potential benefits for a single year, the probabilistic aggregation of benefits over time, which is also essential in benefit-cost analysis, is not incorporated in the model. In addition, neglecting intrazonal flows and their disruption might underestimate the actual damage from an earthquake since intrazonal flows will

[18] With only a few exceptions, the bridges closer to the epicenter have more disrupted than others in the scenario analysis.

also be paralyzed with varying degrees, which might eventually have an influence on interzonal flows as well.

While this chapter attempted to establish the methodology to perform the scenario analysis of the earthquake and its economic impact, additional analyses need to be done in the future in some areas. Most of all, individual links and bridges need to be considered for the sensitivity analysis for the overall economic damage. By so doing, each link and bridge can be ranked along with the economic significance. Such a list of bridges and links is able to determine not only the retrofit priority before the earthquake, but the restoration priority afterwards.

References

Applied Technology Council. 1991. *ATC-25: Seismic Vulnerability and Impact of Disruption on Lifelines in the Conterminous United States*. Applied Technology Council, CA; Federal Emergency Management Agency.

Brookshire, D.S., S.E. Chang, H. Cochrane, R.A. Olson, A. Rose, and J. Steenson. 1997. "Direct and Indirect Economic Losses from Earthquake Damage," *Earthquake Spectra,* 13 (4): 683-701.

Chang, S.E. 1998. "Direct Economic Impacts," in M. Shinozuka, A. Rose, and R.T. Eguchi eds. *Engineering and Socioeconomic Impacts of Earthquakes: An Analysis of Electricity Lifeline Disruptions in the New Madrid Area*. Buffalo, NY; Multidisciplinary Center for Earthquake Engineering Research: 75-94.

Chang, S.E., H.A. Seligson, and R.T. Eguchi. 1995. "Estimation of Economic Losses Caused by Disruption of Lifeline Service," Proceedings of the Fourth U.S. Conference on Lifeline Earthquake Engineering, ASCE, San Francisco, CA: 48-55.

Chang, S.E., M. Shinozuka, and J.E. Moore II. 2000. "Probabilistic Earthquake Scenarios: Extending Risk Analysis Methodologies to Spatially Distributed Systems," *Earthquake Spectra,* 16: 557-72.

Cho, S., P. Gordon, H.W. Richardson, J.E. Moore, and M. Shinozuka. 2000. "Analyzing Transportation Reconstruction Network Strategies: A Full Cost Approach," *Review of Urban and Regional Development Studies,* 12 (3): 212-27.

Cochrane, H. 1974. "Predicting the Economic Impact of Earthquakes," in H.C. Cochrane, J.E. Haas, M.J. Bowden, and R.W. Kates eds. *Social Science Perspectives on the Coming San Francisco Earthquake*, Natural Hazards Research Paper No.25. Boulder, CO; NHRAIC.

Cochrane, H.C. 1997. "Forecasting the Economic Impact of a Midwest Earthquake," in B.G. Jones ed. *Economic Consequences of Earthquakes: Preparing for the Unexpected.* Buffalo, NY; NCEER: 223-47.

Cole, S. 1998. "Decision Support for Calamity Preparedness: Socioeconomic and Interregional Impacts," in M. Shinozuka, A. Rose, and R.T. Eguchi eds. *Engineering and Socioeconomic Impacts of Earthquakes: An Analysis of Electricity Lifeline Disruptions in the New Madrid Area.* Buffalo, NY; Multidisciplinary Center for Earthquake Engineering Research: 125-153.

Davis, H.C. and E.L. Salkin. 1984. "Alternative Approaches to the Estimation of Economic Impacts Resulting from Supply Constraints," *Annals of Regional Science,* 18: 25-34.

Dendrinos, D. and M. Sonis. 1987. "The Onset of Turbulence in Discrete Relative Multiple Spatial Dynamics," *Applied Mathematics and Computation,* 22: 25-44.

Dendrinos, D. and M. Sonis. 1988. "Nonlinear Discrete Relative Population Dynamics of the U.S. Regions," *Applied Mathematics and Computation,* 25: 265-285.

Dendrinos, D. and M. Sonis. 1990. *Chaos and Socio-Spatial Dynamics.* New York, NY; Spring-Verlag.

Ham, H. 2001. *An Integrated Interregional Input-Output and Transportation Network Model for Assessing Economic Impacts of Unexpected Events*, Unpublished Ph.D. dissertation, University of Illinois, Urbana, IL.

Hewings, G.J.D., M. Sonis, F. Cuello, and F. Mansouri. 1996. "The Role of Regional Interaction in Regional Growth: Competition and Complementary in the U.S. Regional System," *Australian Journal of Regional Studies*, 2 (2): 133-149.

Hwang, H., Y.H. Chiu, and J.B. Liu. 2001. "Seismic Fragility Analysis of Reinforced Concrete Bridges," Paper submitted to the 8th International Conference on Structural Safety and Reliability, Newport Beach, CA.

Hwang, H., J.B. Jernigan, and Y. Lin. 2000. "Evaluation of Seismic Damage to Memphis Bridges and Highway Systems," *Journal of Bridge Engineering*, 5 (4): 322-30.

Jackson, R.W. and M. Sonis. 2001. "On the Spatial Decomposition of Forecasts," *Geographical Analysis*, 33: 58-75.

Kim, T.J., D.E. Boyce, H. Ham, and Y. Okuyama. 1999. "Implementation of an Integrated Transportation Network and Input-Out Model for Assessing Economic Impacts of Unexpected Events: A Solution Algorithm and Issues," Paper presented in the 21st Northeastern Regional Science Association Annual Conference, Ithaca, NY.

Kim, T.J., H. Ham, and D.E. Boyce. 2002. "Economic Impacts of Transportation Network Changes: Implementation of a Combined Transportation Network and Input-Output Model," *Papers in Regional Science*, 81: 223-46.

Magalhães, A., M. Sonis, and G.J.D. Hewings. 1999. "Regional Competition and Complementarity Reflected in Relative Regional Dynamics and Growth of GSP: A Comparative Analysis of the Northeast of Brazil and the Midwest States of the U.S.," Regional Economics Applications Laboratory, Discussion Paper (99-T-8), Urbana, IL.

Moses, L.N. 1955. "The Stability of Interregional Trading Patterns and Input-Output Analysis," *American Economic Review*, 45 (5): 803-32.

Nazara, S., G.J.D. Hewings, and M. Sonis. 2001. "Hierarchical Spatial Interaction: An Exploratory Analysis," Regional Economics Applications Laboratory, Discussion Paper (01-T-12), Urbana, IL.

Okuyama, Y. 1997. "A Review of Interregional/Multiregional Input-Output Modeling," Regional Economics Applications Laboratory, Discussion Paper (97-T-2), Urbana, IL.

Okuyama, Y., G.J.D. Hewings, T.J. Kim, D.E. Boyce, H. Ham, and J. Sohn. 1998. "Economic Impacts of An Earthquake in the New Madrid Seismic Zone: A Multiregional Analysis," Paper presented in the 45th North American Meeting of Regional Science Association International, Santa Fe, NM.

Okuyama, Y., G.J.D. Hewings, T.J. Kim, D.E. Boyce, H. Ham, and J. Sohn. 1999. "Economic Impacts of an Earthquake in the New Madrid Seismic Zone: An Overview," Paper presented in the 21st Northeastern Regional Science Association Annual Conference, Ithaca, NY.

Quigley, J.M. 1998. "Earthquake! The Use of Economics, Engineering, and Statistical Information to Invest in Seismic Safety," *Journal of Applied Regional Science*, 3: 9-20.

Rose, A.Z. 1981. "Utility Lifelines and Economic Activity in the Context of Earthquakes," in J. Isenberg ed. *Social and Economic Impact of Earthquakes on Utility Lifelines*. New York, NY; American Society of Civil Engineers: 107-20.

Rose, A. and J. Benavides. 1998. "Regional Economic Impacts," in M. Shinozuka, A. Rose, and R.T. Eguci eds. *Engineering and Socioeconomic Impacts of Earthquakes: An Analysis of Electricity Lifeline Disruptions in the New Madrid Area*. Multidisciplinary Buffalo, NY; Center for Earthquake Engineering Research: 95-123.

Rose, A., J. Benavides, S.E. Chang, P. Szczesniak, and D. Lim. 1997. "The Regional Economic Impact of an Earthquake: Direct and Indirect Effects of Electricity Lifeline Disruptions," *Journal of Regional Science*, 37 (3): 437-58.

Sonis, M. and D. Dendrinos. 1987. "A Discrete Relative Growth Model: Switching, Role Reversal and Turbulence," in P. Eichborn and P. Friedrich eds. *International Perspectives on Regional Decentralization*.

Tierney, K.J. and J.M. Dahlhamer. 1998. "Earthquake Vulnerability and Emergency Preparedness among Businesses," in M. Shinozuka, A. Rose, and R.T. Eguci, eds. *Engineering and Socioeconomic Impacts of Earthquakes: An Analysis of Electricity Lifeline Disruptions in the New Madrid Area*, Buffalo, NY; Multidisciplinary Center for Earthquake Engineering Research: 53-73.

13 Benefit Cost Analysis for Renewal Planning of Existing Electric Power Equipment

Yoshiharu Shumuta

Central Research Institute of Electric Power Industry, Japan
e-mail: shumuta@criepi.denken.or.jp

13.1 Introduction

Recent earthquake disasters have caused major damage to electric power systems, leading to significant economic disruption. For example, the restoration expenditure of the electric power facilities due to the 1995 Hyogoken-Nanbu Earthquake amounted to no less than 226 billion yen (Kansai Electric Power Company, 1995). This expenditure would be equivalent to about 58% of the total maintenance cost (about 390 billion yen) of the Kansai Electric Power Company in the 1994 fiscal year (The Federation of Electric Power Companies Japan, 1996). Particularly, most of the damage by the 1995 Hyogoken-Nanbu earthquake was concentrated on existing old equipment, which did not meet the present seismic design standard. Therefore, the seismic upgrading of such old equipment was recognized as a serious subject after the 1995 Hyogoken-Nanbu earthquake.

In order to perform a reasonable seismic upgrading of the existing old equipment, the technologies for evaluating their seismic risk are fundamentally needed. These evaluation technologies are divided into two groups: (1) seismic hazard and (2) seismic fragility.

The seismic hazard part implies the evaluation of the earthquake occurrence probability and of its intensity including the Peak Ground Acceleration (PGA) distributed in a target region. Though there are many existing research studies, Cornell (1968) and Kameda *et al.* (1997) performed the pioneering and contributive research in this field. The Headquarters for Earthquake Research Promotion (2001) has officially been announcing the earthquake occurrence probabilities caused by major active faults or plate boundaries in and around Japan using its own web site. As a result, it is easier to get seismic hazard information for each local region in Japan.

The seismic fragility part focuses on the evaluation of the seismic-resisting capacity of electric power equipment. Ang *et al.* (1992) and Huo and Hwang (1995) proposed the seismic fragility curves for substations and its equipment in the USA. Oikawa *et al.* (2001) also proposed the seismic fragility curves for electric power transmission facilities including substation equipment and transmission towers using actual damage records from the 1995 Hyogoken-Nanbu earthquake. The Public Interest Energy Research (PIER) program has been publishing some reports associated with the seismic performance of existing substation equipment before and after retrofitting (for example, California Energy Commission (2000) and Gilani *et al.* (1999)).

In the seismic upgrading process, if a large number of equipment in a target power system needs to be seismically upgraded, it is difficult to simultaneously carry out it because of the limited budget. This suggests that more attention should be paid to the system performance of the damaged power system in order to minimize its functional loss and its economic loss due to an earthquake. System performance measures are useful in the context for evaluating the degree of system improvement afforded by the various levels of seismic upgrade of equipment. This aids the upgrade prioritization under budget constraints. System performance measures may also be effective when discussing what levels of risk or potential loss are acceptable or unacceptable.

Some research studies have been performed to evaluate the seismic performance of electric power systems that include Yamada and Noda (1988), Ang *et al.* (1992), and Shumuta (1995, and 1996). However, there are few studies associated with the prioritization of the seismic upgrade for a large number of equipment. Matuda *et al.* (1991) proposed an empirical prioritized index of the seismic upgrade for every substation. The proposed index is called '*Composite Earthquake Index*' (CEI) and it enables us to quantify the composite weight of vulnerability and the consequence of each substation to the entire power system. The CEI was applied to a seismic upgrade planning for 200 substations located in the PG&E territories, California. However, the CEI could not evaluate the degree of the system performance improvement by means of upgrading its component. Shumuta (1998) proposed a method to prioritize the substation equipment based on a system performance evaluation and a cost effectiveness analysis against an earthquake scenario.

On the other hand, substation equipment is exposed to not only seismic hazards but also other hazards including typhoons and continuous environmental factors associated with its progressive deterioration during its life cycle. In the actual renewal planning, these multi-hazards are usually considered as important risk factors to maintain the supply reliability of the entire power system. Particularly, practitioners in Japan are highly concerned about the progressive deterioration of equipment. However, a low frequency but large-scale earthquake was not taken into consideration as an important hazard. During the actual facility planning, the seismic upgrade and the renewal are treated as an independent subject, respectively, and the seismic countermeasure is regarded as a lower priority under recent tighter budget constraints in Japan. As a result, a lot of old equipment to be seismically upgraded still remains in the existing power systems.

This chapter presents the benefit-cost analysis for the renewal planning of substation equipment based on a life cycle costing approach. Chang *et al.* (1998) performed the pioneering work related to the life cycle cost analysis including the system performance evaluation for an infrastructure system considering natural disasters, especially an earthquake. They demonstrated how the proposed life cycle costing framework could be applied to a water delivery system for a seismic upgrading program. As for the electric power system, the benefit cost analysis based on the system performance evaluation or the lifecycle costing approach usually neglects the damage cost associated with natural disasters. For example, Billinton (1994) evaluated the relationship between the system reliability under normal conditions and the total cost including the facility investment cost, operation cost and customer interruption cost using actual monitoring data in Canada. Neudorf *et al.* (1995) performed a benefit cost analysis based on the lifecycle costing approach for the renewal of an old power plant located at PG&E, California. These studies did not analyze the benefit cost associated with a natural disaster because of its low frequency of occurrence.

The proposed framework for the benefit cost analysis is similar to that of Chang *et al.* (1998). However, this chapter does not consider the maintenance cost. This is because the maintenance of substation equipment is expected to be performed at regular intervals based on the preventive maintenance policy. This suggests that the change in the maintenance cost depending on the renewal priority is extremely low. In contrast, the revenue loss due to multi-hazards is considered in this chapter. The power lifeline is "private" in the sense that it is business bent on making a profit while the water system demonstrated by Chang *et al.* (1998) is a "public lifeline" supported by the government. Thus, this chapter considers the revenue loss due to a seismic event in the benefit-cost analysis.

In the numerical examples, the proposed model is applied to the actual secondary power system located in the western part of Japan. Particular attention is paid to issues of the acceptable seismic performance level and the cost sensitivity against uncertain factors included in the seismic risk assessment. To demonstrate the availability of the proposed model, the analytical results from the proposed model are compared to that of an existing model. Section 13.2 introduces the existing renewal model adapted for the target power system. Section 13.3 provides the formulation of the proposed model. It extends and improves the basic framework outlined by Shumuta (2001). Section 13.4 illustrates how the proposed model leads to different policies/decisions than the existing renewal planning in the numerical examples. Section 13.5 contains the conclusion and identifies areas for further research.

13.2 Existing Renewal Planning Model

13.2.1 General Framework

The existing renewal planning including the renewal-prioritization of substation equipment is divided into two sections; comprehensive and partial renewal. The comprehensive renewal aims at the large-scale replacement of substation equipment. For example, when the Circuit Breaker (CB) and Line Switch (LS) are replaced by a Gas Insulated Switch (GIS), many CB and LS installed in the same substation are simultaneously replaced. In this case, the renewal priority of equipment is determined for every substation. On the other hand, the partial renewal mainly focuses on the minor replacement of the deteriorating parts of equipment. Its renewal priority is also determined for equipment. In this chapter, only the comprehensive renewal planning is focused on.

In order to prioritize the comprehensive renewal of substation equipment, the power industry would usually like to consider not only the degree of the system performance improvement but also that of the environment & maintainability improvement in the target region. Equation (13.1) is the existing objective function, F_l, for the comprehensive renewal planning proposed by Hoshiya *et al.* (1989). A power industry in Japan has applied this objective function in the actual renewal planning. The renewal priority becomes higher with an increase in the value of F_l.

$$F_l = \sqrt{\alpha \cdot (SPI_l)^2 + \beta \cdot (EMI_l)^2} \tag{13.1}$$

where F_l is the objective function of substation l; SPI_l is the substation performance improvement index by the comprehensive renewal of substation l; EMI_l is the environment & maintainability improvement index by that of substation l ; and α and β indicate the weight parameters. Note that Hoshiya *et al.* (1989) did not mention how to determine the weight parameters, α and β. Thus, in actual renewal planning, these weight parameters are subjectively determined by the user.

13.2.2 Substation Performance Improvement Index (SPI_l)

SPI_l indicates the degree of substation performance improvement. It is defined as follows:

$$SPI_l = \frac{E(PF^{pre})_l - E(PF^{post})_l}{RC_l} \tag{13.2}$$

where $E(PF^{pre})_l$ and $E(PF^{post})_l$ indicate the expected power energy losses before and after the renewal of substation l, respectively; and RC_l is the total renewal cost of substation l.

The expected power energy loss of each substation varies depending on its damage probability associated with the equipment deterioration and on the power energy loss level caused by damage to each piece of equipment. Thus, $E(PF^{pre})_l$ and $E(PF^{post})_l$ are defined as:

$$E(PF^{pre})_l = \sum_{j=1}^{n(l)} Pf^{pre}{}_j \cdot DW_j \tag{13.3.a}$$

$$E(PF^{post})_l = \sum_{j=1}^{n(l)} Pf^{post}{}_j \cdot DW_j \tag{13.3.b}$$

where $n(l)$ is the total number of target equipment in substation l; $Pf^{pre}{}_j$ and $Pf^{Post}{}_j$ indicate the annual damage probabilities of equipment j before and after its renewal, respectively; and DW_j is the power energy loss caused by the damage to equipment j. In Hoshiya *et al.*(1989), DW_j is evaluated by a connectivity analysis with a substation network model assuming that only one or two equipment accidents simultaneously occurs in a substation.

Some alternatives to estimate $Pf^{pre}{}_j$ and $Pf^{post}{}_j$ have been discussed in Hoshiya *et al.* (1989); as a result, they are only concerned with parameterizing the equipment age and its total operating times. This can be expressed as:

$$Pf^{pre}{}_j \;\; / \, Pf^{post}{}_j \Rightarrow Pf_j = P_j(t_d, n_d) = a_j \cdot \exp(b_j \cdot t_d + c_j \cdot n_d) \tag{13.4}$$

where $P_j(t_d, n_d)$ is the damage probability of equipment j with its age t_d and its total operating times n_d; a_j, b_j, c_j indicate the recurrent coefficients for the damage probability function; here, table 13.1 shows the damage probability functions of the different equipment proposed by Hoshiya *et al.* (1989). They proposed these empirical functions using statistical data associated with the equipment accident due to its deterioration.

The total renewal cost of substation l, $RC_{l,}$, is defined as:

$$RC_l = \sum_{j=1}^{n(l)} rc_j \tag{13.5}$$

where, rc_j is the renewal cost of equipment j.

Table 13.1 Damage Probability Function Proposed by Hoshiya *et al.* (1989)

Equipment j	Annual damage probability function
ABB:	$Pf_j = P_j(t_d, n_d) = 0.00007462 \times \exp(0.09714t_d + 0.001037n_d)$
OCB	$Pf_j = P_j(t_d, n_d) = 0.0001307 \times \exp(0.08186t_d + 0.000367n_d)$
GCB	$Pf_j = P_j(t_d, n_d) = 0.00004723 \times \exp(0.100107t_d + 0.000266n_d)$
TrB	$Pf_j = P_j(t_d, n_d) = 0.0002253 \times \exp(0.10194t_d)$
LS	$Pf_j = P_j(t_d, n_d) = 0.00001524 \times \exp(0.09594t_d)$

Note: t_d: equipment age, n_d: operating times; ABB: Air Blast circuit Breaker; OCB: Oil Circuit Breaker; GCB: Gas Circuit Breaker; TrB: Transformer; LS: Line Switch.

13.2.3 Environment and Maintainability Improvement Index (EMI_l)

In addition to the substation performance improvement, the degree of environment and maintainability improvement are evaluated by:

$$EMI_l = \sum_{j=1}^{n(l)} \sum_{k=1}^{nf} w_k \cdot DI_{jk} \tag{13.6}$$

where w_k is the weight of the risk item k; DI_{jk} is the risk point of equipment j in risk item k; and nf is the total number of risk items.

Table 13.2 shows the risk items, their weights and their risk points shown in Hoshiya *et al.* (1989). The risk items are classified into four viewpoints; (i) Equipment damage, (ii) Local impact, (iii) Maintainability, and (iv) Complaints from customers. Equipment damage pertains to the possible degree of damage to the equipment. Local impact considers the consequences of power failures to customers around the target substation. Maintainability pertains to the difficulty in the maintenance activities. Complaints from customers pertain to the complaint level from customers. Note that since the weights and the risk points in table 13.2 were given as the initial condition in Hoshiya *et al.* (1989) and these values may vary depending on the users.

In table 13.2, particular attention should be paid to the risk item associated with the salt stain density. As for natural disaster risk, the salt, which is mainly carried by typhoons from the sea to equipment, is only categorized in table 13.2. The salt causes serious damage to equipment. The salt stain density of equipment becomes high as its location is close to the seaside. On the other hand, the seismic risk is not categorized though it causes serious damage to substation equipment. This is because the earthquake has seldom struck equipment during its own life cycle while a typhoon strikes in Japan every year.

Table 13.2 Risk Items Taken into Consideration in Actual Renewal Planning for Substation Equipment

View Point	Equipment Damage				Local Impact				Maintenance Difficulty		Complaint from Customers	
Risk Item (k)	CB accident rate		Salt Stain Density		Land Use Condition		Neighborhood facing Substation		Installation Condition	Security	Maintenance Activity	Mechanical Noise
Weight (w_k)	13		10		1		4		11	12	9	9
Risk Point (DI_{jk})	10 times of the average in 6 years	10	Extra high density (seaside)	10	Commercial	10	Skyscraper, High density residential	10	Classified into 3 grades (10, 5 and 0)	Classified into 3 grades (10, 5 and 0)	Classified into 3 grades (10, 5 and 0)	Classified into 3 grades (10, 5 and 0)
	5times	8	High	8	Residential	9	Residential	7				
	2times	5	Medium	5	Semi-industrial	5	Road	5				
	None	0	Low	0	Industrial	2	Others	0				

Note: Average of CB accident rate [0.2 (over 77kv,22 and 23kV), 0.1 (6kv)]

13.3 Proposed Renewal Planning Model

13.3.1 General Framework

Figure 13.1 shows the conceptual figure of the benefit cost analysis for the optimum renewal planning, which presents the trade-off relationship between Investment cost and Damage cost against multi-hazards. The ordinate indicates cost such as present value and the abscissa indicates the performance level of the entire power system. The increment of the Investment cost reduces the Damage cost and improves the performance level under multi-hazards. The proposed model based on this concept enables us to find the optimum condition for minimizing the total cost and to quantitatively discussing an acceptable performance level under multi-hazards. The renewal priority and renewal schedule for old equipment are also provided by the optimum condition.

In the renewal planning, since the Investment Cost is considered as a renewal cost, the total life cycle cost is defined as:

$$E(TC^{NT}) = RC^{NT} + E(DC^{NT}) \Rightarrow \min \tag{13.7}$$

where $E(TC^{NT})$ is the expected total cost in the analytical time period, NT; RC^{NT} is the renewal cost in NT; and $E(DC^{NT})$ is the expected total damage cost caused by multi-hazards in NT.

In the actual comprehensive renewal planning mentioned in section 13.2, the salt stain risk and the deterioration risk are treated as the scheduling multi-risks. Furthermore, this chapter emphasizes that the seismic risk should also be considered as a scheduling risk in the comprehensive renewal planning.

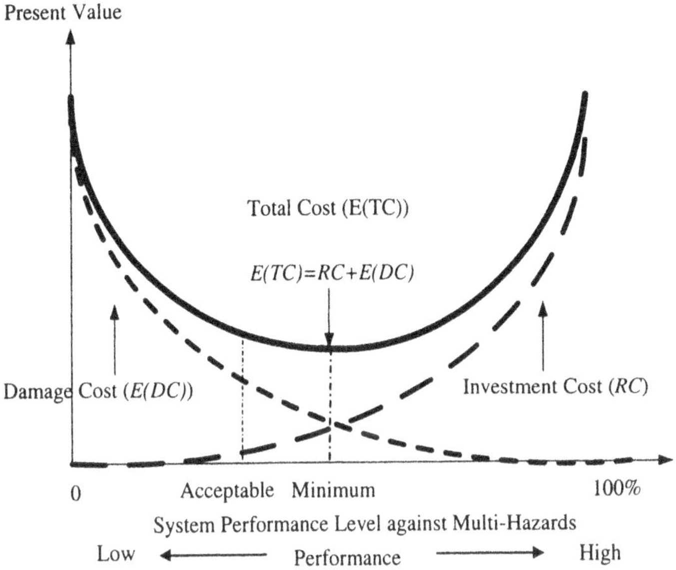

Figure 13.1 Conceptual Figure of Benefit Cost Analysis

Figure 13.2 shows multi-hazards, multi-risks, and the components of the total cost considered in this chapter. Earthquakes and typhoons are defined as hazards causing seismic risk and salt stain risk, respectively. On the other hand, it is difficult to specify the hazard causing deterioration risk. For example, according to the Japan Electrical Manufactures' Association (1999), equipment quality is gradually degraded by thermal energy, electric field, mechanical stress, and many environmental factors including corrosion. However, the deterioration mechanism due to the above factors is not yet clearly understood. Therefore, in this chapter, equipment age and its operating times are only parameterized to evaluate the deterioration risk based on Hoshiya *et al.* (1989).

The expected damage cost can be divided into two major categories associated with the damage to the supply side and that to the demand side. The damage cost associated with the supply side consists of the repair and revenue loss costs. The damage cost pertaining to the demand side indicates the power interruption cost. In this chapter, the total cost is further divided into the two types shown in figure 13.2. The total cost associated with the power industry is defined as the sum of the investment cost and the damage cost pertaining to the supply side. Similarly, the total cost associated with the power market consists of the investment and all the damage cost contents related to the supply and demand sides. Thus, the total damage cost is defined as:

$$E(DC^{NT}) = \begin{cases} E(TREC^{NT}) + E(TREL^{NT}) & \text{(associated with power industry)} \\ E(TREC^{NT}) + E(TREL^{NT}) + E(TSC^{NT}) & \text{(associated with power market)} \end{cases} \quad (13.8)$$

where $E(TREC^{NT})$ is the expected total repair cost in NT; $E(TREL^{NT})$ is the expected total revenue loss in NT; and $E(TSC^{NT})$ is the expected total power interruption cost in NT.

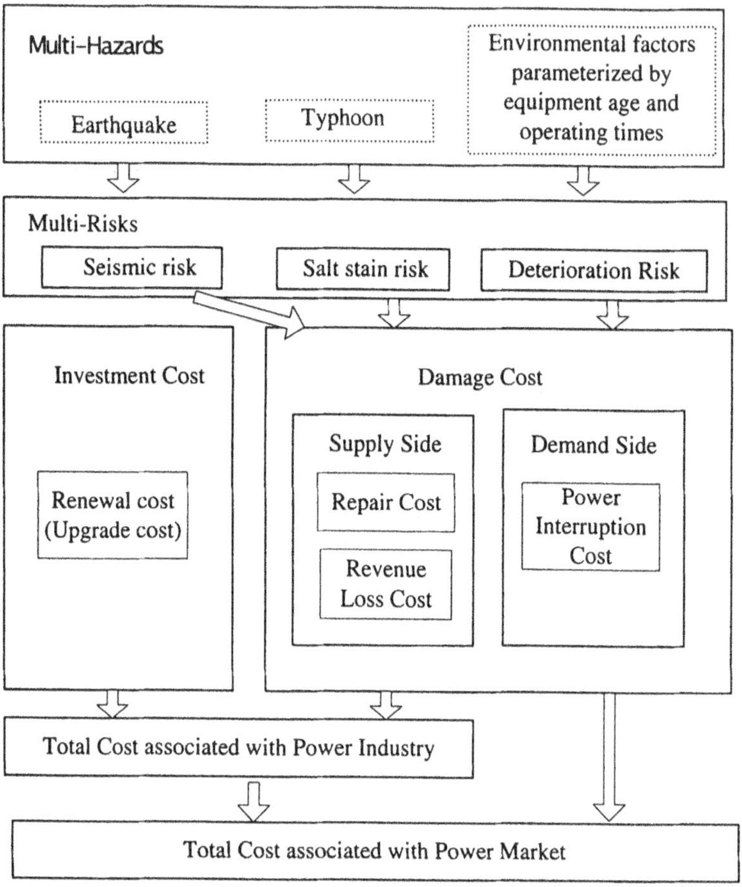

Figure 13.2 Total Cost Framework

13.3.2 Renewal Cost

The discount rate including the inflation rate and the interest rate in a year is taken into account in order to evaluate the present value of the total renewal cost in NT. The total renewal cost is defined as:

$$RC^{NT} = \sum_{t=1}^{NT} \sum_{j=1}^{NR(t)} (1+r_1)^{-t} \cdot rc_j \tag{13.9}$$

where $NR(t)$ is the total number of renewal equipment at time t; r_l is the annual discount rate; and rc_j is the renewal cost of equipment j.

13.3.3 Repair Cost

The expected total repair cost associated with the deterioration risk, the salt stain risk and the seismic risk is expressed as:

$$E(TREC^{NT}) = E(TREC_d^{NT}) + E(TREC_s^{NT}) + E(TREC_e^{NT}) \tag{13.10}$$

where $E(TREC_d^{NT})$, $E(TREC_s^{NT})$ and $E(TREC_e^{NT})$ indicate the total repair cost associated with the deterioration risk, the salt stain risk, and the seismic risk, respectively. Note that in equation (13.10), it is assumed that the coincident probabilities, which indicate the correlations associated with the occurrence of the three hazards, is very small and could be neglected though it was defined by Wen (1990).

In order to evaluate $E(TREC_d^{NT})$, it is assumed that substation equipment progressively deteriorates with increasing age and operating times. On the basis of this assumption, the expected repair cost associated with the deterioration risk is defined as:

$$E(TREC_d^{NT}) = \sum_{t=1}^{NT} \sum_{j=1}^{n} (1+r_1)^{-t} \cdot P_j^t(t_d, n_d) \cdot rpc_j \tag{13.11}$$

where n is the total number of target equipment for renewal; $P_j^t(t_d, n_d)$ is the damage probability of equipment j at time t with its age t_d and its operating times n_d; and rpc_j is the repair cost of equipment j; in the numerical examples, equation (13.4) is used to evaluate $P_j^t(t_d, n_d)$.

The salt stain risk is caused by typhoons and seasonal wind. According to the Committee of Salt-Resistance Design for Substation Equipment (1980), over 80 % of the damage due to salt stain is caused by a typhoon. In this chapter, a typhoon is treated as the only hazard for the salt stain risk. Thus, $E(TREC_s^{NT})$ is defined as:

$$E(TREC_s^{NT}) = \sum_{t=1}^{NT} \sum_{j=1}^{n} (1+r_1)^{-t} \cdot Po^t(s) \cdot P_j^t(r_s, d_s / s) \cdot rpc_j \tag{13.12}$$

where $Po^t(s)$ is the annual occurrence probability of typhoon s at time t; and $P_j^t(r_s, d_s / s)$ is the conditional damage probability of equipment j at time t with salt resisting capacity r_s and salt stain density d_s assuming a typhoon s has occurred.

The National Astronomical Observatory of Japan (1999) has been developing a database associated with natural hazards including typhoons. Using this database, $Po^t(s)$ are estimated by the Poisson process in this study. On the other hand, the Committee of Salt-Resistance Design for Substation Equipment (1980) investigated the salt stain density distribution and the annual average accident rate of substations due to the salt in Japan. These data are useful to evaluate $P_j^t(r_s, d_s / s)$ including the salt resisting capacity r_s and salt stain density d_s. Appendix 13.A.1 shows the basic formulation of $P_j^t(r_s, d_s / s)$ used for the numerical examples.

$E(TREC_e^{NT})$ is similarly evaluated. It is defined as:

$$E(TREC_e^{NT}) = \sum_{t=1}^{NT} \sum_{j=1}^{n} Po^t(e) \cdot (1+r_1)^{-t} \cdot P_j^t(r_e, a_e / e) \cdot rpc_j \tag{13.13}$$

where $Po^t(e)$ is the annual occurrence probability of earthquake e at time t; and $P_j^t(r_e, a_e/e)$ is the damage probability of equipment j at time t with seismic resisting capacity r_e and seismic force a_e assuming earthquake e has occurred.

The occurrence probability of an earthquake is often modeled as the Poisson process. However, the Headquarters for Earthquake Research Promotion (2001), which has officially announced the occurrence probabilities of major earthquakes in and around Japan, compared the Poisson process with other typical probability distribution models including the logarithmic normal, the Gamma, the Weibull, the Gumbel and the Brownian Passage Time distributions. As a result, it was concluded that the Brownian Passage Time (BPT) distribution firstly adapted by Matthews *et al.* (2002) is the most suitable probability model to estimate the occurrence of an earthquake. This chapter is supposed to employ the officially announced data associated with seismic risk in order to propose a practical model for the benefit cost analysis. Therefore, it is assumed that $Po^t(e)$ follows the BPT distribution.

Many approaches to evaluate $P_j^t(r_e, a_e/e)$ have been previously discussed. Showing the detailed model is beyond the scope of this chapter (please refer to existing references including Ang *et al.* (1992), Huo and Hwang (1995), Oikawa *et al.* (2001), the Public Interest Energy Research (PIER) reports, etc.). In this chapter, the seismic fragility functions of substation equipment proposed by Oikawa *et al.* (2001) are employed to evaluate $P_j^t(r_e, a_e/e)$ in the numerical examples.

13.3.4 Revenue Loss Cost

The total expected revenue loss also consists of three categories and is expressed as:

$$E(TREL^{NT}) = E(TREL_d^{NT}) + E(TREL_s^{NT}) + E(TREL_e^{NT}) \tag{13.14}$$

where $E(TREL_d^{NT})$, $E(TREL_s^{NT})$ and $E(TREL_e^{NT})$ indicate the total revenue loss associated with the deterioration risk, the salt stain risk, and the seismic risk, respectively.

$E(TREL_d^{NT})$ is defined as:

$$E(TREL_d^{NT}) = \sum_{t=1}^{NT} (1+r_i)^{-t} \cdot E(PEL(\mathbf{P}^t(\mathbf{x}_d))) \times PC/Unit^t \tag{13.15}$$

where $PC/Unit^t$ is the average unit power price at time t; and $E(PEL(\mathbf{P}^t(\mathbf{x}_d))$ is the expected power energy loss of the entire target power system with the equipment damage probability vector $\mathbf{P}^t(\mathbf{x}_d)$ at time t;. $\mathbf{P}^t(\mathbf{x}_d)$ represents the damage probability of equipment associated with the deterioration defined by $P_j^t(t_d, n_d)$ (for j=1,...,n).

Shumuta (1998) proposed models to evaluate the expected power energy loss of the primary and the secondary power system due to an earthquake scenario based on the Monte Carlo Simulation (MCS). However, compared with the damage probability of equipment associated with the earthquake, that associated with the deterioration is very small (*e.g.*, $P_j^t(t_d,n_d) < 10^{-3}$). This suggests that it is difficult to apply the MCS to evaluate $E(PEL(\mathbf{P}^t(\mathbf{x}_d)))$ because a large trial number is needed for the MCS. On the other hand, even if a minor accident such as a single accident associated with equipment occurs at a substation, since the power system has redundancy, its damage hardly affects the performance of the entire power system. For this reason, to evaluate $E(PEL(\mathbf{P}^t(\mathbf{x_d})))$, it is assumed that equipment damage associated with the deterioration in a substation does not affect the performance of other substations and its damage independently occurs. Under this assumptions, $E(PEL(\mathbf{P}^t(\mathbf{x_d})))$ is defined as:

$$E(PEL(\mathbf{P^t}(\mathbf{x_d}))) = \sum_{l=1}^{NS} f_l(\mathbf{P^t}(\mathbf{x_d}))\cdot g_l(\mathbf{P^t}(\mathbf{x_d})) =$$

$$= \sum_{l=1}^{NS} f_l(P_1^t(\mathbf{x_d}),\cdot\cdot P_j^t(\mathbf{x_d}),\cdots P_{n(l)}^t(\mathbf{x_d}))\ \cdot g_l(P_1^t(\mathbf{x_d}),\cdot\cdot P_j^t(\mathbf{x_d}),\cdots P_{n(l)}^t(\mathbf{x_d})) \cong$$

$$\cong \sum_{l=1}^{NS}(\sum_{de=1}^{n(l)}(d_1\cdot\cdot d_j\cdot\cdot d_{n(l)}\times f_l(\delta_1\cdot\cdot\delta_j\cdot\cdot\delta_{n(l)})\cdot g_l(\delta_1\cdot\cdot\delta_j\cdot\cdot\delta_{n(l)}))$$

$$\text{s.t.}\qquad \delta_j = \begin{Bmatrix} 1 & (j = de) \\ 0 & (j \neq de) \end{Bmatrix},\ d_j = \begin{Bmatrix} P_j^t(x_d) & (\delta_j = 1) \\ 1 - P_j^t(x_d) & (\delta_j = 0) \end{Bmatrix} \tag{13.16}$$

where $f_l\ (P^t(\mathbf{x_d}))$ is the power failure function for substation l at time t with equipment damage probability vector $\mathbf{P^t}(\mathbf{x_d})$; $P_j^t(\mathbf{x_d})$ indicates $P_j^t(t_d,n_d)$; here the detail formulation of $f_l(\delta_1\cdot\cdot\delta_j\cdot\cdot\delta_{n(l)})$ should be referred to Shumuta (1996, and 1998); $g_l(\mathbf{P^t}(\mathbf{x_d}))$ is the restoration time function for substation l at rime t ; NS is the total number of substations of the target power system; and de is the damaged equipment.

As for the restoration time function, $g_l(\delta_1\cdot\cdot\delta_j\cdot\cdot\delta_{n(l)})$, some data are available in Japan. For example, The Agency of Natural Resources and Energy (1998) has officially announced statistics related to the accident and its restoration time of electric power equipment on its own website. The power industries have also been accumulating confidential data associated with equipment accidents and its restoration time. In this chapter, the restoration time for the single accident of substation equipment is assumed as a constant value for every equipment type based on the Agency of Natural Resources and Energy (1998).

$E(TREL_s^{NT})$ is evaluated as:

$$E(TREL_s^{NT}) = \sum_{t=1}^{NT}(1+r_1)^{-t}\cdot Po^t(s)\cdot E(PEL(\mathbf{P^t}(\mathbf{x_s})))\ \times PC/Unit^t \tag{13.17}$$

where $E(PEL(\mathbf{P^t}(\mathbf{x_s})))$ is the expected power energy loss of the entire target power system with the equipment damage probability vector $\mathbf{P^t}(\mathbf{x_s})$ associated with the salt assuming typhoon s has occurred; here it is assumed that $\mathbf{P^t}(\mathbf{x_s})$ represents $(P_j^t(r_s\ d_s/s)$; for $j=1...n)$. In general, since the damage probability of equipment associated with the salt stain risk is also very low, the same model defined by equation (13.16) is employed to evaluate $E(PEL(\mathbf{P}^t(\mathbf{x_s})))$.

Similarly, $E(TREL_e^{NT})$ is expressed as:

$$E(TREL_e^{NT}) = \sum_{t=1}^{NT}(1+r_1)^{-t}\cdot Po^t(e)\cdot E(PEL(\mathbf{P^t}(\mathbf{x_e})))\ \times PC/Unit^t \tag{13.18}$$

where $E(PEL(\mathbf{P^t}(\mathbf{x_e})))$ is the expected power energy loss of the entire power system with equipment damage probability vector $\mathbf{P^t}(\mathbf{x_e})$ assuming earthquake e has occurred; here $\mathbf{P^t}(\mathbf{x_e})$ represents $(P_j^t(r_e,a_e/e)$; for $j=1...n)$.

To evaluate $E(PEL(\mathbf{P}^t(\mathbf{x}_e)))$, an existing model proposed by Shumuta (1998) could be employed. A summary of this model is illustrated in Appendix 13.A.2. Note that since an earthquake causes not only damage to the substation equipment but also to the other supply and demand side facilities including transmission tower, household, commercial, and industrial factories, the revenue loss is also caused by the damage to them. However, since the proposed model is expected to be applied in the upgrade planning of the substation equipment, evaluating the revenue loss associated with the damage to such non-equipment facilities is beyond the scope of this chapter. Thus, this chapter only considers the revenue loss associated with the damage to substation equipment.

13.3.5 Power Interruption Cost

The Power interruption cost has been employed as an index to evaluate the economic loss to customers caused by the power energy loss. It is expressed as:

$$E(TSC^{NT}) = E(TSC_d^{NT}) + E(TSC_s^{NT}) + E(TSC_e^{NT}) =$$
$$= \sum_{t=1}^{NT}\sum_{l=1}^{NS}(1+r_1)^{-t}\cdot\Big(E\big(PEL_l\big(\mathbf{P}^t(\mathbf{x}_d)\big)\big) + Po^t(s)\cdot E\big(PEL_l\big(\mathbf{P}^t(\mathbf{x}_s)\big)\big) + Po^t(e)\cdot E\big(PEL_l\big(\mathbf{P}^t(\mathbf{x}_e)\big)\big)\Big)\times$$
$$\times PO_l/Unit \qquad (13.19)$$

where $E(TSC_d^{NT})$, $E(TSC_s^{NT})$ and $E(TSC_e^{NT})$ indicate the total power interruption cost caused by the deterioration risk, the salt stain risk, and the seismic risk, respectively; and $PO_l/Unit$ is the average unit power interruption cost of the local area distributed by substation l.

Various approaches to evaluate the power interruption cost have been developed. Kurihara *et al.* (1998) provided a research review associated with the power interruption cost. According to Kurihara *et al.* (1998), the method for the evaluation of the power interruption cost is divided into four basic approaches: (1) the contingent valuation method (CVM), (2) the direct costing method, (3) the proxy method, and (4) Market-based method.

The contingent valuation method establishes a monetary value of outage for incremental changes in the levels of services. This is quantified either through the customer's willingness to pay (WTP) to avoid having an interruption, or the willingness to accept (WTA) compensation for having had one. Valuation based on WTP and WTA are valuable measures and serve as upper and lower bounds, respectively, in the cost of the outage assessment.

The direct costing method is the most obvious approach for the determination of customer interruption costs for given outage conditions. The respondent is given a worksheet and asked to identify the impacts and evaluate the costs associated with particular outage scenarios. This approach provides consistent results in those situations where most losses tend to be tangible, directly identifiable, and quantifiable.

The proxy method (or called indirect costing method) is based on the economic principle of substitution in which the valuation of a replacement is used as a measure of worth of the original item. For example, the cost of the power generator for emergency, or expenditures associated with the power industry is regarded as a substitute valuable. It is useful when social considerations and inconvenient effects are expected to comprise a significant part of the overall interruption costs, such as in the residential sector.

The market-based costing method evaluates the short term or long-term power interruption costs using econometrics approach. For example, the demand curve for customers, who have a load interception contract with the electric power industry, can be estimated on the basis of the data associated with the power failure in a target period. If the demand curve is given, the consumer surplus, which indicates the power interruption cost, can be evaluated.

As for Japan, there are few studies related to the power interruption costs. For example, Nishino *et al.* (1982) investigated the unit power interruption cost in Japan using the proxy method and the CVM. The Institute of Energy Economics (1994) analyzed the impacts of the Japan economy due to the power failure based on the Market-based method. Nakamura and Yamashiro (1999), and Fujita *et al.* (2000) investigated the average unit power interruption cost of the local customers in Hokkaido, Japan, by means of CVM. However, investigating the unit power interruption cost is beyond the scope of this chapter. Therefore, in the numerical examples, the average unit power interruption costs estimated by Fujita *et al.* (2000) are used to estimate $PO_l/Unit$.

13.3.6 System Performance

Evaluating the system performance is useful to determine the minimum requirement of the supply power reliability as a constraint in the cost benefit analysis or to understand the cost effectiveness of the upgrade. Thus, the system performance under the multi-hazards in the total analytical time period, NT, is defined as:

$$SPL^{NT} = \left(1 - \frac{\sum_{t=1}^{NT} \left\{ \sum_{l=1}^{NS} \left(f_l(\mathbf{P^t}(\mathbf{x_d})) + Po^t(s) \cdot f_l(\mathbf{P^t}(\mathbf{x_s})) \right) + Po^t(e) \cdot f_g(\mathbf{P^t}(\mathbf{x_e})) \right\}}{\sum_{t=1}^{NT} DP^t} \right) \times 100 \tag{13.20}$$

where SPL^{NT} is the system performance level of the entire target power system in NT; DP^t is the target demand power of the entire target power system under the normal conditions for time period t; $f_l(\mathbf{P^t}(\mathbf{x_d}))$, $f_l(\mathbf{P^t}(\mathbf{x_s}))$ and $f_g(\mathbf{P^t}(\mathbf{x_e}))$ are the power failure functions of substation l and entire power system g shown in 13.3.4 and Appendix 13.A.2. $Po^t(s)$ and $Po^t(e)$ indicate the annual occurrence probability of typhoon s and earthquake e at time t.

13.3.7 Renewal Priority

To determine the renewal priority of substation l, the cost benefit index is evaluated for every substation through an iterative procedure. It is defined as:

max $\{CB_l(m);\, l = 1,\ldots,NS\}$ $\Rightarrow$ top priority substation for renewal at iterative step m;

$$CB_l(m) = \frac{E(DC(m-1)) - E(DC_l(m))}{\sum_{j=1}^{NR(l)} rc_j} \tag{13.21}$$

where $CB_l(m)$ is the cost benefit index of substation l for iterative step m; $E(DC(m\text{-}1))$ is the annual average total damage cost of the entire power system at iterative step m-1; here $(E(DC(m\text{-}1))$ indicates $E(DC^{NT})$ for $NT=1$. Particularly, $E(DC(0))$ indicates $E(DC^{l})$ with no comprehensive renewal substation; $E(DC_l(m))$ also indicates $E(DC^{l})$ with renewing substation l from the renewal condition of the entire power system for iterative step m-1; after renewing substation l with max $(CB_l(m))$, $E(DC(m))$ is evaluated step by step; and $NR(l)$ is the total number of renewal equipment in substation l; and rc_j is the renewal cost of equipment j. Note that the renewal priority of a substation becomes higher with the maximum cost benefit index value at the earlier iterative step.

13.3.8 Renewal Scheduling

In order to optimize the renewal schedule of substations in the analytical time period, the dynamic programming method is applied. The objective function is defined as:

$$f^t(RC^t) = \min\left\{g^t(rc^t) + f^{t-1}(RC^t - rc^t)\right\} \tag{13.22}$$

$$\text{s.t.} \quad \begin{cases} 0 \le rc^t \le MaxRC \\ f^1(RC^1) = \min\left\{g^1(rc^1)\right\} \\ 2 \le t \le NT \end{cases}$$

where $f^t(RC^t)$ is the cost function for time period t with the total renewal cost (or total number of renewal substations) RC^t; rc^t is the renewal cost at time t; $g^t(rc^t)$ is the total cost when rc^t is invested at time t; and $MaxRC$ is the maximum renewal cost (or maximum number of renewal substations) for a one year period.

As an initial condition of this model, the renewal priority of each substation was previously calculated by equation (13.21). Thus, alternatives of the renewal substations at time t, which are needed to determine rc^t in equation (13.22), are automatically determined in order of the renewal priority of the substation. It follows an iterative procedure that starts with any initial alternatives, $g^1(rc^1)$, as $f^1(RC^1)$, using these on the right-hand side of equation (13.22). Now the right-hand side is fully known, so the results of applying the operator can be calculated for every rc^2. The result a new function value is called, $f^2(RC^2)$. They are then used on the right-hand side to get the next iterate function value, $f^3(RC^3)$, and so on. After carrying out this iterative process until all substations are renewed or the analytical time period reaches NT, every substation is allocated as the renewal substation for the appropriate time.

13.4 Numerical Examples

Numerical examples are developed for an actual electric power system to demonstrate the applicability of the proposed model. In the numerical examples, the comprehensive renewal indicates that circuit breakers (CB) and line switches (LS) are replaced with Gas Insulated Switches (GIS) for each substation.

13.4.1 Characteristics of the Target Power System

Figure 13.3 shows a target power system located in Japan. It consists of 125 substations including 9 primary substations and 115 secondary substations. The primary substation includes the substations with the voltage levels of 154, 275 and 500kV. The target power system is divided into 9 local power system groups depending on the primary substations. Each local group has one primary substation as the power source. On the other hand, the secondary substation is a 77kV substation. Each secondary substation has its own distribution district as shown in figure 13.3. The power system data were provided by a power industry distributing the power to this area.

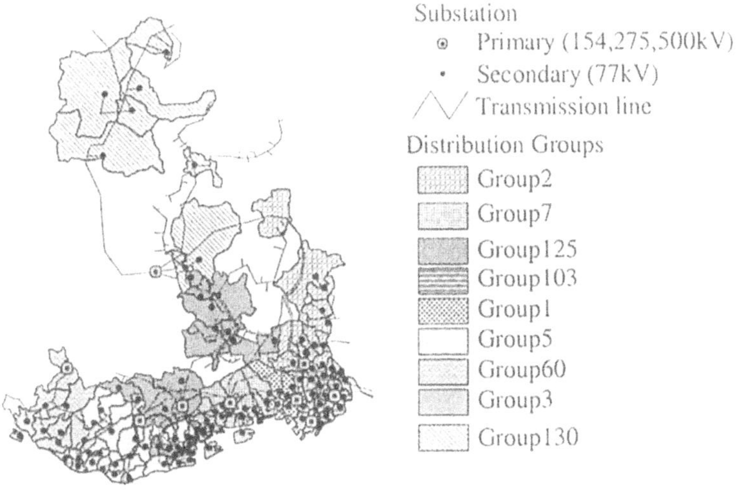

Figure 13.3 Target Power System

Table 13.3 shows the statistics of the Target Power System (TPS). To emphasize the local characteristics of the TPS, an actual power system located in Tianjin, China (CPS) was also listed in table 13.3. The TPS has the features of a typical power system in a developed region. For example, D/T is helpful as an index for comparing the development level of the local region and the power consumer's profile. D/T in the TPS is about 3 times bigger than the CPS. D/T becomes higher with an increase in the population density. Similarly, T_{max} is usually employed to specify the load profile. T_{max} for the TPS is about three fourths of that for the CPS. These results suggest that the TPS has different levels of power consumers including not only normal households but also heavy power consumers such as big factories or commercial facilities. On the contrary, the power consumption for each customer in China is still small and averaged. These features illustrate that the target area is the more developed region in Asian countries. Note that in numerical examples, it is assumed that the target demand power, DP^t, defined in equation (13.20) is 4929 MW as a constant value.

Table 13.3 Comparison of Statistics between the Target Power System (TPS) and an Actual Power System Located in Tianjin, China (CPS)

No.	Contents	Target Power System (TPS)	China Power System (CPS)
(1)	Compared Time	1998	1997
(2)	Supply Area (km^2)	2669	11,920
(3)	Substation & Switching Station	125	206
(4)	Transmission Line Length (km)	1,355	3,140
(5)	Distribution Line Length (km)	19,917	14,738
(6)	Power Sales per Year (billion kWh)	21.5	16.4
(7)	Maximum Load (MW)	4,929	2,880
(8)	Population (thousand)	3,710	8,980
(9)	Power Price (US$/kWh)	0.1731	0.0613
(10)	Population Density (persons/ km^2)	1,390	753
(11)	Load per Person (kW/ person)	1.267	0.3207
(12)	Energy per Person (kWh/ person)	5795	1827
(13)	Load Density (MW/ km^2)	1.76	0.24
(14)	D/T=(5)/(4)	14.70	4.69
(15)	T_{max} (h)=(6)/(7)	4362	5694
(16)	Profit per Year (billion $)=(6) ×(9)	3.72	1.01

13.4.2 Modeling of the TPS

Figure 13.4 shows a power system model of the target power system. The target power system is modeled as 9 system groups including 203 nodes and 213 links. The nodes are divided into two types; substation and dummy nodes. The substation node has a sub-network consisting of sub-nodes for substation equipment including the transformers, circuit breakers, and line switches. Nodes 1-23, 25-84, 86-128, 130, and 197 show substation nodes. In the substation nodes, nodes 1,2,3,5,7,60,103,125 and 130 are defined as primary substations. The total number of sub-nodes in the target power system reaches 7,252 in total. The dummy node is modeled as the junction of the transmission lines.

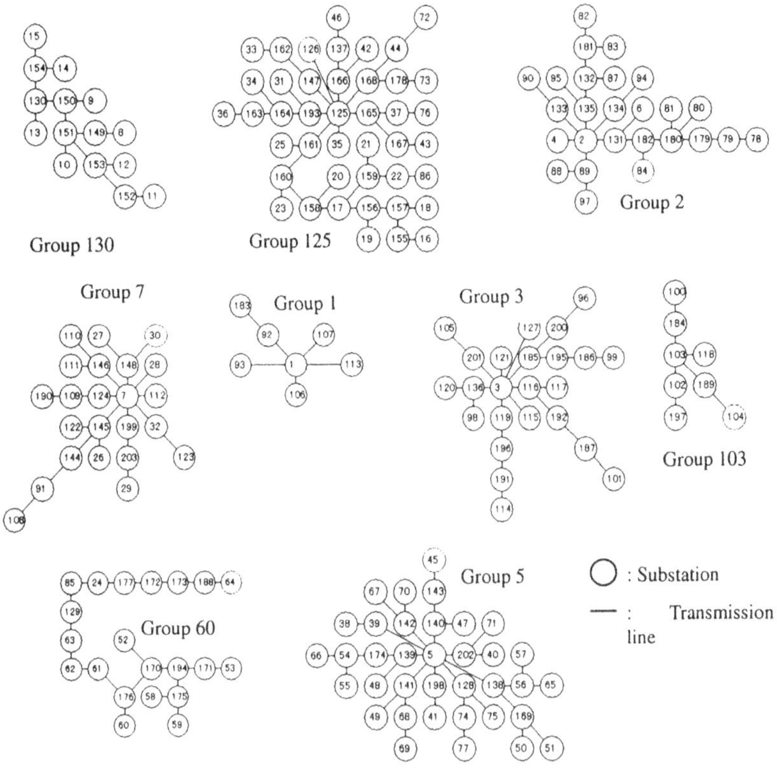

Figure 13.4 Analytical Model for the Target Power System

13.4.3 Multi-risk Conditions of the TPS

(1) Deterioration risk

Table 13.4 shows the statistics of the equipment age installed in the TPS and the CPS. In table 13.4, the equipment of the TPS is relatively older than that of the CPS. This result demonstrates that the TPS has much older equipment and its deterioration risk is a more serious problem compared with the CPS. The damage probability of equipment associated with the deterioration, $P_j^I(t_d, n_d)$, was estimated based on the functions in table 13.1 and the provided data from the power industry.

Table 13.4 Statistics of the Equipment Age Installed in the TPS and the CPS

Contents	TPS			CPS		
Target equipment	Average age (Year)	Over 20years (%)	Over 10years (%)	Average age (Year)	Over 20years (%)	Over 10years (%)
Switches	18.41	44.71	75.38	9.87	15.26	52.63
Circuit Breakers	17.32	41.92	72.12	8.56	14.69	50.86

(2) Salt stain risk

According to the Committee of Salt-Resistance Design for Substation Equipment (1980), the target area is divided into five levels of salt stain density regions; less than 0.01 mg/cm^2 (none), 0.01to 0.03 mg/cm^2 (low), 0.03 to 0.06 mg/cm^2 (medium), 0.06 to 0.12 mg/cm^2 (high), and more than 0.12 mg/cm^2 (extra high). The salt stain risk becomes higher for with areas closer to the seashore. To evaluate the damage probability of equipment associated with the salt stain risk, $P_j^t(r_s, d_s/s)$, the accident rate of equipment for every salt density, $\lambda_j(m)$, is needed. $\lambda_j(m)$ was provided by the Committee of Salt-Resistance Design for Substation Equipment (1980). The annual occurrence probability of typhoon *s* at time period *t*, $Po^t(s)$, was assumed to be 0.955 based on National Astronomical Observation of Japan (1999).

(3) Seismic risk

As mentioned before, the damage probability of equipment associated with an earthquake, $P_j^t(r_e, a_e/e)$, was estimated based on the work of Oikawa *et al.* (2001). According to Oikawa *et al.* (2001), $P_j^t(r_e, a_e/e)$ follows the logarithmic normal distribution with median ,m_j, and standard deviation , β_j , shown in table 13.5. On the other hand, figure 13.5 shows the Peak Ground Acceleration (PGA) in the target area. The PGA is estimated based on an earthquake scenario due to the fault slide shown in the same figure.

Table 13.6 shows the earthquake scenario caused by the fault slide. In order to evaluate the PGA, an attenuation function proposed by Kanai *et al.* (1966) was applied and some confidential data associated with ground conditions was employed. According to the Headquarters for Earthquake Research Promotion, this fault slide has a recurrent cycle of 1000 to 2000 years. The last earthquake due to the fault slide occurred in 1596. These local data associated with the target fault was employed to evaluate $Po^t(e)$ which follows the BPT distribution.

Table 13.5 Parameters for seismic fragility functions of substation equipment (Oikawa *et al.*, 2001)

	Higher than 187kV							All voltage class
	Transformer	Circuit Breaker	Line switch	Lightning Arrester	Instrument Transformer	Power Capacity	Shut Reactor	Transmission Tower
Median (m_j) (GAL)	340	620	480	550	550	790	610	770
β_j	0.29	0.29	0.29	0.29	0.29	0.29	0.29	0.32
	Lower than 187kV							All voltage class
	Transformer	Circuit Breaker	Line switch	Lightning Arrester	Instrument Transformer	Power Capacity	Shut Reactor	Transmission Tower
Median (m_j) (GAL)	500	1010	780	830	830	790	610	600
β_j	0.29	0.29	0.29	0.29	0.29	0.29	0.29	0.29

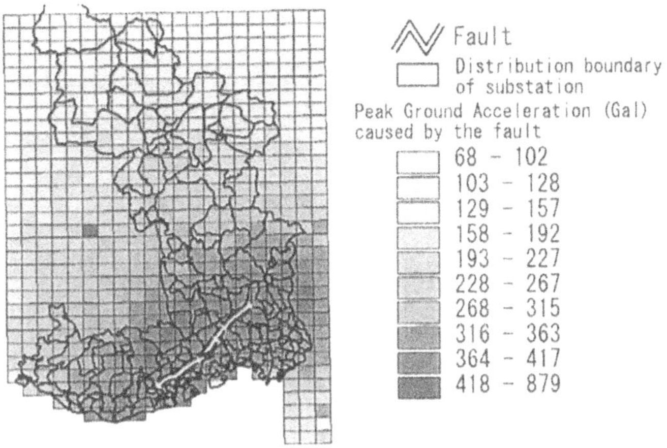

Figure 13.5 Distribution of PGA Caused by the Earthquake Scenario (fault slide)

Table 13.6 Earthquake Scenario

Latitude	Longitude	Epicenter depth (km)	Fault length (km)	Direction of fault slip	Magnitude
34.74	135.25	10	20	N54°E	7.0

13.4.4 Power Interruption Cost

One of the difficult tasks is to evaluate the power interruption cost for each distribution district. As mentioned before, investigating the average unit power interruption cost for every distribution district is out of the scope of this chapter. Thus, in the numerical examples, we employed the research result by Fujita *et al.* (2000). Fujita *et al.* (2000) classified the power consumers into three types depending on the consumer types, *i.e.*, residential, commercial, and industrial consumers. Residential consumers indicate the power consumers less than 50kw. Commercial consumers are from 50kw to 500kw. Industrial consumers indicate more than 500kw. Table 13.7 shows the average unit power interruption cost for every customer type evaluated by Fujita *et al.* (2000). To estimate the average unit power interruption cost for every distribution district, $PO_l/Unit$, we developed a GIS based database associated with the number of establishments and households for every distribution district based on the census data of Japan published by the Statistics Bureau & Statistics Center (1996). Based on the number of establishments and households, $PO_l/Unit$ is defined as:

$$PO_l/Unit = \sum_{ct=1}^{3} \frac{NC_l(ct)}{NC_l(1) + NC_l(2) + NC_l(3)} \times PIC(ct)/Unit \qquad (13.23)$$

where ct is the consumer type; here $ct=1$ is residential; $ct=2$ is commercial; $ct=3$ is industrial; $NC_l(ct)$ is the number of ct consumers in the distribution district of substation l; and $PIC(ct)/Unit$ is the average unit power interruption cost of ct consumers.

Table 13.7 Power Interruption Costs by Customer Type (Fujita *et al.*, 2000)

Customer type	Power interruption cost (yen/kwh)
Residential	1,350
Commercial	6,200
Industrial	2,300

13.4.5 Other Cost Parameters

In the numerical examples, the renewal costs of equipment, rc_j, and the repair cost of equipment rpc_j were assumed based on the Federation of the Electric Power Company of Japan (1992). This guideline is widely used in the Japanese power industries to objectively estimate the maintenance cost. The annual discount rate r_l is assumed as 0.0 based on the recent economic condition in Japan.

13.4.6 Comparison of Existing and Proposed Models

Figure 13.6 shows the comparison of the cost effectiveness, which indicates the renewal cost versus the damage cost, between the existing and proposed models. As mentioned before, the existing model can not consider the effect of seismic risk though it includes the evaluation models pertaining to the deterioration and salt stain risks.

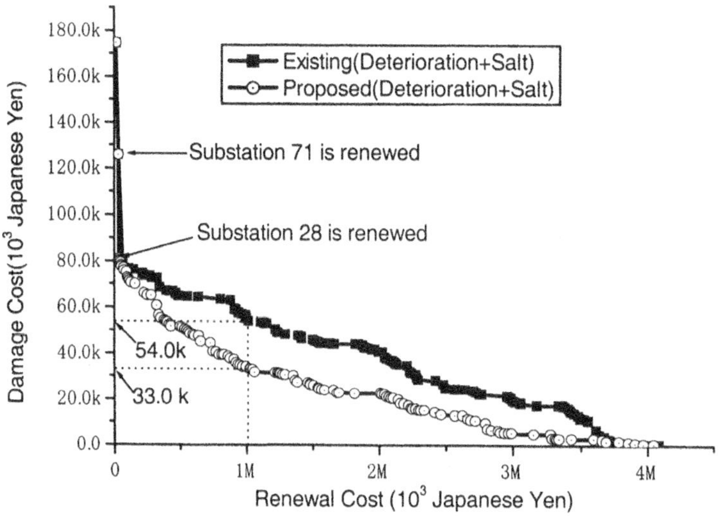

Figure 13.6 Comparison of Cost Effectiveness between the Existing and Proposed Models

Table 13.8 Renewal Priority based on the Existing and Proposed Models (Deterioration and Salt Stain Risks Considered)

Priority	SubNo (Existing)	SubNo (Proposed)	Priority	SubNo (Existing)	SubNo (Proposed)	Priority	SubNo (Existing)	SubNo (Proposed)	Priority	SubNo (Existing)	SubNo (Proposed)
1	71	71	32	29	105	63	91	53	94	42	56
2	28	28	33	77	99	64	92	77	95	35	62
3	123	76	34	70	45	65	53	74	96	126	4
4	62	106	35	33	58	66	49	72	97	83	97
5	111	120	36	114	50	67	41	35	98	87	118
6	39	107	37	76	88	68	99	36	99	82	92
7	43	117	38	37	110	69	89	68	100	86	17
8	124	75	39	1	21	70	45	52	101	6	89
9	54	100	40	109	34	71	105	73	102	84	20
10	55	66	41	106	9	72	102	124	103	81	112
11	113	70	42	120	197	73	94	33	104	80	37
12	122	94	43	74	87	74	95	10	105	22	84
13	31	82	44	72	27	75	103	26	106	15	60
14	127	126	45	73	29	76	97	125	107	17	30
15	36	41	46	40	102	77	98	2	108	18	48
16	34	80	47	69	90	78	44	12	109	78	5
17	119	39	48	128	95	79	93	86	110	23	103
18	32	13	49	65	67	80	116	43	111	19	11
19	56	123	50	58	14	81	90	51	112	20	25
20	3	113	51	57	83	82	197	65	113	25	46
21	110	122	52	7	127	83	88	78	114	21	16
22	68	54	53	48	23	84	2	111	115	79	121
23	38	109	54	27	47	85	96	101	116	4	3
24	115	96	55	26	69	86	100	81	117	13	116
25	75	38	56	52	7	87	104	6	118	14	18
26	112	104	57	51	114	88	101	119	119	9	1
27	108	49	58	30	98	89	118	8	120	10	93
28	66	32	59	47	40	90	117	108	121	12	31
29	107	91	60	60	57	91	125	79	122	8	128
30	67	59	61	50	115	92	5	55	123	16	44
31	121	42	62	59	22	93	46	15	124	11	19

Note: Priority: Renewal Priority; SubNo(Existing): Substation Number detemined by the Exsitng Model; SubNo(Proposed): Substation Number detemined by the Proposed Model.

On the other hand, the proposed model cannot consider the maintainability and the effect of the customer's complaint defined in the existing one to prioritize the renewal substation. Thus, in order to compare both models under the same conditions, the maintainability and the complaints from customers defined in table 13.2, and the seismic risk were neglected. In addition, to evaluate the damage and renewal costs, the analytical time period, *NT*, is assumed to be one year. Furthermore, to evaluate the objective function, *F*, in the existing model, the weights defined in equation (13.1) were assumed to be α=173.6 and β=11.56, which were the differences between max and min values of SPI_l and EMI_l, respectively.

In figure 13.6, the renewal cost increases with the number of renewal substations. The substations in the entire power system were renewed in order of the renewal priorities. Table 13.8 shows the renewal orders for every substation determined by the existing and proposed models. For example, when about 1 billion Japanese yen is invested based on the renewal orders in table 13.8, the damage costs associated with the existing and proposed models become about 54 and 33 million Japanese yen, respectively. The damage cost associated with the existing model is about 1.3 times higher than that associated with the proposed model. This result suggests that the proposed model enables us to determine the renewal priority of substations more cost effectively than the existing one under the assumed conditions. In actuality, according to some experts in the target power industry, complaints from customers have seldom occurred and there are slight differences in the maintainability of substations. These expert opinions also support the applicability of the proposed model.

13.4.7 Sensitivity Analysis

Figure 13.7 shows the comparison of the annual system performance levels (*SPL*) with the change in the elapsed time from the latest earthquake event in the target area. *SPL* indicates SPL^{NT} defined by equation (13.20) for *NT*=1 and is evaluated under the three hazard conditions: 1) the Earthquake (M7.0); 2) environmental factors associated with the deterioration; and 3) the salt stain caused by the typhoon. The abscissa shows the elapsed time from the occurrence of the latest earthquake. The ordinate indicates the annual system performance. As mentioned in section 13.3, it was assumed that the occurrence probability of the earthquake was in accord with the Brownian Passage Time (BPT) distribution. Therefore, the *SPL* for the earthquake becomes lower with an increase in the elapsed time. On the other hand, the *SPLs* for the other two hazards always become constant regardless of the elapsed time. This is because it is assumed that three multi-hazards independently occur and there are no relationships among the elapsed time and the other two hazards. In the target area, it has been 406 years since the last earthquake occurred (at the time of 2002). Figure 13.7 suggests that at the present condition, the seismic risk is the lowest level in the three risks. However, after 194 years, at which the elapsed time is 600 years, the seismic risk is higher than the salt stain risk. Furthermore, after 774 years, at which the elapsed time reaches 1180 years, the seismic risk becomes the highest of the three risks. To determine the renewal priority of existing equipment, the existing renewal planning usually neglects the seismic risk. This result supports this renewal policy if the acceptable performance level is lower than the *SPL* under the salt stain risk. However, if the elapsed time exceeds about 1180 years, the earthquake causes the highest functional loss of the three hazards. In this case, the seismic risk should be considered for the renewal planning to maintain an acceptable performance level of the target power system.

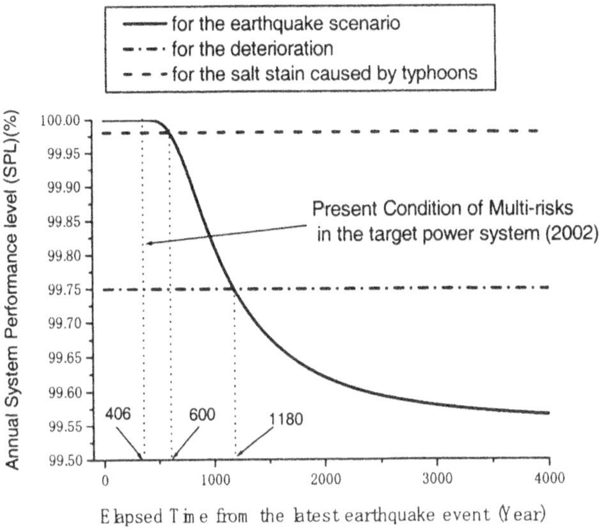

Figure 13.7 Comparison of System Performance Levels (SPLs) under the Multi-hazards with Change in the Elapsed Time from the Latest Earthquake Event

The Occurrence Probability of the Earthquake (OPE) includes significant uncertain factors because its estimation is based on the limited data. Therefore, it is necessary to carry out the sensitivity analysis between the total costs versus the OPE. Figure 13.8 shows the sensitivity of the minimum total cost and its optimum renewal cost associated with the power market on the change in the OPE. Note that the optimum renewal cost indicates the renewal cost to minimize the total cost associated with the power market. The minimum total cost and its optimum renewal cost were evaluated in total analytical time period, *NT*. In figure 13.8, *NT* is assumed to be one year.

Particular attention should be paid to the optimum renewal cost on the change in the OPE from 0.08 to 0.2. During the change in the OPE between them, the optimum renewal cost quickly increases. To discuss the reason, figure 13.8 also includes the tradeoff relationships between the damage cost and the renewal cost with the following OPE conditions; 0.001.0.01.0.1 and 1.0. For the condition when OPE = 0.1, there is little change in the total cost regardless of the increase in the renewal cost because the damage cost and the renewal cost are well balanced, compared with those of the other OPE conditions. This result suggests that if the OPE becomes around 0.1, the optimum renewal cost to minimize the total cost is dramatically changed. As a result, the sensitivity of the optimum renewal cost versus the OPE increases from 0.08 to 0.2. It illustrates that the proposed model enables us to find a critical boundary for the OPE from an economic point of view.

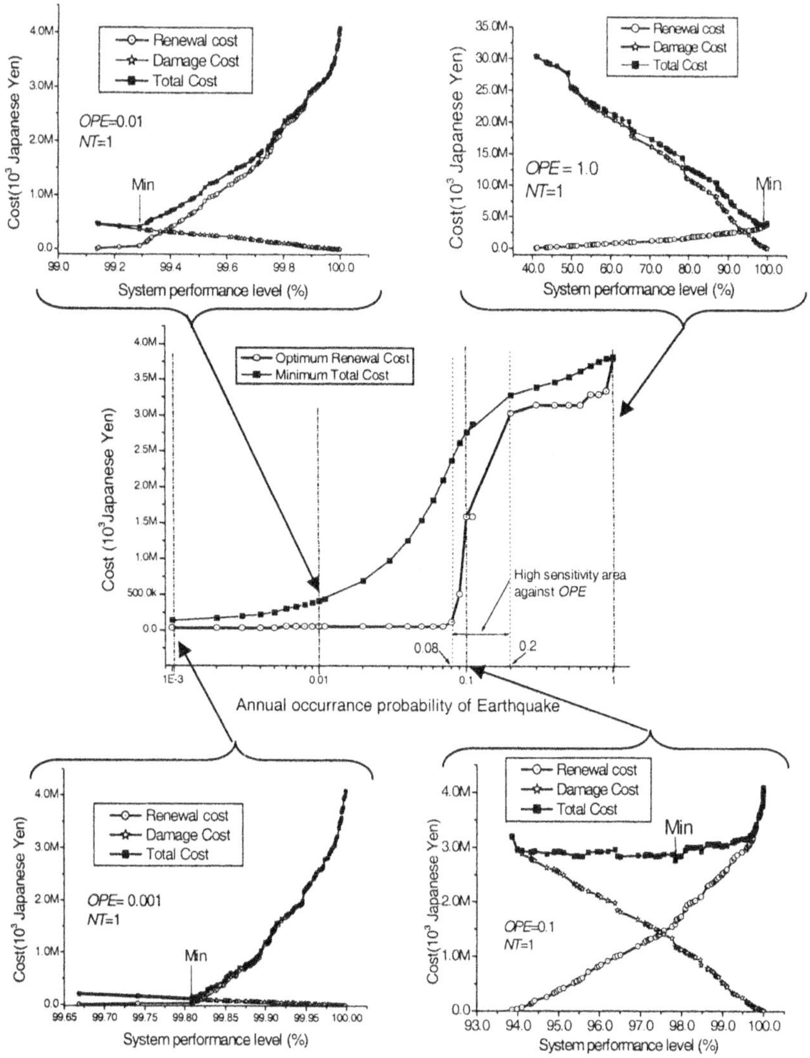

Figure 13.8 Sensitivity of Minimum Total Cost and Optimum Renewal Cost on Change in the Annual Occurrence Probability of Earthquake (Tradeoff Relationships between Renewal Cost and Damage Cost on Change in the Occurrence Probability of Earthquake)

Figures 13.9 and 13.10 show the tradeoff relationship between the renewal cost and the damage cost for the three hazards in 25 years. The total analytical time period, *NT*, is assumed to be 25 years. In figure 13.9, it is assumed that the maximum number of comprehensive renewal substations per year (*MaxRC*) is five. Figure 13.9 just focuses on the cost components associated with the power industry defined in figure 13.2. Figure 13.10 considers, on the other hand, all cost components associated with the power market. As mentioned before, the differences between figures 13.9 and 13.10 are whether the power interruption cost is considered or not as one of the contents of the damage cost. Figure 13.9, which does not include the power interruption cost, suggests that in order to minimize the total cost, no renewal cost should be invested in 25 years. In other words, from an economic point of view for the power industry, the present system performance level is too reliable to upgrade any substations for 25 years. In contrast, figure 13.10, which indicates the power interruption cost, shows that 300 million Japanese yen should be invested as the renewal cost to minimize the total cost associated with the power market.

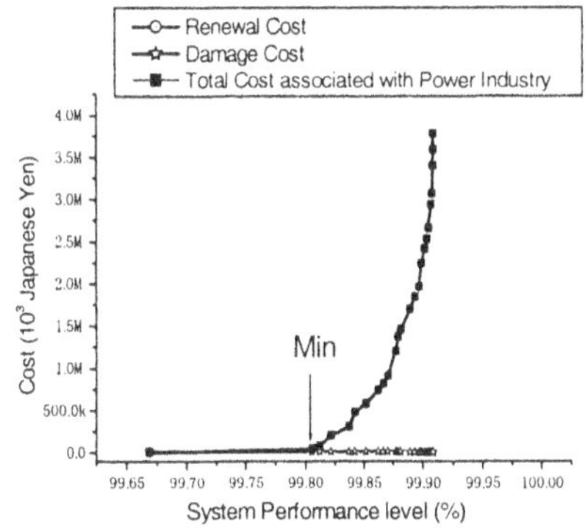

Figure 13.9 Tradeoff Relationship between Renewal Cost and Damage Cost Associated with Power Industry in 25 years

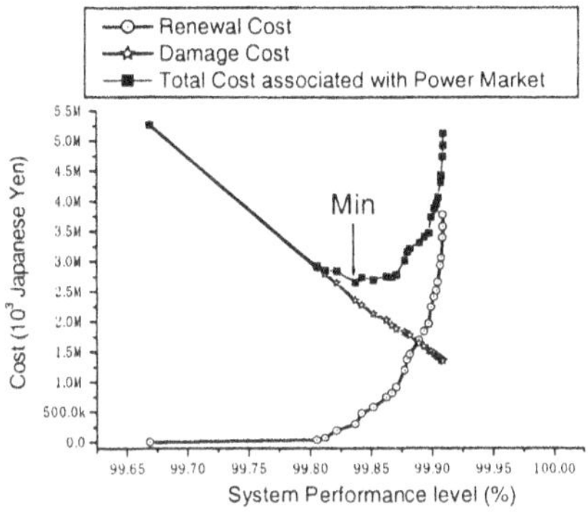

Figure 13.10 Tradeoff Relationship between Renewal Cost and Damage Cost Associated with Power Market in 25 years

13.5 Conclusions

This chapter presented a benefit cost analysis framework to support the comprehensive renewal planning of substation equipment. The proposed model was applied to an actual power system in Japan and compared with the existing model. The conclusions are as follows:

(1) The proposed model enables us to determine the renewal priority more cost effectively and more objectively than the existing one against multi-hazards including earthquakes, typhoons, and other environmental factors associated with equipment deterioration.

(2) In order to discuss the reasonable system performance level of the entire power system, a framework was provided from the standpoint of the comparison of the system performance during multi-hazards. As a result, it was demonstrated that the seismic risk might be higher than the other scheduled risks according to the local condition though the seismic risk has not considered as an important risk

factor in the conventional model.

(3) The sensitivity of the optimal renewal cost on the change in the occurrence probability of an earthquake was investigated. As a result, it was demonstrated that the proposed model enables us to identify a critical boundary of uncertain factors for the seismic risk assessment from an economic point of view.

(4) The target power system has no need to upgrade all substations at the present time in order to minimize the total cost associated with the power industry. On the other hand, to minimize the total cost associated with the power market, it should upgrade them appropriately selected by the proposed model.

Findings from this study indicate a number of areas for further research that will be important in developing applications to more complex systems. First, the discount rate should be improved. In order to discuss the long-term renewal scheduling, the discount rate is a significant parameter in the proposed model. The methodology to determine the discount rate is needed to improve the accuracy of the cost evaluation.

To expand the applicability of the proposed model, the revenue loss model should be improved. If an earthquake occurs, many facilities associated with not only the supply side but also the demand sides are simultaneously damaged. Both side damages may cause a power sales income loss to the power industry. In order to discuss a reasonable supply sides' seismic performance level which can balance the countermeasures costs before the earthquake and the business interruption losses after the earthquake, the revenue loss associated with the damage to the demand side should also be evaluated.

A maintainability assessment model should be developed. This chapter does not mention the cost assessment pertaining to the maintainability subjectively considered in the existing model. To improve the proposed model, the cost assessment models for the maintainability need to be developed.

A unit power interruption cost should be improved. In Japan, there are few studies related to the unit power interruption cost, especially during and after an earthquake event. To improve the accuracy of the total cost associated with the power market, further studies will be performed.

References

Agency of Natural Resources and Energy. 1998. *Electricity Safety Statistics* (in Japanese).

Ang, A.H-S., J. Pires, R. Schinzinger, R. Vallaverde, and E. Yoshida. 1992. "Seismic Reliability Of Electric Power Transmission System –Application to the 1989 Loma Prieta Earthquake," Technical Report to the NSF and NCEER, Department of Civil Engineering, University of California, Irvine.

Billinton, R. 1994. "Evaluation Of Reliability Worth In An Electric Power System," *Reliability Engineering And System Safety,* 46: 15-23.

California Energy Commission. 2000. "Electric System Seismic Safety and Reliability," Consultant Report, Public Interest Energy Research.

Chang, S.E., M. Shinozuka, and D.B. Ballantyne. 1998. "Life Cycle Cost Analysis with Natural Hazard Risk: A Framework and Issues for Water Systems," *Optimum Performance of Civil Infrastructure Systems*, ASCE: 58-73.

Committee of Salt-Resistance Design For Substation Equipment. 1980. "Salt-Resistant Design," *Electric Cooperative Study*, 34 (3) (In Japanese).

Cornell, C.A. 1968. "Engineering Seismic Risk Analysis," *Bulletin of the Seismological Society of America*, 58 (5): 1583-1606.

Ellsworth, W.L., M.V. Matthews, R.M. Nadeau, S.P. Nishenko, P.A. Reasenberg, and R.W. Simpson. 1999. "A Physically Based Earthquake Recurrence Model for Estimation of Long-Term Earthquake Probabilities," USGS Open-File Report: 99-522.

Federal Emergency Management Agency. 1999. "8.5 Electric Power System," HAZUS99-SR1, Technical Manual, Washington, DC.

Federation of Electric Power Companies Japan. 1996. *Handbook of Electric Power Industry.* Japan (in Japanese).

Federation of the Electric Power Companies of Japan. 1992. *The Standard Maintenance Expenses Calculation Guideline.* Japan (in Japanese).

Fujita, S., K. Nakamura, and S. Yamashiro. 2000. "An Economic Estimation of Distribution System Reliability Using Power Interruption Costs," *J.IEIE Jpn*, 20 (6): 457-462.

Geographical Survey Institute. 1996. *Detail Digital Information.* (10m Grid Land Use).

Gilani, A.S., A.S. Whittaker, G.L. Fenves, and E. Fujisaki. 1999. "Seismic Evaluation and Retrofit of 230-Kv Porcelain Transformer Bushings," Report No.PEER 99/14, Pacific Earthquake Engineering Research Center, University of California, Berkeley, CA.

Headquarters for Earthquake Research Promotion. 2001. *Evaluation Method and Its Application for the Probability of Long-Term Earthquake Occurrence.* Ministry of Education, Culture, Sports, Science and Technology, Japan (in Japanese).

Hoshiya, M. *et al.* 1989. "Study on Life Cycle Estimation of Substation Equipment," Technical Report, Association for the Development of Earthquake Prediction, Japan (in Japanese).

Huo. J.R. and H. Hwang. 1995. "Seismic Fragility Analysis of Equipment and Structure in a Memphis Electric Substation," Technical Report NCEER-95-0014, Buffalo, NY.

Institute of Energy Economics. 1994. "Social Activity in Tokyo during Large-Scale Power Outage, Economic Effect Survey," Research Report IEE SR-255, Japan (in Japanese).

Japan Electrical Manufacture's Association. 1999. *Study on Long-Term Reliability of Substation Facilities*, Japan (in Japanese).

Japan Electrical Manufacturers' Association. 1999. *Discussion of Reliability of Substation Equipment in Long-Term Usage*, Japan (in Japanese).

Kameda, H., Y. Ishikawa, T. Okumura, and M. Nakajima. 1997. "Probabilistic Scenario Earthquakes -Definition and Engineering Applications," *Journal of Structural Mechanics and Earthquake Engineering*: 75-87 (in Japanese).

Kanai, K., S. Hirano, K. Yoshizawa, and T. Asada. 1966. "Observation of Strong Earthquake Motions in Matsushiro Area. Part 1," Bulletin of the Earthquake Research Institute, 44: 1269-1296.

Kansai Electric Power Company. 1995. *Restoration Report of the Hanshin-Awaji Earthquake.* Japan (in Japanese).

Kurihara, I., K. Okada, H. Asano, N. Watanabe, H. Matuoka, and M. Inoue. 1998. "Supply Reliability Evaluation Considering Power Outage and Supply Costs," CRIEPI Report T98505, Japan (in Japanese).

Matthews, M.V., W.L. Ellsworth, and P.A. Reasenberg. 2002. "A Browinian Model for Recurrent Earthquakes," *Bulletin of the Seismological Society of America*, 92 (6): 2233-2250.

Matuda, E.N., W.U. Savage, K.K. Williams, and G.C. Laguens. 1991. "Earthquake Evaluation of a Substation Network," Proceedings of 3rd U.S. Conference TCLEE/ASCE: 295-317.

Nakamura, K. and S. Yamashiro. 1999. "A Survey Study on Estimation of Customer Interruption Costs," *T.IEE Japan*, 119-B (2): 284-290 (in Japanese).

National Astronomical Observatory of Japan. 1999. *Rikanenpyo, 99.* Tokyo, Japan; Maruzen. Co., Ltd.

Neudorf, E.G., M. Stephenson, and M. Logan. 1995. "Cost-Benefit Analysis of Power System Reliability: Two Utility Case Studies," *IEEE Trans. PWRS*, 110 (3).

Nishino, Y., S. Ueki, and F. Makino. 1982. "Estimation of Outage Cost in Japan," CRIEPI Report, No.582007, Japan.

Oikawa, T., S. Fukushima, H. Takase, T. Uchiyama, and K. Muramatsu. 2001. "Seismic Reliability Evaluation of Electric Power Transmission Systems and Its Effect on Core Damage Frequency," *Transaction*, SMiRT16, #1569.

Seismic Countermeasure Committee for Electric Power Facility. 1995. "Working Report on Seismic Countermeasure for Substation Facility," The Federation of Electric Power Company, Japan (in Japanese).
Shumuta, Y. 1996. "A Method for Seismic Reliability Analysis of Secondary Transmission Systems," *Advances in Structural Optimization*: 168-179.
Shumuta, Y. 1998. "Study on Seismic Retrofit Planning Method of Substation Facilities on the Basis of the Seismic Risk Assessment of Electric Power System," CRIEPI report U33, Japan (in Japanese).
Shumuta, Y. 2001. "Basic Study on Rational Seismic Performance Level of Electric Power System Based on Multi-Risk Assessment," Proceedings of the Second Asian Symposium of Risk Assessment and Management (SASRAM), Kobe, Japan.
Shumuta, Y., K. Takenaka, N. Momonoi, and K. Ishida. 1995. "A Method for Seismic Risk Reliability Analysis of Electric Power Primary System," *Journal of Structure Mechanics and Earthquake Engineering*, 507/I-30: 243-253 (in Japanese).
Statistics Bureau & Statistics Center. 1996. *Census Data of Japan*. Tokoy, Japan; Ministry of Public Management, Home Affairs, Posts, and Telecommunications.
Wen, Y.K. 1990. *Structural Load Modeling and Combination for Performance and Safety Evaluation*. Elservier Science Publisher.
Yamada, Z. and S. Noda. 1988. "Damage Estimation of Electric Power System during Large Earthquake," *Journal of Japan Science for Natural Disaster Science*, 7 (1): 10-25 (in Japanese).

Appendix

13.A.1 Formulation of $P_j^{\prime}(r_s, d_s/s)$

The damage probability of equipment associated with the salt stain risk is defined as:

$$P_j^{\prime}(r_s, d_s / s) \equiv P_j(r,d) = \int_0^{\infty} f_{D\,j}(d)\left[\int_0^{d} f_{R\,j}(r)dr\right]dd = \int_0^{\infty} f_{D\,j}(d)F_{R\,j}(d)dd \qquad (13.A.1)$$

where $P_j^{\prime}(r_s, d_s/s)$ is the conditional damage probability of equipment j at time t with salt resisting capacity r_s and salt stain density d_s assuming typhoon s has occurred; $f_{D\,j}(d)$ is the probability density function of equipment j pertaining to salt stain density d; $f_{R\,j}(r)$ is the probability density function of equipment j pertaining to salt resisting capacity r; and $F_{R\,j}(d)$ is the probability distribution function of $f_{R\,j}(r)$ under salt stain density d. Note that in equation (13.A.1), it is assumed that the salt stain density has the same probability distribution regardless of the typhoon size in equation (13.A.1).

In this chapter, $f_{D\,j}(d)$ was evaluated using a salt stain distribution map provided by a power industry. It is expressed as:

$$f_{Dj}(d) = \begin{Bmatrix} 1: \text{(if equipment } j \text{ is installed in a place with salt stain density } d) \\ 0: \text{(othewise)} \end{Bmatrix} \qquad (13.A.2)$$

$F_{R\,j}(d)$ are evaluated as:

$$F_{Rj}(d) = 1 - \exp\left\{-\int_0^{d} \lambda_j(d)dd\right\} \qquad (13.A.3)$$

where $\lambda_j(d) = a_d / NF_{l(j)}$ is the accident rate of equipment j with salt stain density d; a_d is the average accident rate of the substation with salt stain density d; and $NF_{l(j)}$ is the total number of equipment in the substation l installed equipment j.

13.A.2 Model for Power Energy Loss of Power System against Seismic Event

The estimation of power energy loss against seismic event is based on the Monte Carlo Simulation (MCS). The basic formulation is expressed as:

$$E(PEL(\mathbf{P^t}(\mathbf{x_e}))) = f_g(\mathbf{P^t}(\mathbf{x_e}))\cdot g_g(\mathbf{P^t}(\mathbf{x_e}))\cdot PLU_g \equiv$$

$$\equiv f_g(P_j^t(\mathbf{x_e}); j=1,2,\cdots,n)\cdot g_g(P_j^t(\mathbf{x_e}); j=1,2,\cdots,n)\cdot PLU_g =$$

$$= \frac{1}{NM}\cdot\sum_{m=1}^{NM} f_g(\delta_{1m}\cdots,\delta_{jm},\cdots,\delta_{nm})\cdot g_g(\alpha_m)\cdot PLU_g \qquad (13.A.4)$$

s.t. $\delta_{jm} = 1\ (ZD_{jm} \geq 1;\ j \text{ is damaged})$

$\delta_{jm} = 0\ (ZD_{jm} < 1;\ j \text{ is undamaged})$

$$\alpha_m = \sum_{j=1}^{n}\delta_{jm}$$

$$ZD_{jm} = \frac{P_j^t(k)}{R_{jm}} \quad (j=1,2,\cdots,n;\ \ m=1,2,\cdots,NM)\ \ (0.0 \leq R_{jm} \leq 1.0) \qquad (13.A.5)$$

where $E(PEL(\mathbf{P^t}(\mathbf{x_e})))$ is the expected power energy loss with seismic damage probability vector $\mathbf{P^t}(\mathbf{x_e})$of equipment assuming earthquake e has occurred at time t; here $\mathbf{P^t}(\mathbf{x_e})$ represents $(P_j^t(\mathbf{x_e})=(Pj(r_e,a_e/e);$ for $j=1...n)$ defined by equation (13.13); $f_g(\mathbf{P^t}(\mathbf{x_e}))$ is the power failure function of target power system g; $g_g(\mathbf{P^t}(\mathbf{x_e}))$ is the restoration time function for g; PLU_g is the coefficient of restoration efficiency; note here that the above power failure function and the restoration time model are static, which has no time parameters to simulate the restoration process. Thus, to evaluate the power energy loss using functions, $f_g(P_1^t(\mathbf{x_e}),..P_j^t(\mathbf{x_e}),..P_n^t(\mathbf{x_e}))$ and $g_g(P_1^t(\mathbf{x_e}),..P_j^t(\mathbf{x_e}),..P_n^t(\mathbf{x_e}))$, the reduction level of the power energy loss due to the change in the system performance during the restoration process is considered using PLU_g in equation (13.A.4). According to Shumuta (1998), PLU_g varies depending on the network condition of the local power system. To consider the effect of the restoration process, PLU_g is assumed as $(0.0 \leq PLU_g \leq 1.0)$; in this chapter, it is assumed that PLU_g is 0.426 based on existing data associated with the restoration time of the target power system after recent earthquakes; m is the trial in the Monte Carlo Simulation (MCS); NM is the total number of trials in the MCS; δ_{jm} is the index function for j at m which takes 0 or 1; R_{jm} is the uniform random value which represents the probable level of resisting capacity of j at m $(0.0 \leq R_{jm} \leq 1.0)$; and ZD_{jm} is the damage function of j at m. To assume δ_{jm} for every trial in the MCS, a uniform random number R_{jm} in the range from 0.0 to 1.0 is produced for the equipment.

The total restoration time of the power system is usually in proportion to the total number of damaged equipment, α_m, defined in equation (13.A.4). The author investigated the relationship between the total restoration time and the total number of damaged equipment caused by recent earthquakes. As a result, the restoration time function, $g_g(\alpha_m)$, was empirically defined as:

$$g_g(\alpha_m) = 1.69 + 0.20\times\alpha_m \qquad (13.A.6)$$

13.A.3 Notation

The following symbols are used in this chapter:

a_e = seismic force;

a_j, b_j, c_j=recurrent coefficients for damage probability function, Pf_j;

$CB_l(m)$ = cost benefit index of substation l for iterative step m;

Ct = customer type; cf=1: residential, cf=2:commercial, ct=3:industrial;

de= damaged equipment;

DI_{jk}=risk point of equipment j in risk item k;

DP^t=demand power of the entire target power system under the normal condition at time t;

d_s = salt stain density;

DW_j= power energy loss caused by the damage to equipment j;

e = earthquake;

$E(DC^{NT})$ = expected total damage cost caused by multi-hazards in NT;

$E(DC(m-1))$ = annual average total damage cost of the entire power system at iterative step $m-1$;

$E(DC_l(m)) = E(DC^l)$ with renewing substation l from the renewing condition of the entire power system for iterative step $m-1$;

EMI_l = environment and maintenanability improvement index by the comprehensive renewal of substation l;

$E(PEL(\mathbf{P}^t(\mathbf{x}_d)))$, $E(PEL(\mathbf{P}^t(\mathbf{x}_d)))$ and $E(PEL(\mathbf{P}^t(\mathbf{x}_e)))$ = expected power energy loss of the entire target power system at time t with equipment damage probability vector $\mathbf{P}^t(\mathbf{x}_e)$;

$E(PEL(\mathbf{x}_s))$ = expected power energy loss of the entire target power system with the equipment damage condition vector $\mathbf{x}_d$ associated with the salt assuming typhoon s has occurred;

$E(PEL(\mathbf{x}_e))$=expected power energy loss of the entire target power system with equipment damage condition vector $\mathbf{x}_e$ assuming earthquake e has occurred;

$E(PF^{pre})_l$, $E(PF^{post})_l$ = expected power energy losses before and after the comprehensive renewal of substation l, respectively;

$E(TC^{NT})$ = expected total cost in the analytical time period NT;

$E(TREC^{NT})$ = expected total repair cost in NT;

$E(TREC_d^{NT})$, $E(TREC_s^{NT})$ and $E(TREC_e^{NT})$ = total repair cost associated with the deterioration risk, the salt stain risk, and the seismic risk, respectively;

$E(TREL^{NT})$ = expected total revenue loss in NT;

$E(TREL_d^{NT})$, $E(TREL_s^{NT})$ and $E(TREL_e^{NT})$ =total revenue loss associated with the deterioration risk, the salt stain risk, and the seismic risk respectively;

$E(TSC^{NT})$= expected total power interruption cost in NT;

$E(TSC_d^{NT})$, $E(TSC_s^{NT})$ and $E(TSC_e^{NT})$ =total power interruption cost caused by the deterioration risk, the salt stain risk, and the seismic risk, respectively;

F_l = objective function of substation l to determine the renewal priority in the existing model;

$f_{Di}(d)$ = probability density function of equipment j pertaining to salt stain density d;

$f_l(\mathbf{P}^t(\mathbf{x}_d))$ and $f_l(\mathbf{P}^t(\mathbf{x}_s))$ = power failure function for substation l with $\mathbf{P}^t(\mathbf{x}_d)$ and $\mathbf{P}^t(\mathbf{x}_s)$);

$f_g(\mathbf{P}^t(\mathbf{x}_e))$ = power failure function for target power system g with $\mathbf{P}^t(\mathbf{x}_e)$;

$f_{Rj}(r)$ = probability density function equipment j pertaining to salt resisting capacity r;

$F_{Rj}(d)$ = probability distribution function of $f_{Rj}(r)$ under salt stain density d;

$f^t(RC^t)$ = cost function for time period t with the total renewal cost (or total number of renewal substations);

$g_l(\mathbf{P}^t(\mathbf{x}_d))$ and $g_l(\mathbf{P}^t(\mathbf{x}_s))$ = restoration time function for substation $l$with $\mathbf{P}^t(\mathbf{x}_d)$ and $\mathbf{P}^t(\mathbf{x}_s)$);

$g_g(\mathbf{P}^t(\mathbf{x}_e))$ = restoration time function for target power system g with $\mathbf{P}^t(\mathbf{x}_e)$;

$g^t(rc^t)$ = total cost when rc^t is invested at time t

j = equipment;

k = risk item

l = substation;

$MaxRC$ = maximum renewal cost (or maximum number of renewal substations) for a one year period;

m = trial in MCS;

n = total number of target equipment for renewal;

nd = operation times of equipment j;
$NC_l(ct)$ = number of ct customers in the distribution district of substation l;
$NE(l)$ = total number of equipment in substation l;
nf = total number of risk items;
$NF_{l(j)}$ = total number of equipment in the substation l installed equipment j;
$n(l)$=total number of target equipment in substation l;
NM = total number of trials in Monte Carlo Simulation;
$NR(t)$ = total number of renewal equipment at time t;
NS = total number of substations of whole power system;
NT = total analytical time period;
$PC/Unit^t$ = average unit power price at time t;
$Pf^{pre}{}_j$ and $Pf^{Post}{}_j$ = annual damage probabilities of equipment j before and after its renewal;
$PIC(ct)$/Unit = average unit power interruption cost of ct customers;
$P_j{}^t(t_d,n_d)$ = damage probability of equipment j at time t with its age t_d and its total operating times n_d;
$P_j^t(r_s, d_s/s)$ = conditional damage probability of equipment j at time t with salt resisting capacity r_s and salt stain density d_s assuming typhoon s has occurred;
$P_j^t(r_e, a_e/e)$ = damage probability of equipment j at time t with seismic resisting capacity r_e and seismic force a_e assuming earthquake e has occurred;
$P_j(t_d, n_d)$ = damage probability of equipment j with its age t_d and its operating times n_d;
$P_j^t(\mathbf{x_d})$, $P_j^t(\mathbf{x_s})$, and $P_j^t(\mathbf{x_e})$ = $P_j^t(t_d,n_d)$, $P_j^t(r_s, d_s/s)$ and $P_j^t(r_e, a_e/e)$, respectively;
PLU_g = coefficient of restoration efficiency for target power system g;
$Po^t(s)$ = annual occurrence probability of typhoon s at time t;
$Po^t(e)$ = annual occurrence probability of earthquake e at time t;
$PO_l/Unit$ = average unit power interruption cost of the local area distributed by substation l;
$\mathbf{P}^t(\mathbf{x_d})$, $\mathbf{P}^t(\mathbf{x_s})$, and $\mathbf{P}^t(\mathbf{x_e})$ = equipment damage probability vector at time t associated with the deterioration risk , the salt stain risk and the seismic risk,respectively;
rc_j = renewal cost of equipment j;
rc^t = renewal cost at time t
RC^{NT} = renewal cost in NT;
RC_l = total renewal cost of substation l;
r_l = annual discount rate.
R_{jm} = uniform random value which represents the probable level of resisting capacity of j at m;
rpc_j = repair cost of equipment j;
r_s = salt resisting capacity;
s = typhoon;
SPI_l = substation performance improvement index by the comprehensive renewal of substation l;
SPL^{NT} = system performance level of the entire target power system in NT;
t = time or time period;
t_d = equipment age
w_k = weight of risk item k;
$\mathbf{x_d}$,$\mathbf{x_s}$, $\mathbf{x_e}$ = damage condition vector of equipment associated with the deterioration risk, salt stain risk and seismic risk, respectively;
ZD_{jm} = damage function of j at m;
$\lambda_j(d)$ = accident rate of equipment j with salt stain density d;
α_d = average accident rate of the substation with salt stain density d;
α,β = weight parameters of objectivetive function F_l;
δ_j, δ_{jm} = index function for j which takes 0 or 1;
α_m - total number of damaged equipment at trail m in the Monte Carlo Simulation.

14 Evaluating the Disaster Resilience of Power Networks and Grids*

Masanobu Shinozuka[1] and Stephanie E. Chang[2]

[1] Department of Civil and Environmental Engineering, University of California, USA
e-mail: shino@uci.edu
[2] School of Community and Regional Planning, and Institute for Resources, Environment, and Sustainability, University of British Columbia, Canada
e-mail: stephanie.chang@ubc.ca

14.1 Introduction

Electric power is essential for virtually every urban and economic function. Failures of electric power networks and grids – whether from natural disaster, accident, or man-made disaster such as terrorism – can cause severe and widespread societal and economic disruption. In the 1994 Northridge earthquake that struck Los Angeles, some 2.5 million customers lost electric power. For the first time in its history, the entire city of Los Angeles was blacked out. Power outages were experienced in many areas of the western U.S. outside the earthquake region and even as far away as Canada (Hall, ed., 1995). On August 14, 2003, an unprecedentedly widespread blackout rippled out from Akron, Ohio, across the northeastern U.S. and parts of Canada, affecting an area with a population of some 50 million (U.S.-Canada Power System Outage Task Force, 2003). In September of 2003, a power outage that began in Switzerland cascaded over a large region of Italy. Examples such as these indicate the importance of being able to anticipate potential power system failures and identify effective mitigation strategies.

Modeling the impacts of electric power disruption is, however, a highly complex problem. Many of the inherent challenges relate to the need to integrate across disciplines – not only civil, mechanical, and electrical engineering, but also economics and other social science disciplines. For example, one must assess how damages to individual pieces of electric power equipment affect power flow across the network. One must model how a damaged network would be repaired and electric power restored over space and time. Additionally, one must capture how loss of electric power would affect households, businesses, and other units of society.

While numerous studies have addressed portions of the problem, very few have attempted to model regional impacts. For example, Nojima and Sugito (2003) modeled power restoration times based on empirical data from Japanese earthquakes

* This work was done under National Science Foundation Grant EEC9701471 through the Multidisciplinary Center for Earthquake Engineering Research and was also supported by National Science Foundation Grant CMS-0112665. The authors acknowledge technical support provided by Mr. R. Tognazzini and Mr. J. Mochizuki of LADWP and Professor T.C. Cheng and Dr. X. Jin, University of Southern California, and Dr. X.J. Dong, University of California, Irvine.

using geographical information systems (GIS); however, they did not evaluate power outage and impacts in an actual urban area. The current study builds on a long-standing research program led by the first author and carried out by a group of researchers. Their results (*e.g.*, Shinozuka *et al.*, 1994; Tanaka *et al.*, 1997; Shinozuka *et al.*, 1998) laid a foundation for the development of a method by which utility transmission network performance can be rationally analyzed, taking into account power flow in a seismically damaged network. Related research developed methods for assessing how power system failures would impact regional economies (Rose *et al.*, 1997).

This chapter introduces the concept of disaster resilience of infrastructure systems in general and of utility power systems in particular. Resilience is defined in terms of robustness of the system function and rapidity in its restoration if the function is impaired in a disaster. The method of estimating resilience in its technical, organizational and socio-economic dimensions with a high level of interaction is then developed. While the study focuses on earthquake disasters, the method is also applicable to the evaluation of resilience of infrastructure systems subjected to other natural, accidental and man-made disasters including terrorist attacks. Based on our experience in the analysis of the seismic performance of the Los Angeles Department of Water and Power (LADWP) system after the 1994 Northridge earthquake, we derived a useful set of data and gained significant insight into the robustness and rapidity for restoration of electric power networks and grids during and after a catastrophic event. For this reason, LADWP's power transmission system is used as a testbed for model development and analysis.

The modeling effort presented in this chapter is unique in scope, addressing not only the modeling of equipment damage but also the associated network, spatial, and socio-economic impacts. Focusing on analyzing power system performance in earthquakes, it makes several notable advances, including: (1) integration of WECC (Western Electricity Coordination Council) database with EPRI's (Electric Power Research Institute's) IPFLOW (Interactive Power FLOW) computer code for systematic power flow analysis using LADWP's power transmission network as an example, (2) development of fragility curves for transformers, circuit breakers and disconnect switches in a transmission network on the basis of damage data from the 1994 Northridge and other California earthquakes, (3) experimental and analytical study on equipment retrofit in order to examine the extent of the enhancement such retrofit can produce in the fragility characteristics, (4) estimation of the likelihood for power outage and attendant economic loss and societal disruption in the form of risk curves, (5) simulation of system restoration curves, (6) development of preliminary performance criteria for power systems, and (7) evaluation of the effect of a disabling event within a power grid on the performance of the grid, which could possibly be as disastrous as the August 14, 2003 power blackout. These advances contribute toward an analytical framework that is based on the concept of disaster resilience.

The remainder of this chapter is organized as follows: Section 14.2 presents the conceptual framework for disaster resilience, introducing the key concepts of "robustness" and "rapidity" for utility infrastructure systems. The analytical approach for implementing this framework begins with quantification of the hazard. In the Los Angeles case study, the seismic hazard is represented through the use of probabilistic scenario earthquakes, described in Section 14.3. Earthquakes cause physical damage to electric power transmission equipment. Section 14.4 summarizes how damageability for both unretrofitted and seismically retrofitted equipment is modeled through the use of empirically-derived fragility curves and analysis supported by shaking table experiments. Section 14.5 then presents a methodology for assessing how equipment failures impact power flow throughout the regional network. Section 14.6 proposes an approach for graphically summarizing the risk of regional power loss in the form of "risk curves." Section 14.7 introduces the concept of system performance criteria and demonstrates how it can be used in the context of risk curves to assess earthquake retrofit strategies. Section 14.8 extends this approach to consider impacts on the regional economy. Section 14.9 discusses how the temporal dimension of impacts can be included through the modeling of system restoration. While the preceding analysis focuses on a single urban area, Section 14.10 indicates a modeling approach and example results for catastrophic failures of electric power grids that cover large, multi-state regions. Finally, Section 14.11 offers conclusions and suggestions for future research.

14.2 Resilience Framework

Resilience is an important concept for the disaster management of infrastructure systems. Two key dimensions of resilience can be referred to as robustness and rapidity in restoration. These can be expressed utilizing a restoration curve typically having characteristics as shown in Figure 14.1.

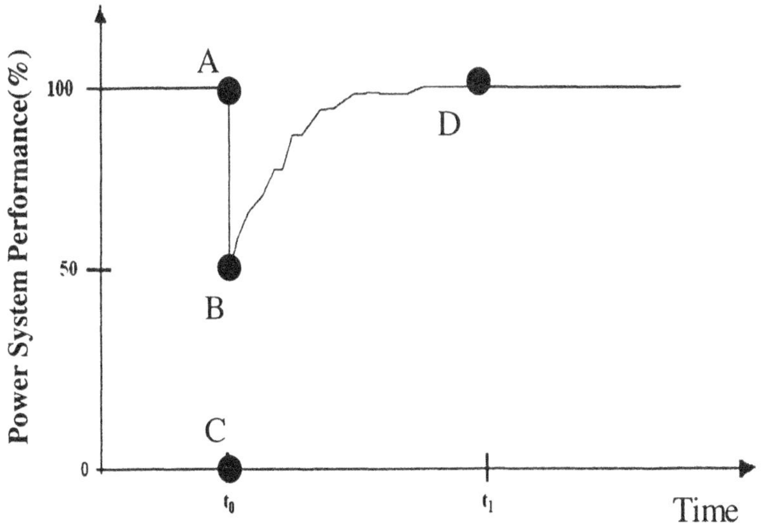

t_0 = Time at which an earthquake occurs
t_1 = Time at which power supply is restored 100%

Figure 14.1 Power Supply as a Function of Time

The curve plots system performance as a function of time. The reduction in performance from 100% at point A (time t_0) to 50% (in this example) at point B results from damaging seismic impact to the system. The restoration curve starting from the initial distress point B to the complete recovery point D (back to 100% at time t_1), exhibits the process of restoration. Hence, the performance percentage corresponding to B (or $B-C$, with C associated with zero performance) represents robustness, and the elapsed time for the total restoration ($t_1 - t_0$) can be used to quantify rapidity, for example:

$$\text{Robustness} = B - C \qquad \text{(in percentage)} \tag{14.1}$$

$$\text{Rapidity} = \frac{A-B}{t_1 - t_0} \qquad \text{(recovery in percentage /time)} \tag{14.2}$$

though equation 14.2 may admittedly be too simplistic.

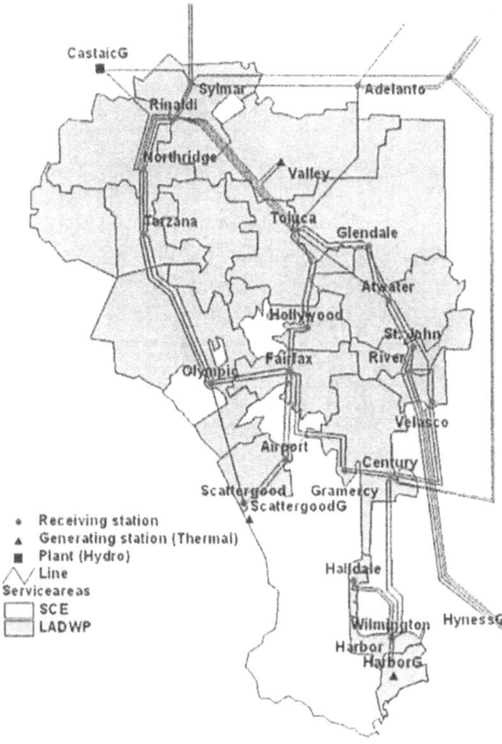

Figure 14.2 Transmission Network and Service Areas of LADWP's Power System

Throughout this chapter, systems analysis of LADWP's transmission network system (figures 14.2 and 14.3) under actual and simulated earthquakes is performed as a testbed. The analysis was based on an inventory database of the network, together with available fragility information of power equipment and Monte Carlo simulation techniques. This is a unique study where an actual database from WECC is used in conjunction with the computer code IPFLOW (EPRI, 1998). In this analysis, transmission lines and generating stations are assumed to be seismically not vulnerable. This assumption, while reasonable, is obviously not always the case, as demonstrated clearly in the 1999 Chi-Chi (Taiwan) earthquake where high voltage transmission towers constructed in sloping mountainous areas and a number of generating stations sustained some damage. Another important issue is the behavior of electric power distribution systems. Each distribution system in general provides service within a corresponding service area. Distribution systems are known to be vulnerable to earthquakes. Their seismic behavior is extremely complex and almost defies any analysis from the flow analysis point of view. This matter is left for future study, where statistical models should be developed to evaluate the risk of outage, and integrated into total system risk assessment procedures.

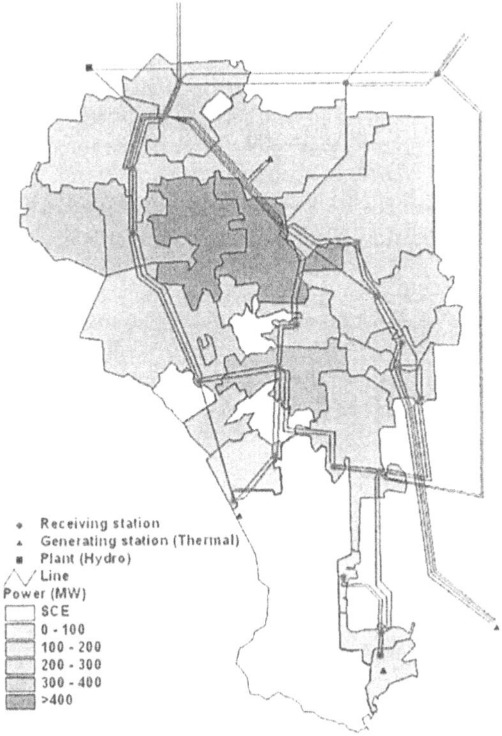

Figure 14.3 Electric Power Output for Service Areas under Intact Condition

14.3 Scenario Earthquakes

For electric power and other urban infrastructure systems, evaluating potential impacts of damage is complicated by the fact that the networks are spatially distributed across a wide area. Risk analysis must account for how the system performs given that the hazard (*e.g.*, earthquake ground motion) is not only spatially variant across a wide area but also, for any given disaster, spatially correlated. Hence, traditional probabilistic methods that can readily be applied for site-specific facilities such as individual buildings cannot be used for these spatially distributed networks.

This chapter therefore analyzes system functionality and impacts in the context of scenarios of individual earthquake events, and then combines the scenario results probabilistically to gain a complete understanding of the seismic performance of LADWP's power system. In total, 47 scenario earthquakes for the Los Angeles region are selected and simulated, as discussed later in detail. These scenarios were developed by Chang *et al.* (2000), applying a loss estimation software tool, EPEDAT, based on K. Campbell's attenuation law (Campbell, 1994), which was used to generate regional ground motion patterns for given earthquake epicenter, magnitude, and depth (USC-EPEDAT, 1999). The 47 events include 13 maximum credible earthquakes (MCEs) on various faults in the Los Angeles region and 34 other events of magnitude 6.0 or higher. These scenario earthquakes are associated with annual "equivalent probabilities" of occurrence so that collectively, they represent the full range of the regional seismic hazard (Chang *et al.*, 2000).

Analysis using these probabilistic earthquake scenarios allows the estimation of "risk curves" that graphically summarize system risk in terms of the likelihood of experiencing different levels of performance degradation in disasters. Risk curves can be developed for performance parameters associated with different dimensions of resilience, including the technical (*e.g.*, power supply in each service area), societal (*e.g.*, rate of households without power supply), organizational (*e.g.*, repair and restoration efficiency), and economic (*e.g.*, regional output or employment loss). Examples of risk curves are provided later in this chapter.

14.4 Development of Fragility Curves

Failure of transmission equipment is one of the most common and significant types of damage to electric power systems due to earthquakes. In particular, transformers are crucial components in an electric power network, and yet their redundancy is not as high as other equipment due mainly to their large size and high cost. Therefore, it is very important to make a pre-event evaluation of the seismic performance of transformers on the basis of damage information obtained from past earthquakes as well as numerical analysis and laboratory tests. In this respect, substantial effort was expended to acquire damage information for transmission level transformers related to the 1994 Northridge earthquake obtained from LADWP's Northridge Earthquake Damage Report. The damage information is transformed into empirical fragility curves expressed in the form of two-parameter (median and log-standard deviation) lognormal distribution, as a function of peak ground acceleration (PGA), which represents the intensity of the seismic ground motion at each site. Use of PGA for this purpose is considered reasonable: The PGA value at the location of each receiving station is determined from PGA contours (Wald, 1999). The maximum likelihood method (Shinozuka *et al.*, 2000) is employed to estimate the two parameters of the fragility curve, resulting in c=0.45g (median) and ζ=0.42 (log-standard deviation). With these two parameters, the fragility curve for the unretrofitted transformers in the transmission network has been developed and is shown in figure 14.4 with the label Case 1 (Dong, 2002).

There exist a number of possible methods for seismic retrofit of transformers, including base-isolation techniques (Murota, 2002, 2003; Saadeghvaziri *et al.*, 2001). The improvements to the seismic performance using these methods are measured by the enhancement index (C_r/C)-1 where C_r and C are the median values of fragility curves for the retrofitted and unretrofitted transformers, respectively. The change in the log-standard deviation is disregarded for the reasons that their influence on the final result is secondary.

The methods of retrofit examined tend to be quite effective, and therefore in the following analysis, 50% and 100% enhancement index values are assumed to be achievable and used for the sake of gaining some useful insight, though 100% is arguably very optimistic. The enhanced fragility curves are also plotted in figure 14.4 (Dong, 2002). Case 2 refers to 50% enhancement and Case 3 to 100% enhancement.

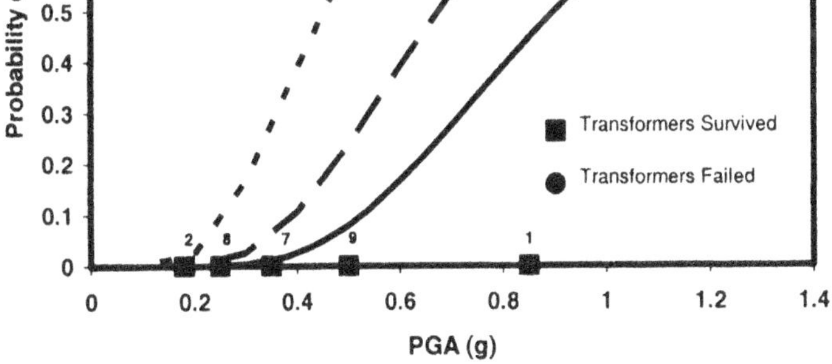

Figure 14.4 Fragility Curves for Transformers in Transmission Network

14.5 Seismic Performance of a Power System

To demonstrate our method of systems analysis, LADWP's network which is part of the WECC's power grid (based on the data available at http://www.wecc.biz/main.html) is analyzed. WECC's grid covers 14 western states in the U.S., two Canadian provinces, and northern Baja California in Mexico. The iterative analysis method to predict the seismic performance of the electrical power network is described below (Dong, 2002; Tanaka *et al.*, 1997).

(1) For each of the 47 scenario earthquakes, generate the PGA value at each receiving station.
(2) For each scenario earthquake, simulate equipment damage using PGA and fragility curves with and without retrofitting.
(3) Perform Monte Carlo Simulation to generate a sample of damaged transmission networks for each scenario earthquake.
(4) Isolate every line that is disconnected to the nodes (at least one end) due to equipment damage.
(5) Calculate power flow using IPFLOW code under the following failure criteria:

- Imbalance of power over the entire network

$$1.05 < \frac{total \quad supply}{total \quad demand} < 1.1 \tag{14.3}$$

- Abnormal voltage at each node

$$\left| \frac{V_{intact} - V_{damage}}{V_{intact}} \right| > 0.1 \tag{14.4}$$

V_{intact} = Nodal voltage under normal operating condition
$V_{damaged}$ = Nodal voltage under damaged condition

- Frequency change (IPFLOW does not check this criterion)
- Loss of connectivity.

(6) Remove any node under abnormal pressure and isolate lines initiating or terminating at the node.
(7) Check the power balance.
(8) If power flow is balanced, compute reduction in average power supply (total and for each service area). If not, network black-out occurs and computation stops here. Steps (4) through (8) are iterative.
(9) Compute relative number of households without power.
(10) Compute loss of GRP (Gross Regional Product) due to power outage.
(11) Develop seismic risk curves (annual probability that system performance will be reduced more than a specified level due to earthquake).
(12) Examine system performance relative to performance criteria with and without seismic rehabilitation.
(13) Determine effectiveness of rehabilitation.

By applying the systems analysis procedure, LADWP's power flow analysis is performed 20 times on the network for each scenario earthquake. Each simulation result represents a unique state of network damage. Figure 14.5 shows the PGA distribution associated with the Northridge earthquake in LADWP's service areas. Figure 14.6 shows the ratio of the average power supply of the damaged network to that associated with the intact network for each service area when only transformers are considered vulnerable. The average is taken over the entire sample size equal to 20. The extent to which the retrofit of transformers contributes to improvement of system performance is evident if we compare the power supply ratio under Case 1 (not enhanced), Case 2 (50% enhanced) and Case 3 (100% enhanced) in figure 14.6.

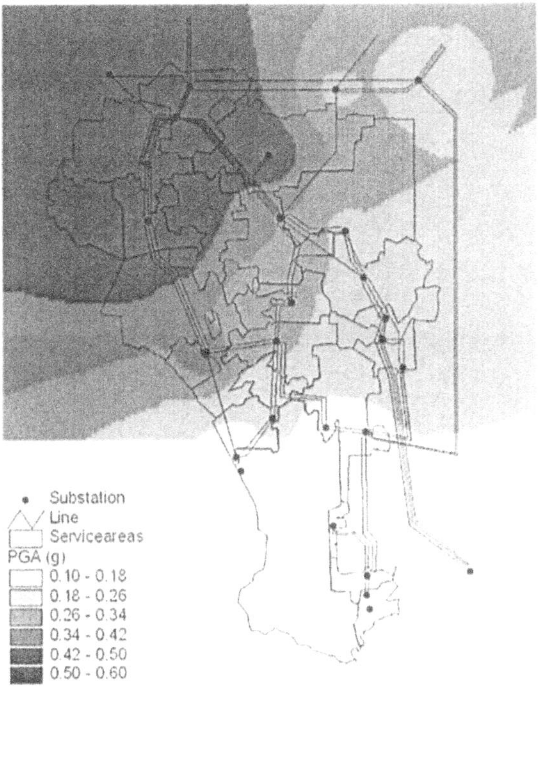

Figure 14.5 PGA under Northridge Earthquake

Aside from transformers, there is a set of transmission level equipment in the power network which is also very vulnerable to earthquakes, particularly circuit breakers and disconnect switches. Their fragility curves are constructed utilizing the equipment damage database developed by Anagnos (1999) resulting from California earthquakes. In addition, buses (and its supporting structures) are critical components, although not included in this analysis. These fragility curves are empirically developed based on Northridge damage data. The results (not shown here in the interest of space) indicate that if transformers, circuit breakers and disconnect switches are all considered vulnerable under earthquakes, the power output is significantly reduced throughout LADWP's service areas. This result matches the actual outcome of the 1994 Northridge earthquake. It is noted that the blackout was quickly remedied in the case of Northridge earthquake due to effective emergency network management. This represents LADWP's organizational effectiveness in emergence response.

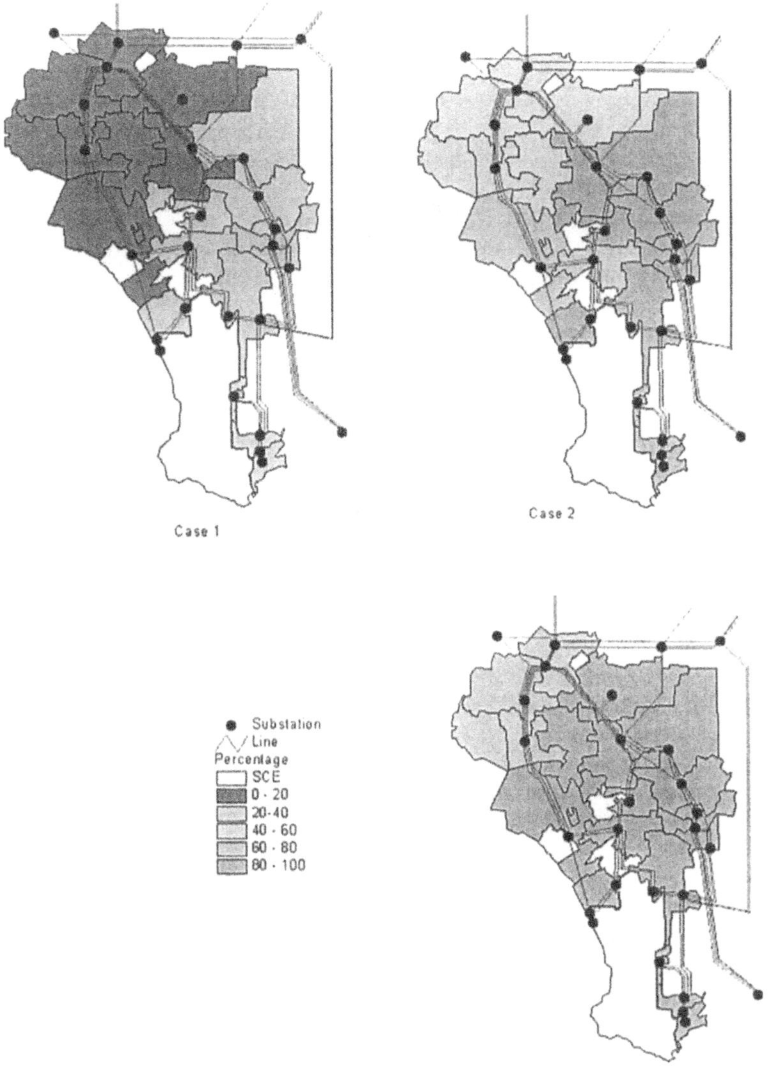

Figure 14.6 Relative Average Power Output with Only Transformers Vulnerable

14.6 Risk Curves of LADWP's Power System

The locations of LADWP's and SCE's receiving stations relative to faults in and around the Los Angeles area indicate that many are near active faults (figure 14.7) and have a high possibility of suffering damage due to an earthquake. In evaluating the risk and costs associated with potential future earthquakes, the performance of the power system in 47 (deterministic) earthquake scenarios was evaluated, as noted earlier. Based on these scenarios, together with their hazard-consistent probabilities, seismic risk for power supply (to the entire service area) of LADWP's power system due to earthquake-induced performance degradation is evaluated in terms of "risk curves." Figure 14.8 shows the risk curves in terms of the severity of power loss for Case 1 (no fragility enhancement) and for Cases 2 and 3 (enhancement index 50% and 100%, respectively). One can develop similar risk curves for each service area. The significant improvement due to seismic retrofit is clearly seen from figure 14.8.

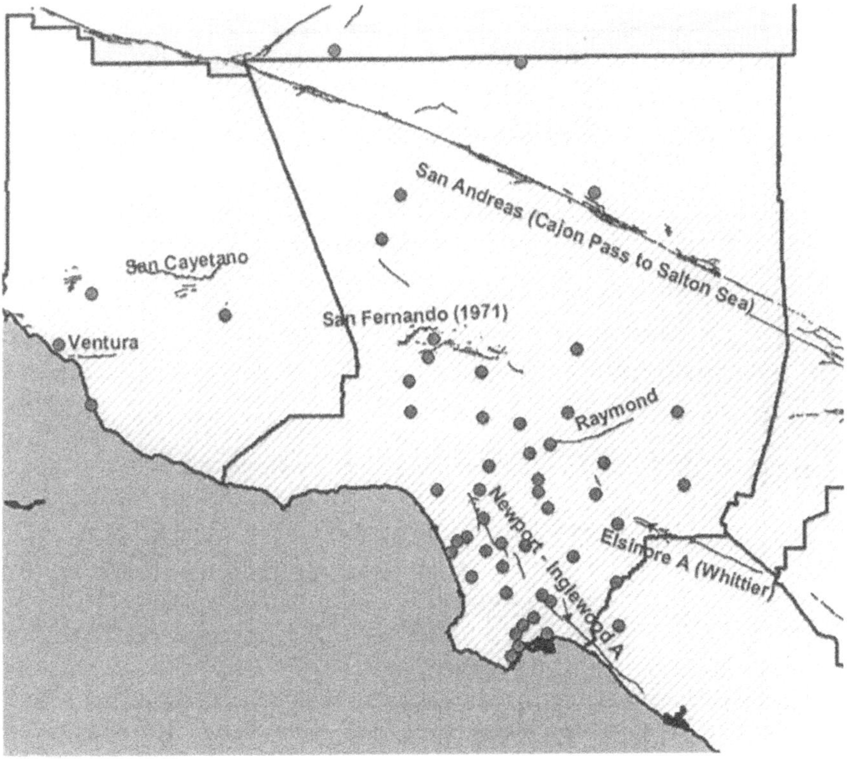

Figure 14.7 Locations of Seismic Faults and 52 Receiving Stations

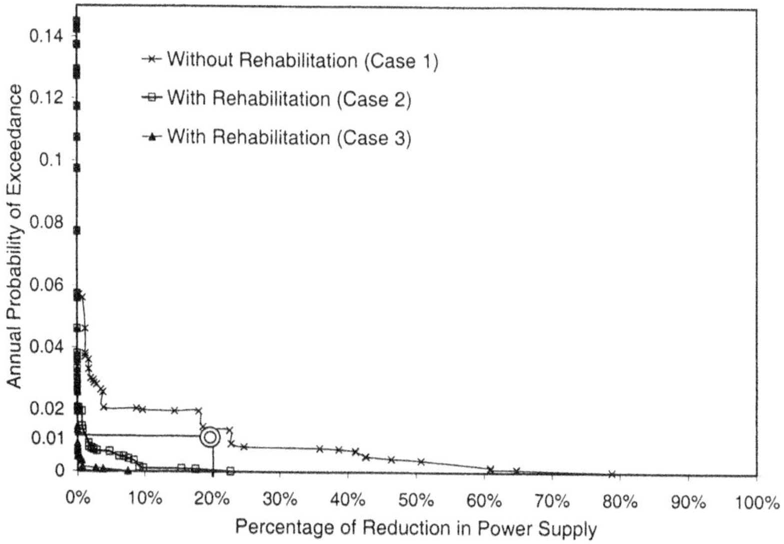

Figure 14.8 Risk Curves for Power Supply Reduction

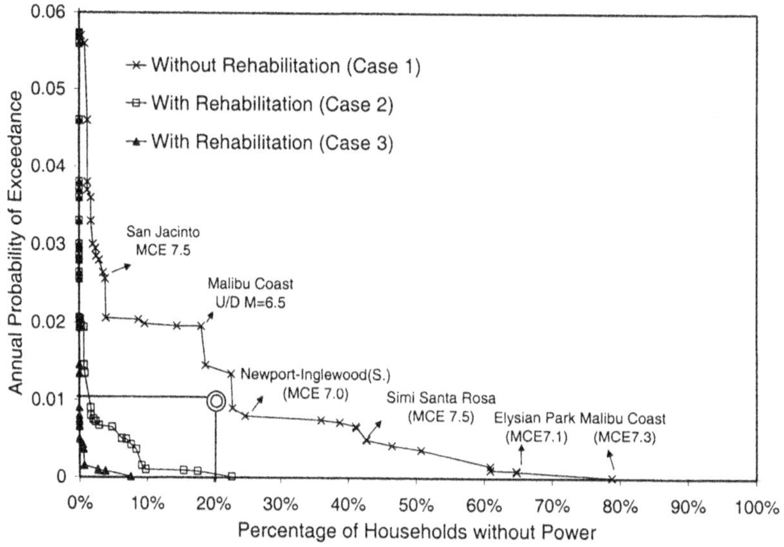

Figure 14.9 Risk Curves for Household Power Outage

The percentage of households with and without power supply immediately after an earthquake is a risk of societal concern, and the associated risk curves are plotted in figure 14.9. These risk curves indicate the ratio of households without power immediately after earthquake, and are computed utilizing census data on the spatial distribution of households across LADWP's service areas. The ratio of households with power, equation 14.5, and the ratio of households without power, equation 14.6, can be defined as follows:

$$\text{Ratio of households with power} = \frac{\sum_{m=1}^{M} \frac{1}{N} \sum_{n=1}^{N} Rd(m,n) \times Hshld(m)}{\sum_{m=1}^{M} Hshld(m)} \tag{14.5}$$

$$\text{Ratio of households without power} = 1 - \frac{\sum_{m=1}^{M} \frac{1}{N} \sum_{n=1}^{N} Rd(m,n) \times Hshld(m)}{\sum_{m=1}^{M} Hshld(m)} \tag{14.6}$$

where:

m = service areas number (1,2,...,M); M=21 in this example
n = simulation number (1,2,...,N); N=20 in this example
$Rd(m,n)$ = power output ratio in service area m under nth simulation
$Hshld(m)$ = total number of households in service area m

The risk curves in this chapter plot the expected annual probability that loss of system power supply will exceed a certain percentage level (figure 14.8) or that the percentage of households without power after an earthquake will exceed a certain percentage level (figure 14.9). Each point in the figure represents one of the scenario events with their occurrence probabilities cumulatively added backward beginning from the earthquake producing the largest percentage so that the risk curve represents a complimentary (or 1 minus) cumulative distribution function of the performance variable (such as percentage of power supply reduction). The risk curve approach is also useful for economic impact analysis (described below), as well as cost-benefit analysis to determine the effectiveness of enhancement technologies (see the curves with solid triangles and squares) (Dong, 2002). Of equal importance is the use of the risk curve in relation to the development of the performance criteria and their verification, as described in the section that follows.

14.7 System Performance Criteria

The performance criteria for power systems listed in tables 14.1 and 14.2 demonstrate a possible format in which the criteria can be given. Table 14.1 lists criteria to be satisfied in pre-event assessment (*e.g.*, through seismic retrofit), and table 14.2, those in post-event emergency response (*e.g.*, through disaster response planning). These tables also include performance criteria for water and acute care hospital systems. This general format for performance criteria for structures and lifelines has been provided by Shinozuka and Dong (2002) and Bruneau *et al.* (2003). In combination, they conceptually establish the degree of community resilience in terms of robustness, rapidity and reliability. Specific values (in percentages for robustness, rapidity in restoration, and reliability) are examples so that the concept can be better understood. The performance criterion for power systems shown in table 14.1 is represented by a point double-circled in figure 14.9 where the robustness corresponds to the annual probability of 0.99 that 80% or more households will have power immediately after any earthquake. The risk curve for Case 1 (without rehabilitation) does not satisfy this criterion, but the risk curves for Case 2 and Case 3 (with rehabilitation) do satisfy it. These criteria can also be used to judge the effectiveness of rehabilitation as carried out in this chapter.

Table 14.1 System Performance Criterion I for Pre-event Assessment and Rehabilitation

	Robustness	**Reliability**
Power	A majority (at least 80%) of households will have continued power supply after earthquake	With a high level of reliability (at least 99% per year)
Water	A majority (at least 80%) of households will have continued water supply after earthquake	With a high level of reliability (at least 99% per year)
Hospital	A majority (at least 95%) of injured or otherwise traumatized individuals will be accommodated in acute care hospitals for medical care	With a high level of reliability (at least 99% per year)

Table 14.2 System Performance Criterion II for Post-Event Response and Recovery

	Rapidity in Restoration	**Reliability**
Power	A majority (at least 95%) of households will have power supply as rapidly as possible within a short period of time (3 days)	With a high level of reliability (at least 90% of earthquake events)
Water	A majority (at least 95%) of households will have water supply as rapidly as possible within a short period of time (3 days)	With a high level of reliability (at least 90% of earthquake events)
Hospital	All the injured and traumatized individuals will be accommodated in acute care hospitals as rapidly as possible within a short period of time (1 day)	With a high level of reliability (at least 90% of earthquake events)

Data collection and modeling for rapidity in restoration are much more difficult to pursue (Shinozuka *et al.*, 2002). Further research is needed to develop analytical models based on past experience so that performance criteria, such as those shown in table 14.2, become meaningful in practice. However, some simulation study was performed and compared with the Northridge repair/restoration data. The result of this chapter provides a potentially successful method of pursuit in this area as demonstrated below in Section 14.9 on "System Restoration."

14.8 Economic Impact

The preceding analysis of systems performance can be readily extended from impacts on households to impacts on the regional economy. Here, direct economic losses are evaluated using a methodology that relates the spatial pattern of electric power outage to the regional distribution of economic activity (see Chang and Seligson, 2003; Chang, 1998).

Direct economic loss, L (dollars), is evaluated for each earthquake simulation and each mitigation condition as follows:

$$L = \sum_{m} \sum_{j} l_j \cdot d_m \cdot e_{m,j} \tag{14.7}$$

where l_j is a loss factor for industry j ($0 \leq l \leq 1$), d_m is a disruption indicator for service area m (d=1 in case of power outage, d=0 in case of no outage), and $e_{m,j}$ is daily industry j economic activity in area m (dollars). The disruption indicators d_m for each electric power service area derive directly from the power supply simulation results described previously.

The loss factors l_j reflect the dependency of each industry on electric power. They were developed empirically on the basis of survey data from the Northridge earthquake. Specifically, a large survey of over 1,100 businesses was conducted by K. Tierney and colleagues at the Disaster Research Center of the University of Delaware (see Webb *et al.*, 2000). Selected data from this survey were used in the current chapter, including information on whether a business lost electric power, for how long, the level of disruptiveness associated with this outage, and whether or not the business closed temporarily in the disaster. Data on other sources of disruption (*e.g.*, building damage, loss of water, etc.) were also used to estimate the net effect of electric power outage. For details on the methodology, see Chang *et al.* (2002) and Chang and Seligson (2003). The loss factors range from a low of 0.39 for mining and construction to a high of 0.60 for manufacturing. These factors pertain to a one-day power outage.

Estimates of industry economic activity by service area, e_{mj}, were based on industry employment data. Employment by industry and zip code were obtained from the Southern California Association of Governments (SCAG) and aggregated, using GIS overlays, to the LADWP service areas. Employment was converted into output using estimates of output per employee in each industry. These productivity estimates were based on California gross state product (GSP) and employment data available from the Bureau of Economic Analysis (BEA).

Loss results are expressed as the percent of gross regional product (GRP) in the LADWP service area that would be lost given electric power outage in each earthquake simulation. Because the duration of power outage is not modeled here, results are assessed in terms of daily GRP loss. This can be interpreted as the loss that would obtain if the outage pattern lasted for one day.

Figure 14.10 shows the results in the form of a risk curve. Each data point on the figure is one of the 47 earthquake scenarios (*e.g.*, magnitude 7.1 Elysian Park earthquake), where results have been averaged over the simulations for that scenario. The figure shows the substantial risk reduction – in terms of economic disruption – that would be afforded by rehabilitation.

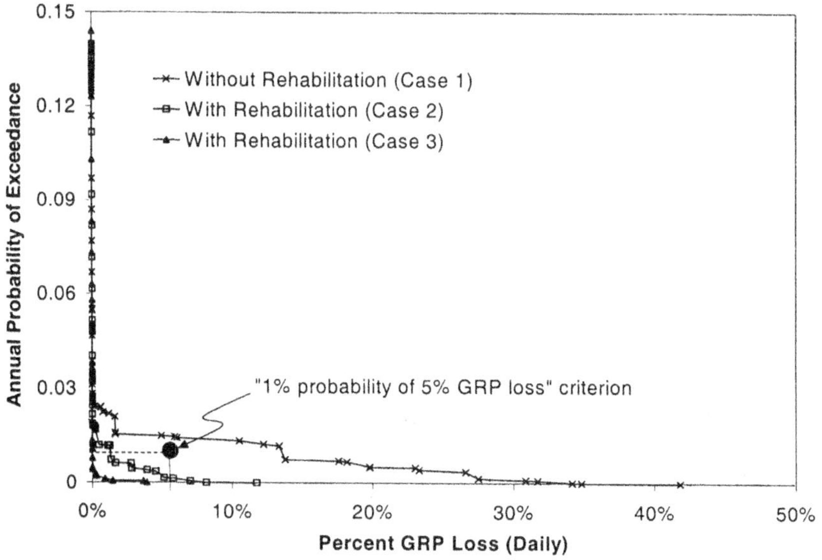

Figure 14.10 Risk Curve for Economic Loss

The solid circle in the figure indicates an example performance criterion for the electric power system in economic terms. In this case, the criterion specifies that economic loss should not exceed 5 percent of GRP with a 1 percent annual probability. The unmitigated system in this instance (Case 1) would not meet the stated performance criterion. However, rehabilitation (either Case 2 or Case 3) would improve the system sufficiently. These risk curves are analogous to those for power loss and households impacted that were developed earlier. They can be used to represent the economic dimension of loss and resiliency, as well as the benefits of seismic retrofit actions.

14.9 System Restoration

It was demonstrated that the restoration for utility power systems tends to be rapid compared with that for water, gas and transportation systems. Figure 14.11 shows the assumed repair or replacement curves for the LADWP system after the Northridge earthquake. The curve plots the probability of damaged equipment repaired or replaced as a function of time (in days). It is postulated that circuit breakers and disconnect switches are more rapidly restored with uniformly distributed probability density over the first one day period, and transformers over the first two days. This not only reflects the relative ease with which each component is repaired /replaced but also the cost of its replacement. The curve for example indicates that a damaged transformer can be replaced or repaired within a half day with a probability of 25%. This is merely an assumption on which we initiate and gain numerical insight for the restoration simulation. In reality, a transformer probably cannot be replaced or repaired with such rapidity unless the degree of damage is light. We follow the fragility simulation and power flow analysis as outlined above and add an extra layer of Monte Carlo simulation where damaged components are restored in accordance with the restoration probability function assumed in figure 14.11. Resulting simulation for restoration (in % of households) is represented by four points in figure 14.12 which underestimates the speed of restoration (in % of customers) actually observed in the aftermath of the Northridge Earthquake. The difference between household- and customer-based percentage restorations is assumed negligible here. Figure 14.12 also includes simulated restoration curves for two other less damaging scenario earthquakes. Note their shapes are in essence the same as the curve BD in figure 14.1. We note that power system restoration procedure carries out repair or replacement of damaged components in the order reflecting the priority established by the utility for the purpose of accelerating the entire network restoration. If such a procedure were taken into consideration, the simulation performed here might have more closely agreed with the empirical curve.

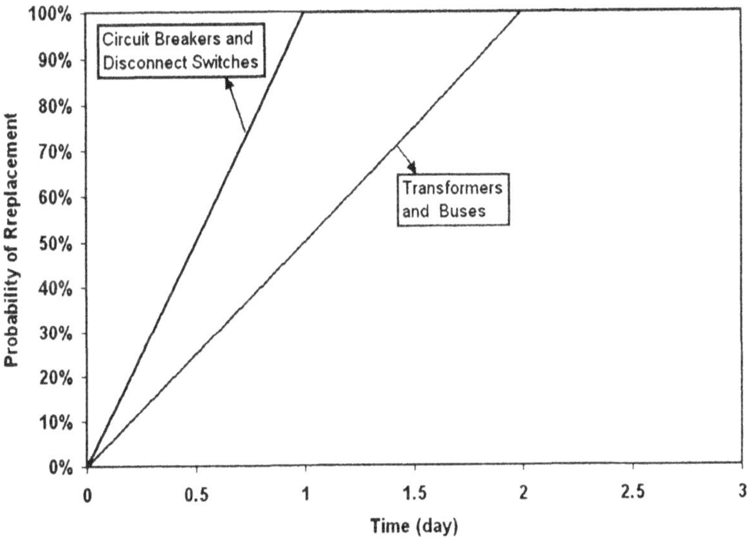

Figure 14.11 Restoration Probability for Transformers, Circuit Breakers, and Disconnect Switches

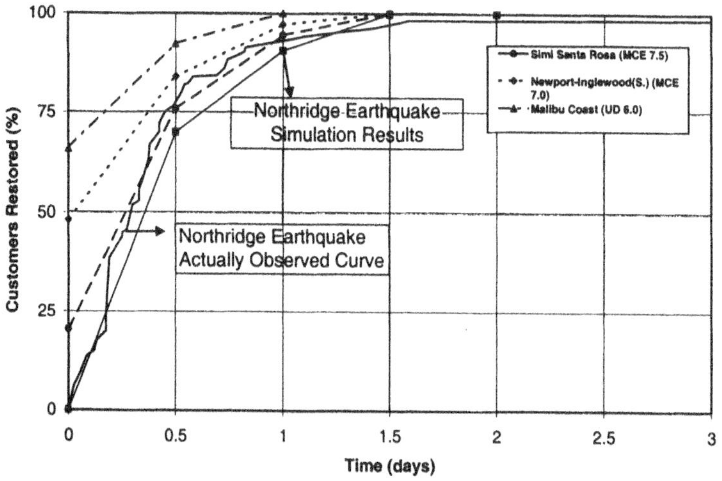

Figure 14.12 LADWP's Power System Customers Restored versus Time

14.10 Effect of a Disabling Event on Performance of Large Systems

A general methodology in this chapter is also developed to analyze the behavior of large electric utility networks or grids under a catastrophic event, such as a severe earthquake or an intentional disablement of system components. In this respect, we are working on the outages recently experienced in the Midwest and northeastern part of the U.S. and southern Canada on August 14, 2003 in order to analyze and simulate what happened, not only to the power grid, but also to transportation, water supply, and other infrastructure systems and associated socio-economic disturbance. Initially, the research was motivated by an accident in August 1996, in which a tree came into contact with a high-voltage aerial power line at a location in the State of Oregon, disabling its transmission capability. The result was a wide-spread blackout in several states and cities, including part of Los Angeles.

We were able to simulate the outage in Los Angeles under the Oregon scenario, making use of grid inventory data and an electric power flow simulation code. The simulation dealt with two scenarios (A and B) involving disabled 500kV transmission lines between Washington State and Idaho State (A) and between Washington State and Oregon State (B) as shown in figure 14.13(a) for Scenario A and figure 14.13(b) for Scenario B. A cross in the WECC Grid in the figures indicates the locations of the interruptions. In Scenario A, power supply to the service area of the LADWP is not affected very much. The power supply to most service areas is maintained at above 90%. Power supply in only one service area in the middle drops between 85% and 90%. The reason is that the disabled 500kV transmission line in Scenario A is a more redundant segment of the grid as far as the LADWP power system is concerned. The effect of a power supply reduction to LADWP service areas comes from power-load balance requirement within the entire WECC system. For Scenario B, however, the disabled transmission line at the border of Washington and Oregon States is a less redundant component, and influences the LADWP power system through two 500kV AC transmission lines and one 500kV DC transmission line. From figure 14.13(b), it is seen that the entire LADWP service area is blacked out when this component is disabled. This is indeed a disaster to the LADWP service area, in spite of various redundancy and tolerance built into the system between the Pacific Northwest and Southern California.

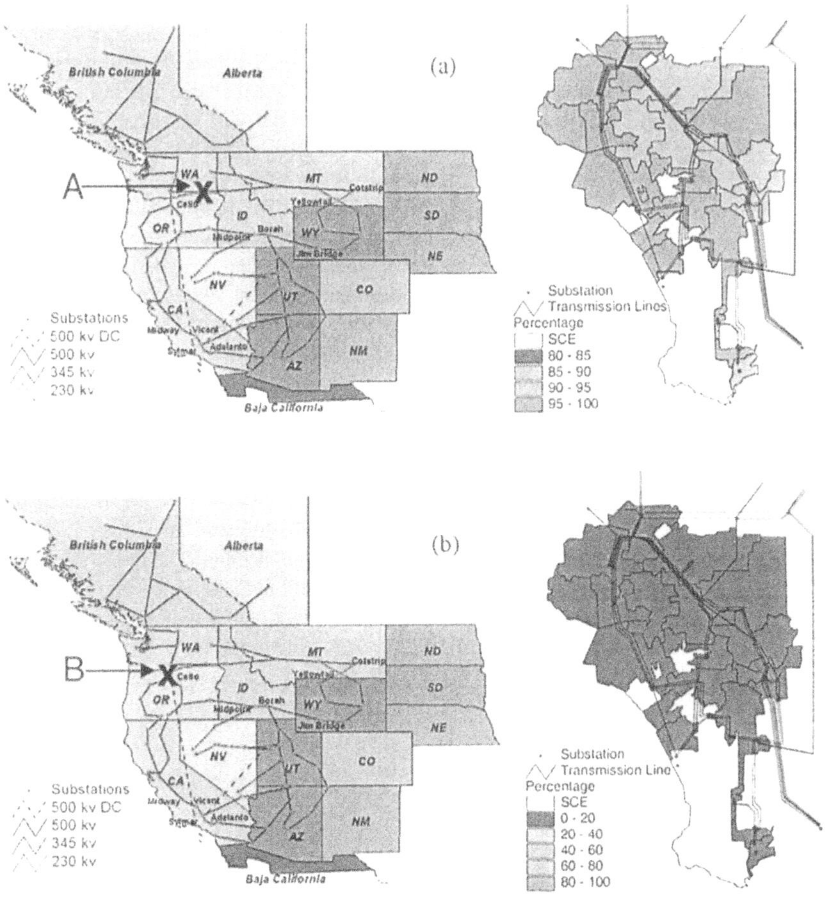

Figure 14.13 (a) Disabled Location A and Reduction in Power Supply, and (b) Disabled Location B and Reduction in Power Supply

This type of accident often creates a chain of events such as power imbalance and abnormal voltage leading to blackout. The research became more seriously significant after the tragic September 11 event from homeland security point of view and has just added another layer of relevance since the August 14, 2003 power outage. This is because disablement can be man-made, accidental or malicious, and cause a devastatingly negative impact on the socio-economic welfare of the directly affected regions immediately and of the nation in the long run.

14.11 Conclusions and Future Research Agenda

This chapter integrated many of the technologies developed by the authors over the years, including GIS inventory data of the LADWP electric transmission system, multiple scenario earthquakes representing the Los Angeles area seismic hazard, fragility characteristics of system components with and without seismic retrofit, systems analysis using WECC's database and EPRI's IPFLOW computer code, loss estimation methods, and preliminary versions of performance criteria definitions and their verification procedures. This integration leads to the capability to evaluate the performance of power systems and the consequences of system interruptions caused by earthquakes. Additional study will be undertaken to further establish performance criteria that can be quantitatively mapped into the response space, in terms of the technological, economic, organizational, and social dimensions of disaster resilience. Future study will also include the development of probabilistic procedures to estimate the reliability of the seismic resilience of the community. Also, integration with other critical systems such as water systems, emergency response organizations, medical care systems (a regionally designated network of acute care hospitals) and highway transportation systems will be carried out to evaluate the community resilience. Finally, the application of SCADA in the inverse analysis of power system performance is of significant future interest. This is to improve the time it takes to achieve restoration of system function degraded by the earthquake, which represents the rapidity aspect of the resilience criteria. An added dimension to our future research is to provide recommendations to prevent, protect, and quickly restore the power network and grid from outages due to natural, technological and malicious causes.

References

Anagnos, T. 1999. "Development of an Electrical Substation Equipment Performance Database for Evaluation of Equipment Fragilities," Final Report for PG & E and PEER, San Jose State University.

Bruneau, M., S.E. Chang, R.T. Eguchi, G.C. Lee, T.D. O'Rourke, A.M. Reinhorn, M. Shinozuka, K. Tierney, W.A. Wallace, and D. von Winterfeldt. 2003. "A Framework to Quantitatively Assess and Enhance the Seismic Resilience of Communities," *Earthquake Spectra*, 19 (4): 733-752.

Campbell, K.W. and Y. Bozorgnia. 1994. "Near-Source Attenuation of Peak Horizontal Acceleration from Worldwide Accelerograms Recorded from 1957 to 1993," Proceedings of the Fifth U.S. National Conference on Earthquake Engineering, Vol. 3, July 10-14, 1994, Chicago, Illinois, Earthquake Engineering Research Institute: 283-292.

Chang, S.E. 1998. "Chapter 6: Direct Economic Impacts," in M. Shinozuka, A. Rose, and R.T. Eguchi eds. *Engineering and Socioeconomic Impacts of Earthquakes: An Analysis of Electricity Lifeline Disruptions in the New Madrid Area*. Buffalo, NY; Multidisciplinary Center for Earthquake Engineering Research: 75-94.

Chang, S.E. and H.A. Seligson. 2003. "Evaluating Mitigation of Urban Infrastructure Systems: Application to the Los Angeles Department of Water and Power," in J.E. Beavers ed. *Advancing Mitigation Technologies and Disaster Response for Lifeline Systems*. Technical Council on Lifeline Earthquake Engineering (TCLEE) Monograph No.25: 474-483.

Chang, S.E., M. Shinozuka, and J. Moore. 2000. "Probabilistic Earthquake Scenarios: Extending Risk Analysis Methodologies to Spatially Distributed Systems," *Earthquake Spectra*, 16 (3): 557-572.

Chang, S.E., W.D. Svekla, and M. Shinozuka. 2002. "Linking Infrastructure and Urban Economy: Simulation of Water Disruption Impacts in Earthquakes," *Environment and Planning B*, 29 (2): 281-301.

Dong, X.J. 2000. "The Seismic Performance Analysis of Electric Power Systems," Ph.D. Dissertation, University of Southern California.

EPRI. 1998. "Interactive Power Flow Version 5.2 Users Manual," November.

Hall, J.F. ed. 1995. "Northridge Earthquake of January 17, 1994: Reconnaissance Report," *Earthquake Spectra*, Supplement C to Vol.11.

Murota, N. 2002. "Earthquake Simulator Testing of Large Scale Electric Transformer Model and Bushing with Base Isolation System," Master's Thesis, University of California, Irvine.

Murota, N. 2003. "Base Isolation for Seismic Protection of Power Transformer," Ph.D. Dissertation, University of California, Irvine.

Nojima, N. and M. Sugito. 2003. "Development of a Probabilistic Assessment Model for Post-Earthquake Residual Capacity of Utility Lifeline Systems," in J.E. Beavers ed. *Advancing Mitigation Technologies and Disaster Response for Lifeline Systems*. Reston, Virginia; American Society of Civil Engineers, Technical Council on Lifeline Earthquake Engineering: 707-716.

Rose, A., J. Benavides, S.E. Chang, P. Szczesniak, and D. Lim. 1997. "The Regional Economic Impact of an Earthquake: Direct and Indirect Effects of Electricity Lifeline Disruptions," *Journal of Regional Science*, 37 (3): 437-458.

Saadeghvaziri, M.A. and M. Feng. 2001. "Experimental and Analytical Study of Base-Isolation for Electric Power Equipment," in *Research Progress and Accomplishments:2000-2001*. Buffalo, NY; Multidisciplinary Center for Earthquake Engineering Research: 29-40.

Shinozuka, M. and X. Dong. 2002. "Seismic Performance Criteria for Lifeline Systems," Proceedings of the Eighth U.S. – Japan Workshop on Earthquake Resistant Design of Lifeline Facilities and Countermeasures Against Soil Liquefaction, Tokyo, Japan, Dec. 15-18, 2002.

Shinozuka, M., A. Rose, and R.T. Eguchi eds. 1998. *Engineering and Socioeconomic Impact of Earthquakes: An Analysis of Electricity Lifeline Disruption in the New Madrid Area*. Buffalo, NY; Multidisciplinary Center for Earthquake Engineering Research.

Shinozuka, M., M. Feng, J. Lee, and T. Naganuma. 2000. "Statistical Analysis of Fragility Curves," *Journal of Engineering Mechanics*, 126 (12): 1224-1231.

Shinozuka, M., T.C. Cheng, X. Jin, X. Dong, and D. Penn. 2002. "System Performance Analysis of Power Networks," Proceedings of the Seventh U.S. National Conference on Earthquake Engineering , Boston, Massachusetts, July 21-25.

Shinozuka, M., Y. Shumuta, S. Tanaka, and H. Koiwa. 1994. "Interaction of Lifeline Systems under Earthquake Conditions," Proceedings of the Second China-Japan-U.S. Trilateral Symposium on Lifeline Earthquake Engineering, April 19-23, Xi'an China: 43-52.

Tanaka S., M. Shinozuka, A. Schiff, and Y. Kawata. 1997. "Lifeline Seismic Performance of Electric Power Systems during the Northridge Earthquake," Proceedings of the Northridge Earthquake Research Conference, Los Angeles, California, August 20-22.

U.S.-Canada Power System Outage Task Force. 2003. "Interim Report: Causes of the August 14th Blackout in the United States and Canada," available at ftp://www.nerc.com/pub/sys/all_updl/ docs/blackout/814BlackoutReport.pdf.

USC-EPEDAT Software. 1999.

Wald, D.J., V. Quitoriano, T.H. Heaton, H. Kanamori, C.W. Scrivner, and C.B. Worden. 1999. "TriNet 'ShakeMaps': Rapid Generation of Peak Ground Motion and Intensity Maps for Earthquakes in Southern California," *Earthquake Spectra,* 15 (3): 537-556.

Webb, G.R., K. Tierney, and J.M. Dahlhamer. 2000. "Businesses and Disasters: Empirical Patterns and Unanswered Questions," *Natural Hazards Review*, 1 (2): 83-90.

Index

Figures are indicated by f, tables by t, and footnotes by n.

C

D

Titles in the Series (continued)

M. M. Fischer and *J. Fröhlich* (Eds.)
Knowledge, Complexity and Innovation Systems
XII, 477 pages. 2001. ISBN 3-540-41969-1

M. M. Fischer, J. Revilla Diez and *F. Snickars*
Metropolitan Innovation Systems
VIII, 270 pages. 2001. ISBN 3-540-41967-5

L. Lundqvist and *L.-G. Mattsson* (Eds.)
National Transport Models
VIII, 202 pages. 2002. ISBN 3-540-42426-1

J. R. Cuadrado-Roura and *M. Parellada* (Eds.)
Regional Convergence in the European Union
VIII, 368 pages. 2002. ISBN 3-540-43242-6

G. J. D. Hewings, M. Sonis and *D. Boyce* (Eds.)
Trade, Networks and Hierarchies
XI, 467 pages. 2002. ISBN 3-540-43087-3

G. Atalik and *M. M. Fischer* (Eds.)
Regional Development Reconsidered
X, 220 pages. 2002. ISBN 3-540-43610-3

Z. J. Acs, H. L. F. de Groot and *P. Nijkamp* (Eds.)
The Emergence of the Knowledge Economy
VII, 388 pages. 2002. ISBN 3-540-43722-3

R. J. Stimson, R. R. Stough and *B. H. Roberts*
Regional Economic Development
X, 397 pages. 2002. ISBN 3-540-43731-2

S. Geertman and *J. Stillwell* (Eds.)
Planning Support Systems in Practice
XII, 578 pages. 2003. ISBN 3-540-43719-3

B. Fingleton (Ed.)
European Regional Growth
VIII, 435 pages. 2003. ISBN 3-540-00366-5

T. Puu
Mathematical Location and Land Use Theory,
2nd Edition
X, 362 pages. 2003. ISBN 3-540-00931-0

J. Bröcker, D. Dohse and *R. Soltwedel* (Eds.)
Innovation Clusters and Interregional Competition
VIII, 409 pages. 2003. ISBN 3-540-00999-X

D. A. Griffith
Spatial Autocorrelation and Spatial Filtering
XIV, 247 pages. 2003. ISBN 3-540-00932-9

J. R. Roy
Spatial Interaction Modelling
X, 239 pages. 2004. ISBN 3-540-20528-4

M. Beuthe, V. Himanen
A. Reggiani and *L. Zamparini* (Eds.)
Transport Developments and Innovations
in an Evolving World
XIV, 346 pages. 2004. ISBN 3-540-00961-2

Druck: betz-druck GmbH, D-64291 Darmstadt
Verarbeitung: Buchbinderei Schäffer, D-67269 Grünstadt

GPSR Compliance

The European Union's (EU) General Product Safety Regulation (GPSR) is a set of rules that requires consumer products to be safe and our obligations to ensure this.

If you have any concerns about our products, you can contact us on ProductSafety@springernature.com

In case Publisher is established outside the EU, the EU authorized representative is:

Springer Nature Customer Service Center GmbH
Europaplatz 3
69115 Heidelberg, Germany

Batch number: 10755466

Printed by Printforce, the Netherlands